Gold Mining in Oregon

Exciting Books From
WEBB RESEARCH GROUP, PUBLISHERS

→Availability and prices subject to change. This is a *partial* list←

Oregon's Seacoast Lighthouses (Includes Nearby Shipwrecks). James A. Gibbs with Bert Webber Every Oregon coast lighthouse! $14.95.

Oregon's Salty Coast From Explorers to the Present Time, James A. Gibbs with Bert Webber. Complete State Park Directory. $12.95.

Terrible Tilly (Tillamook Rock Lighthouse); The Biography of a Lighthouse. (Near Cannon Beach) Bert and Margie Webber. $10.95.

Yaquina Lighthouses on the Oregon Coast. Dorothy Wall with Bert Webber. Both Yaquina Bay and Yaquina Head Lighthouses. $10.95.

Battery Point Light[house] and the Tidal Wave of 1964; Includes St. George Light[house]. Bert and Margie Webber. Visit it! $9.95.

Bayocean: The Oregon Town That Fell Into the Sea. Bert and Margie Webber. Includes Salishan, other erosion damages areas. $12.95.

Lakeport: Ghost Town of the South Oregon Coast. Bert and Margie Webber. Weird shipwreck; includes Coast Guard in WW-II. $12.95.

I'd Rather Be Beachcombing. Bert and Margie Webber. Emphasizes glass fishing floats and where/how to find them. $12.95.

Battleship Oregon: Bulldog of the Navy. Bert Webber $12.95.

Oregon Covered Bridges – Expanded Edition; Bert and Margie Webber. (6 New Bridges!) Photos, how to find all 65 standing bridges. $10.95.

The Search For Oregon's Lost Blue Bucket Mine; Stephen Meek Wagon Train of 1845. Charles Hoffman with Bert Webber. $12.95

Silent Siege-III: Japanese Attacks on North America in World War II; Ships Sunk, Air raids, Bombs Dropped, Civilians Killed. Bert Webber. Hundreds of Japanese attacks on 28 U.S. states – kids killed – and Canada never reported in papers due to agreement between gov't and press not to cause public panic. 304 pages, 462 photos, maps. $28.95.

Gold Mining in Oregon – Documentary; Includes How to Pan for Gold. Edited by Bert Webber. Data from Oregon State Dept. of Geology. All major gold mines located, how operated. 368 pages, pictures, maps – is the "Mother Lode" book about Oregon gold mines! $29.95.

Dredging For Gold – Documentary. Sumpter Valley Dredge State Park (Oregon). Huge dredges work 7-days a week scooping up gold! $8.95.

Indians Along the Oregon Trail; The Tribes of Nebraska, Wyoming, Idaho, Oregon, Washington Identified. Bert Webber, M.L.S. (Encyclopedic) Everything about these native Americans. $17.95

"OUR BOOKS ARE HISTORICALLY ACCURATE AND FUN TO READ"

Shipping costs: Add $2.50 first book, 75¢ each extra book – except *Silent Siege III* and *Gold Mining* shipping 4.95

Books Are Available Through All Independent Book Sellers or By Mail from Publisher

Send stamped, self-addressed envelope to speed copy of latest catalog to you.

WEBB RESEARCH GROUP, PUBLISHERS

P.O. Box 314 Medford, OR 97501 USA

GOLD

MINING IN OREGON

Bert Webber, Editor

How to Pan For Gold

Rare Pictures

Bibliography

Selected Text From *Bulletin 61* of the Oregon Department of Geology and Mineral Industries: *Gold and Silver in Oregon*

Special Index of Mines

Webb Research Group Publishers
Books About the Oregon Country

Printed in Oregon U.S.A.

Published by:
Webb Research Group Publishers
Books About the Oregon Country
P. O. Box 314
Medford, OR 97501

Color photographs:

FRONT COVER
This is Gordon A. "Alaska Al" Tracy
Santa Margarita, California
Photo by Jill McCoy, Morro Bay, California.
The picture was made specifically for this book

BACK COVER:
Photo by Bert Webber
at Oregon Trail Mercantile
Baker City, Oregon

➣TECHNICAL NOTICE: A portion of this volume is a reprint of Oregon State Department of Geology and Mineral Industries *Bulletin 61*. That work contained a number of blank pages which are eliminated here. To maintain the integrity of the Index, the numbered pages remain intact but in so doing, some odd-numbered pages, normally on the right-hand side of the book appear on the left-hand side. Likewise, some even-numbered pages, normally on the left-hand side of the book, appear on the right-hand side. When a former blank page was removed, as page #38, there is a note about this on page #39, etc. We regret any inconvenience.

Library of Congress Cataloging-in-Publication Data:

Webber, Bert
Gold mining in Oregon / Bert Webber.
p. cm.
Includes bibliographical references and index
ISBN-936738-77-4
1. Gold mines and mining – Oregon – History
II. Title.
TN423.07E93 1994 94-9940
622'.3422'09795–dc20 [published in 1995] CIP

Contents

The "Golden" Freeway. Interstate-5 at Exit No. 32 north of Medford. The contractor extracted many thousands of dollars in gold from the gravel that went into this section of the highway but admitted, "we never got anywhere near all of it." —Bert Webber photo

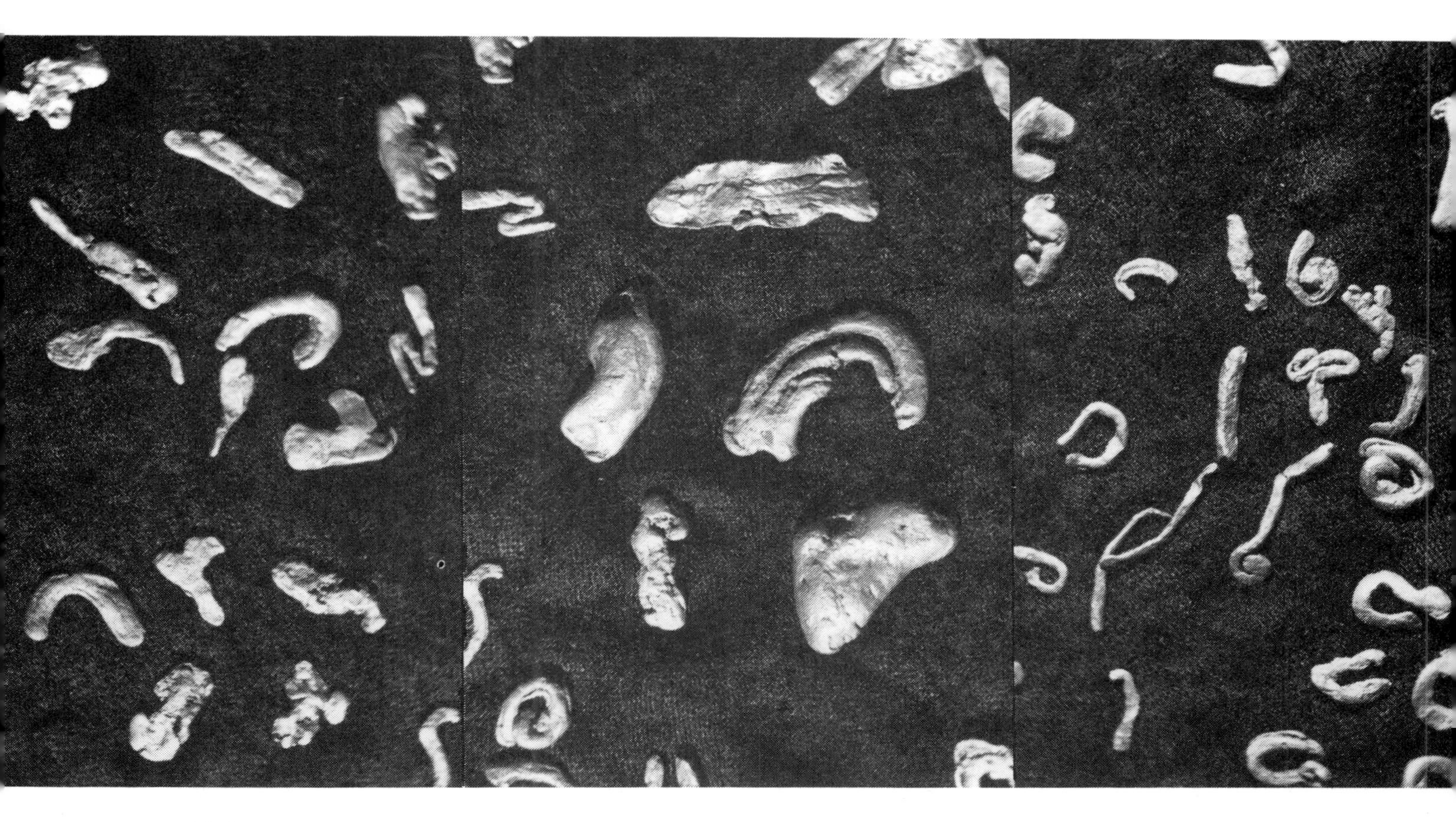

Gold

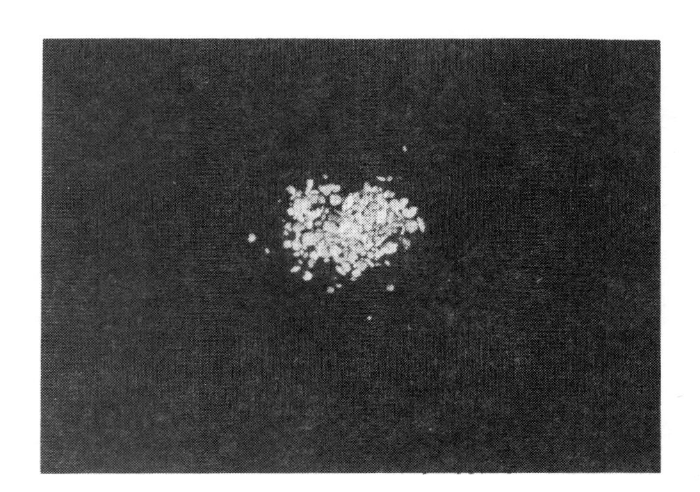

Specimen of gold (top) **that turned up in pans in Baker County. The pictures are about double the size of the originals.** (Center) **A "pinch-of-gold" was standard for trade before coins became common and was worth about $1. A pinch was the amount of dust a miner could hold between a thumb and forefinger. Merchants were careful to see that fingers were thoroughly dry before pinching. At variance also was the size of the man's fingers.** (Lower) **Actual size of nuggets as demonstrated by the rulers.**

Gold shown in top panel and center-lower, courtesy of Oregon Trail Mercantile - Baker City. Pinch-of-gold and gold in lower left, lower right courtesy of Rogue Valley Coin and Jewelry Exchange - Medford

—Bert Webber photos

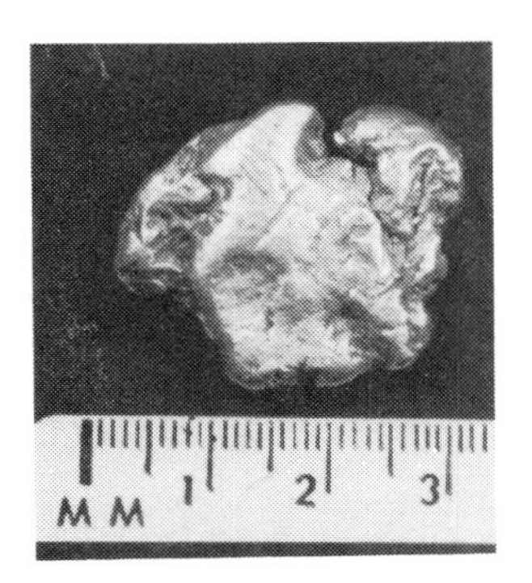

See page 61 for picture of the 80.4 ounce Armstrong Nugget

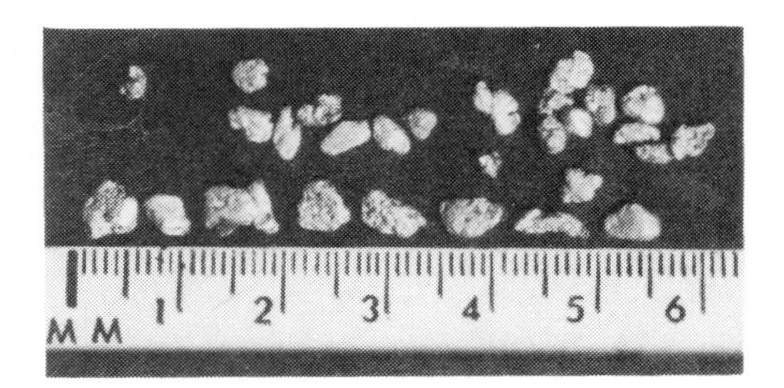

Panning For Gold

Because gold is heavier than most sediment and gravels in a stream, it and other heavy metals called "black sand" (including pyrite, magnetite, ilmenite, chromite, garnet), can be collected in a gold pan when the correct panning techniques are used.

For the most part, vessels for use in gold panning and the methods of panning are unchanged over the years. The bevel-sided dishpan is the traditional utensil for panning for gold. But anything that will hold water, and of course, the dirt and gravel from a likely spot, will do. However, if one expects good results from the effort of getting to a site, then spending the time bending over and wading in cold water, the use of a proper pan is really essential.

There is nothing sophisticated about panning for gold. Just about anyone can to it. The technique, possibly overly simplified, is merely placing dirt believed to contain gold, and water, in a pan then shaking it until the gold, which is heavier than water and most sediment, sinks to the bottom of the pan. Then pick out the gold.

Some "miners" who go out on a weekend looking strictly for gold as a small business, may carry a metal detector. This is a short cut to locating more likely sites than just guesswork.

The "riffle" (Garret) pan, is better than the old fashioned straight pan. This pan has ridges along one side that trap gold as one pours off sediment and water.

Another supplemental "pan" is really a screen with a frame around it. Its purpose is to separate large and small gravel. Holes in the screen should be no less than about 1/4-inch or larger than 3/4 to 1-inch. The screen appears to work best with dry matter as wet material tends to clog the holes. After basic shaking (sifting), the larger pieces can be inspected with a magnifying glass for gold particles. The smaller parts that fell through the holes, maybe directly into the gold pan, are mixed with water then shaken in the usual manner.

All the glitters is not gold.

Pyrite, commonly called "fool's gold," has tricked many a professional as well as weekend-fun gold seeker since the beginning of placer mining.

On close examination, pyrite does not really look like gold. Pyrite has a brassy color and is sometimes tarnished. Because it occurs in crystals, it changes shade as one holds a specimen hand and rotates it in the sun. Gold, on the other hand, is always gold colored. It is soft and can be bent (malleable).

When you see gold colored flecks that either float on the top of the water or are so light weight they easily wash over the edge of the pan, you probably have small pieces or "books of mica. Mica is a transparent and heat resistant material that has been used in doors of wood stoves to view the condition of the fire inside. Mica was extensively used as the insulator for the element in toasters. Mica breaks fairly easily into flat, thin sheets. Because Mica comes in a variety of colors, the goldish color is sometimes misidentified as gold by inexperi-enced panners.

When you are lucky with panning and discover gold in your pan, it may be in any number of shapes. These can be mere flakes, wire-like pieces, feather-shaped crystals, lumps or nuggets. These pieces of gold vary in size from microscopic to as large as a fist.

Great stories have been told of how large the nuggets were, and indeed, there are many large nuggets in museums on exhibit. In Oregon, the very famous Armstrong nugget (80.4 oz.) is seen in the display in the U.S. National Bank in Baker City.

Gold panners are always optimistic. Like Winston Churchill, they "never give up."

Weekend-gold seekers who want to pan for gold need be aware of hundreds of officially staked claims throughout Oregon. But there are "open" areas in Oregon for public panning.

All areas of "State Lands" below the vegetation line on navigable rivers and streams, as well as ocean

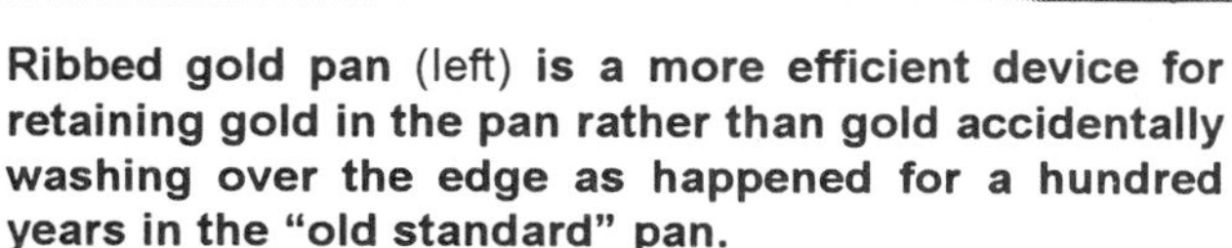

Ribbed gold pan (left) **is a more efficient device for retaining gold in the pan rather than gold accidentally washing over the edge as happened for a hundred years in the "old standard" pan.**
Gold, jewelry and pans courtesy of Oregon Trail Mercantile, Baker City.
—Bert Webber photos

beaches, belong to the State of Oregon. These are open for recreational gold panning.

In addition, panning is permitted on almost all streams and rivers in campgrounds on BLM and U. S. Forest Service land.

Specific locations: *

(1) Quartzville Recreational Corridor in the Western Cascades of the Salem BLM District. Obtain specific information at the BLM office in Salem.

(2) Butte Falls Recreational Area about 45 miles northeast of Medford. Obtain specific information from the BLM office in Medford.

(3) Applegate Ranger District about 25 miles southwest of Medford. Obtain specific information from the U.S. Forest Service office in Medford.

(4) There are three areas set aside for gold panning in the Wallowa-Whitman National Forest in northeast Oregon. Obtain specific information from the U.S. Forest Service in Baker City.

Additional gold panning areas:

•The Eagle Forks Campground, northwest of Richmond in Pine Ranger District at Halfway.

•McCully Forks Campground west of Sumpter, Baker Ranger District

•Deer Creek campground north of Phillips Lake, Baker Ranger District

•Powder River Recreational Area, just below Mason Dam, Baker Ranger District.

⇨ *If one is unsure about the status of the land, investigate with the nearest appropriate State, BLM or U.S. Forest Service offices.*

⇨ *It is wise to get permission before setting foot on private land – even to cross private lane between two known public land areas.*

⇨ *If what you might find along the way is obviously of historical or scientific value, do not disturb it.*

Equipment needed for gold panning:

Gold pan
Garden hand trowel
Tweezers
Small Magnet
Magnifying glass
Small transparent container (pill bottle with screw or snap top; plastic 35mm film canister with snap top)
Hand towel

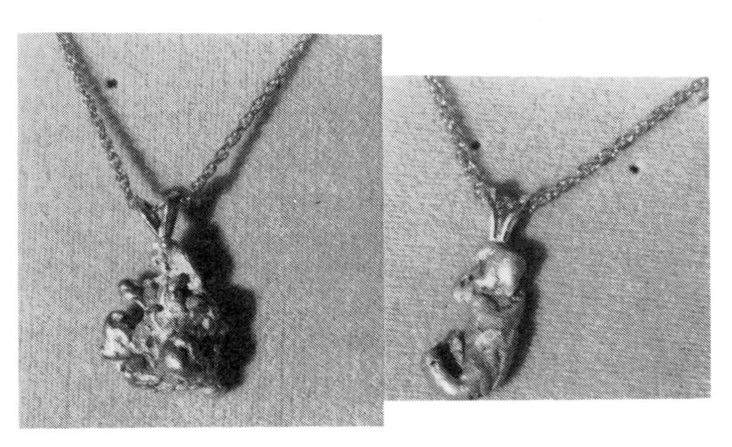

* From time-to-time, federal and state laws and regulation change. Always check with proper authority *first*. Accept the content of this book for gold panning locations as "guidelines" only.

① After digging in the bottom of a shallow creek, below the easily seen loose gravel and preferably to bed rock (the most likely area to find gold, on the down-stream side of large boulders), fill, but do not over-fill a gold pan. Take care with the trowel or shovel due to risk of losing flecks of gold while lifting the tool out of the water. Try not to let water and sediment dribble off the tool for doing so allows any gold to drain right back into the creek. Place in the pan between half to two-thirds full of soil and gravel from the stream bank or channel.

② Experience indicates that the better sites are away from swiftly running water. Place the pan totally under water to break up lumps of clay, then discard the stones. Better to set the stones aside for later inspection with a magnifying glass in one's search for gold.

③ Holding the pan under the water's surface with your hands on opposite sides of the pan. Use hands to break up stubborn clay lumps then rotate, vigorously, the pan halfway back and forth rapidly agitating to wash out the dirt (clay). The shaking and rotating motions should be in all directions. Do not be gentle. This swirls the water and causes heavy matter to sink to the bottom of the pan. Take your time.

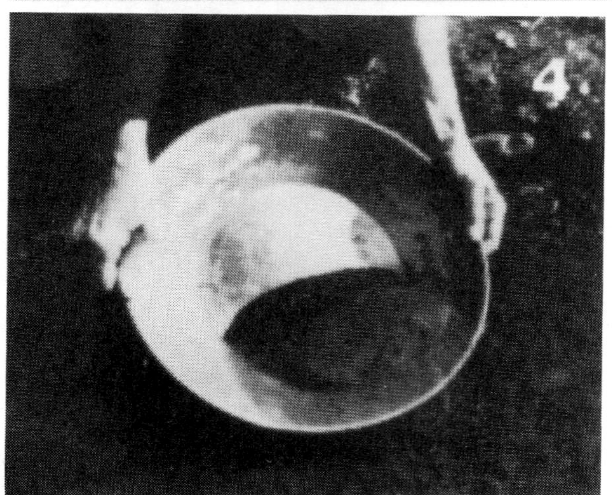

④ Continue holding the pan under the surface of the water but tilt the pan forward away from your body and slightly downward. Continue to rotate and shake the pan to let the light gravel and sand dribble out the front. Push top matter and large chunks of rock out of the pan with your thumbs. Don't rush. Sloppy draining of the pan, with un-broken lumps of clay going over the edge, may return much gold back into the creek. Better, always slurry your pan's drainings into a bucket so the content can be re-washed. Don't hurry. "Haste makes waste" – you lose your gold.

⑤ Take the pan with the residue, probably now reduced to about 4 cupfuls of concentrated sediment, along with some water out of the stream. Rotate the pan in a circular motion but watch carefully what is happening: The water is separating lighter from heavier materials and the gold, if gold is present and you are working your pan correctly, is lagging behind other material in the bottom of the pan. Don't rush!

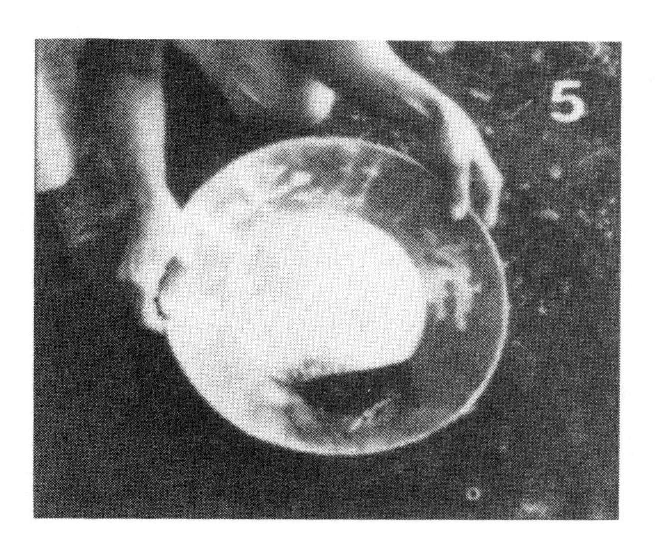

—Gold panning photos courtesy of Oregon Dept. of Geology and Mineral Industries

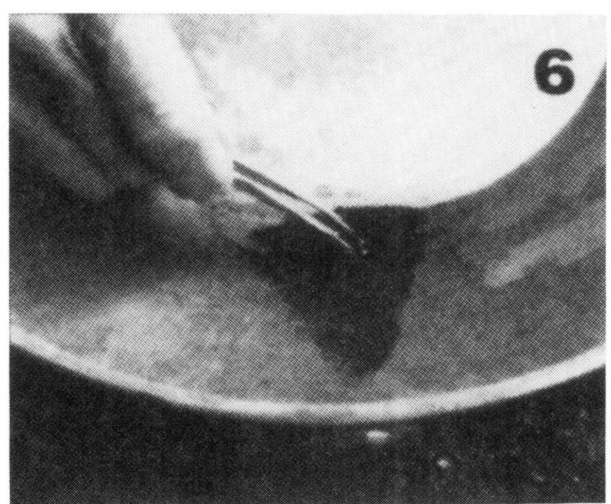

⑥ Stop rotating the pan. If you are fortunate, you will see a few flecks of gold in the dark material in the bottom of the pan. Carefully pour out the water and let the black sand and the gold dry. (It may be good to place the damp black sand with the gold, in a closed-top bucket to be taken home and finished there.) When dry, lift out most of the black sand with a small magnet, then separate the gold from the remainder of the sediment with tweezers. There may be more than gold in your pan! Be alert for silver.

Jim Pickett, Sr and his family pan for gold in Canyon Creek (Grant County) in 1934.
—Photo courtesy of Grant County Museum, Canyon City

(Left) **Beach sand from Curry County may contain gold. Exhibited at Curry County Historical Society, Gold Beach.** —Bert Webber photo

Early Gold Discoveries

It was in 1848 when James W. Marshall, a man from Oregon, discovered gold while working at Sutter's Mill in California. A state park marks the site where the mill was located, where on January 24, 1848, Marshall saw flakes of gold in the South Fork of the American River.*

Marshall took his samples to Sacramento and people laughed at him for claiming to have found gold. But a newspaperman, Samuel Brannan, spread the news through his paper. Within days the rush was on and the rest is history.

The first gold to be found in Oregon was in the summer of 1850† near the confluence of Josephine Creek and the Illinos River. But the discovery that triggered a rush, was that of James Cluggage and John R. Pool, some prospectors who, were working their way north from Sacramento, when they found gold in Daisy Creek, in 1851†, a few hundred yards south of the center of the city of Jacksonville. Within days prospectors founded the shanty town of Jacksonville. The area was pristine wilderness. Walling, writing in 1884, said:

> Nowhere else in America, possibly not in the world have the forces of nature so conspired to beautify and render a region thoroughly as delightful [as here].

Gold had been discovered in Eastern Oregon in 1845† by members of the ill-fated Stephen Meek Wagon Train. Meek was looking for a shortcut across Oregon to avoid crossing the Blue Mountains, when people in his party found gold in what has become known as the "Blue-Bucket Mine." The trouble is, no one recognized the nuggets as gold and they failed to record where they found them.**

Some prospectors were seeking the Blue Bucket Mine when Henry Griffin stumbled on to gold on October 23, 1861. But it was not the Blue Bucket. He was in present Baker County where the boom-town of Auburn quickly sprouted.

In Sumpter, on the north bank of the Powder River, was a valley that would eventually be reaped of gold as a farmer reaps wheat for, in a few years, over $16 million in gold would be the yield from the Sumpter Quadrangle. It was in the Powder River at Sumpter that the Yuba gold dredges worked.*** □

—Bert Webber photos

Visitor from New Jersey, Rick Webber, toured Eastern Oregon in 1994 and stopped at this abandoned mine and trommel. It is a little south of Canyon City on Highway No, 395.

---

* The first recorded discovery of gold in the west was in Placerita Canyon, near Los Angeles, in 1842. The strike was worked very quietly.

** An excellent account is in *The Search For Oregon's Lost Blue Bucket Mine; The Strephen Meek Wagon Train of 1845 - An Oregon Documentary.* Refer to special bibliography on page *xvii*-A.

*** For details of several prominent dredges, including a mammoth size dredge working at the present time, refer to *Dredging For Gold – Documentary.* See special bibliography on page *xvii*-A.

† The dates of 1850 and 1851 for earliest gold discoveries in Oregon are clearly recorded. The date of the Meek Wagon Train, 1845, is clearly recorded. The gold found by members of that train, along that trail, was not positively identified as gold until much later when samples were compared with specimens from California. As the specific site of the Blue Bucket Mine has never been identified, some purists in history like to claim it never existed.

Historical marker at Applegate and S. Oregon streets, commemorates the discovery of gold in what became Jacksonville. The town, now a National Historic Landmark, was over 30 years old before nearby Medford became a happy thought. Trommel (gravel washer) (lower) **from Sterling Mine still in use over 100 years later at a Central Point aggregate firm.**

Gold in Jackson Creek *

Bill Dobbyn and Fred Christean had a mining partnership called Forest Creek Placer Mine. They often dealt with Godward Mercantile Company in Jacksonville. On March 5th, 1938, for example, Fred took in 1-oz. 13P-16 gr and was paid $47.95. Three days later he returned with more dust valued at $81.95 but the proprietor didn't have the cash so he receipted for the gold with the notation, "Pay next time."

Bill Dobbyn told one interviewer years later that Godward would pay higher prices if the miner brought in several ounces at one time, rather than smaller amounts frequently.

Dobbyn and Christean decided to set up a dredge in Jackson Creek late in 1941, and were just getting started when the Japanese attacked Pearl Harbor on December 7th. They operated for a range of about 200-yards from N. Oregon Street upstream to a point where bed rock appeared on the surface. Dobbyn told the authors that they bought the old wood-bottomed dredge, which was about 20 x 30 feet in size, in California. It was hauled to Jacksonville then reassembled at the site.

Very little gold dredging was done however as the War Production Board shut off all non-military essential materials. Steel cable was the major item of shortage, thus they shut down and engaged in other pursuits during the war.

After the war, materials were again available so they returned to Jackson Creek. Their dredge was fitted with a 1-yard capacity bucket. They moved around considerably within the confines of the area, working upstream to the left of the present creek bed then back down. The course of the creek was changed as the dredge moved slowly along as it took lots of water to operate. Dobbyn emphasized however, that the project was not intended to be a year-around job, as Jackson Creek traditionally dries up any time after mid-June.

Some examples of Dobbyn and Christean's labors are reflected in the receipts for gold dust received from the United States Mint in San Francisco.

On April 17, 1946, Bill Dobbyn carried dust to the U. S. National Bank branch in Medford, for which a teller receipted for 40-oz, 19-PW 14 gr gold-amalgam. The bank advanced $1,230 allowing $30/Fine ounce against what the government would pay. On April 22, the Mint issued its report:

BULLION DEPOSIT – MEMO REPORT

in favor of the United States National Bank of Oregon (Deposit No. 1608) with these details:

Dry-land dredge working field on former Wendt property east of Jacksonville 1946-48. Jacksonville Mining Company dug up hundreds of tons of earth, recovered a lot of gold, but expenses denied the operators of much profit.
—Don Wendt Foundation photo

DEPOSITOR: C. & D. Mining Company.
MINE: Jackson Creek (Oregon)
DESCRIPTION: Grains & amalgam
ASSAY No. 2101
WEIGHT BEFORE MELT: 40.97
WEIGHT AFTER MELT: 39.85
ASSAY CERTIFICATE FINENESS: Gold .874- ¾
Silver .014-¼
BREAKDOWN OF FEES:

Melting:	**$1.00**
Refining:	**$1.59**
Handling:	**$3.05**
Total Chg:	**$5.64**

VALUES: Gold $1,220.03
Silver $ 3.09

TOTAL (Less Charges): $1,217.48

From this remittance, the U. S. National Bank deducted $2.08 advanced to Railway Express to move the original amalgam to San Francisco, plus the bank's handling charge (profit) $1.20.

Dobbyn and Christean realized $1,214.20 from this single shipment, which they split 50-50 between them. Dobbyn recalls their usual routine was to carry gold dust to the bank about once a week.

On May 23, 1946, these fellows received another BULLION DEPOSIT – MEMO REPORT from the mint for 51.814 FINE ounces of gold and 6.94 FINE ounces of silver. Their proceeds on this shipment were $1,805.60. On this shipment the breakdown of their amalgam shows: .880 gold; .118 silver; .002 base metals as copper, lead, etc.

When Jackson Creek dried up, the fellows pulled the dredge out and sold it. They set to getting a newer, metal-bottomed dredge which they worked on Foots Creek at some miles distant from Jacksonville. The area chosen didn't do well enough so they gave up gold mining.

During the early 1930's, Dobbyn said he had worked areas other than Jackson Creek and had some luck. "I didn't get rich but I made a decent living," he told the authors. Bill Dobbyn winces just a little when talking about his venture saying:

Man – wouldn't I like to have all that gold back today and sell it at current prices! I got $30 an ounce then and it's well over $300 per ounce now. □

* Extracted from: *Jacksonville, Oregon; Antique Town in a Modern Age.* Webb Research Group. 1994.

Laws of the Miners

★Who owned the mineral rights?
★Was there a controlling law that allowed early prospectors to stake a claim on public land?
★If disputes arose, how were they settled?
★What was the relationship between early miners and Indians?

The rulers of ancient nations controlled the mineral land. The Greek policy was state ownership. Among the Romans, mines belonged to the state. Both the Greeks and the Romans allowed leases. In England, all gold and silver, in hand or still in the ground, on public or private lands, belonged to the crown. The king could grant the right to search for minerals but the right had to be stipulated in the patent. In Germany, in the Middle Ages, the principle of free mining went wherever the miners went and all persons were able to look for useful minerals. Local custom allowed that the discoverer of a deposit was given the right, within some limits, to hold the property on the condition that he make known his discovery promptly after finding it. If his discovery was not published in a prescribed manner, he would lose it.

The earliest mining laws in America are credited to the Spanish. King Charles III, in 1783 set down the principles that applied to all early mining camps of the west. This area included what are now the states of Oregon, California, Arizona and Colorado and probably Wyoming.

The chief facet of the Spanish law was that no land could be claimed by mere speculation. The prospector had to prove there were minerals on it. It is noted that if a group of men were prospecting together, the discoverer of minerals would have the right to a double claim.

The mining law of the United States arose as a result of the discovery of gold in California. Miners insisted on the right to help themselves from the federal domain free from fees or permits. After all, the American government had no idea what was there and with the Mexican withdrawal, there was no governing body.

The military governor had his hands full and did not interfere with the miners therefore to establish some semblance of order, the miners held meetings and organized mining districts. Before 1866 when the Federal Mining Act was passed, there was no restriction on the power of mining districts other than a basic understanding that a mining rule had to be reasonable.

When the Act of 1866 came into being, it "grandfathered" the local rules and regulations if they did not conflict with existing Federal or State laws.

Mining districts were informally established. The first individuals arriving on a new diggings would get together, even if this was only a handful of men, and organize a district on short notice. The geographical limits were defined and one of the men became the official recorder by election. That was all there was to it. They were in business.

As the California districts had been formed and were working before gold was discovered in Oregon, the California rules and regulations usually became the model for Oregon miners. This came about simply because many miners who had been in California, came to Oregon and brought their rules with them.

Oregon's first mining laws were very simple and direct. While some laws were drawn up by lawyers or men of general education, many of the miners were uneducated but had a sense of the practical. They used the simple form in which to state their case in earnest frontier terms. If something wasn't working, the miners who had formed the district got together and added amendments.

At Kerbyville (Josephine County), a group of the miners formed a district and set down these conditions:

> Know all men by these presents: that the miners in council assembled this the 1st day in April 1852, do ordain and adopt the following rules and regulations to govern this camp:
>
> Resolved, 1st, That 50 yards shall constitute a claim in the bed of the creek extending to high water mark on each side.
>
> Resolved, 2d, that 40 feet shall constitute a bank or bar claim on the face extending back to hills or mountain.
>
> Resolved 3d, that all claims not worked, when workable, after 5 days, be forfeited or jumpable.
>
> Resolved 4th, that all disputes arising from mining claims shall be settled by arbitration and the decision shall be final.
>
> E. J. Northcutt
> Chairman

Of course, later laws were in greater detail and provided the terms that had to be met for staking and operating a claim.

History has demonstrated that all prospecting and even working a properly established and recorded claim, did not always "pan out."

In the case of the late E. S. McComas, his diary 1862-67 reveals:

DURING THE NATIONAL FINANCIAL PANIC OF 1907, money was being hoarded to the point where there was not enough to meet daily business needs. James Howard, ex-president of the Bank of Sumpter, got together with Fred Mellis who operated a placer mine. Mellis designed and hand-made gold coins. Each coin contained 2 oz of gold and enough copper to add strength. The local people were convinced that as long as

their mines could produce gold, and these gentlemen could make coins, there was no need for local panic. Presumably these coins have disappeared into private collections. The first of two designs, the one with the copper, was hammered out in November. About the same time a second issue with an alloy of silver appeared. Apparently only about 100 coins of both types were made. Numismatists class such coins as "satirical and political" because of the imprint, "IN GOLD WE TRUST"

Illus. courtesy of Donald H. Kagin, Ph.D., *Private Gold Coins and Patterns of the United States.*

Oregon's Mint

The Provisional Government passed an Act on February 16, 1849 that provided for gold minting of coins. This was because United States coins were in short supply. With a population of about 1,000, Oregon City was designated the site of the for the mint. At the time the legislation for the mint passed, Oregon had been brought into the United States as a Territory by the U. S. Congress. On March 2, the new Governor arrived and immediately declared the coinage act unconstitutional as only the Federal Government possessed the right of coinage.

Nevertheless, Hamilton Alexander made dies for a $5 gold coin and a blacksmith, Thomas Powell, built a press. The first coins were probably struck in March 1949. For a lengthier account see the book *Oregon City By Way of the Barlow Road at the End of the National Historic Oregon Trail.* □

> July 22nd. Busted again. On the first day of July 1864 a company of ten men was organized in La Grande known as the La Grande Mining Company. The object of said company being to mine gold on Grande Ronde River and its tributaries. The capitol stock of said company was five thousand dollars. We expended $800.00 & found that the diggings would not ``pan out.'' The company busted after paying up all of the liabilities. Well, this has been my luck in mining.

The Southern Oregon Indians never liked white men since the introduction of whites as part of the Jedediah Smith party of June 1828. There had been a standoff, over a stolen ax, at a site on the Umpqua River. The Indians retaliated – killed nearly all of the whites.

Hoards of Oregonians, on learning of the gold strike near Sacramento, headed for California through the Umpqua and Rogue River Valleys and encountered Indians along the way. In an effort to keep peace, the Army sent General Joseph Lane, who set up Fort Lane near the present city of Gold Hill in Jackson County. An agreement was reached whereby the Indians would allow the whites to pass through the territory as well as to settle and do some mining in the Rogue Valley. One hastle led to another and the Rogue Indian Wars resulted.

Much of the trouble with Indians was due to the acts of irresponsible whites. In a letter to Governor Charles Stevens, March 10, 1856, are these lines:

> The Indians who were peacefully living on their reservation were often molested by irresponsible whites hanging around the mining camps. These fellows, too lazy and too mean to earn an honest living, would go on the reservation and take anything that could get their hands on. If an Indian interfered, he was shot down. Massacre of Indians incited them to battle as much as a massacre of whites would be followed by an expedition against the Indians.
>
> ...In a number of cases, whiskey sold to the Indians were the indirect cause of trouble. Finally, the miners and settlers determined to exterminate the Indians.

When the Indian Agent, Joel Palmer decided to move the Indians to the Grande Ronde Reservation, so intense was the feeling of the whites against the Indians, that Palmer called on the Army to protect the Indians during the move.

Finally, the *Oregonian* ran an article on December 20, 1856 that concluded:

> ...and as there are no Indians in the mountains, miners do not run the risk as they used to, while out prospecting, of having their scalps taken. □

Special Bibliography

Evans, James R. and Bert Webber. *Flagstaff Hill on the National Historic Oregon Trail*. Webb Research Group. 1992.

Geographic Names System (Oregon). Branch of Geographic Names, U. S. Geological Survey [unfinished project] Dec. 1992.

Hoffman, Charles and Margaret. (Bert Webber, Editor and Indexer) *Hoffman's Rockhound Guide - Expanded Ed.* Webb Research Group. 1993.

Hoffman, Charles S. *The Search For Oregon's Lost Blue Bucket Mine; The Stephen Meek Wagon Train of 1845 - An Oregon Documentary.* Webb Research Group. 1992.

Mayo, Roy F. *Gold Mines of Southwest Oregon*. Nugget Enterprises. 1987.

Oliver, Herman. *Gold and Cattle Country.* Binford and Mort. 1961.

Spreen, Christian A. *A History of Placer Gold Mining in Oregon 1850-1870.* Unpub thesis Univ of Oregon. 1939.

Romanowitz, Charles M. and H. J. Bennett and W. L. Dare. *Gold Placer Mining*. [Info circ. 8462] U.S. Dept of Interior Bureau of Mines. *ca.* 1969.

Webber, Bert. *Dredging for Gold.* Webb Research Group. 1994.

__________. *Oregon Names; How to Say Them and Where They Are Located.* Webb Research Group. 1995.

Webber, Bert and Margie Webber. *Oregon City By Way of the Barlow Road at the End of the National Historic Oregon Trail.* Webb Research Group. 1993.

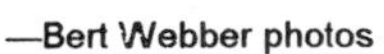
—Bert Webber photos

Oregon Belle Mine
Jacksonville
Mining has resumed

Old Gold Mine Causes

Jacksonville Street Collapse

Patrons in the Jubilee Club Restaurant saw, on October 14, 1982, the pavement "dip" under passing cars then cave in when a heavy truck passed over a few minutes later. By another hour, the street, with a near noiseless *whoosh*, collapsed as bystanders watched in awe. The site was over an abandoned gold mine that had been excavated from a shaft on the northwest corner of 4th and California Streets in the 1930's.

Research photo-journalist Bert Webber entered the "mine" through the hole while Steve DeKorte, Public Works Department, stood guard. Webber, a 6-footer, stands on caved-in earth about 10 feet above the floor of the cavern which was on bed rock. There is about 3 feet between the top of Webber's head and street level.

Jacksonville, during the Great Depression, had dozens of backyard, as well as several under-street working mines. When World War-II started, many of the shafts were filled in but it appears nobody bothered to stuff the caverns or tunnels. Years later, a lady drove her car into her carport when suddenly the car sank and rested on its frame.

When a fire truck stopped at the corner of 5th and C Streets, the street under the rear wheels suddenly collapsed.

Some folks who live there, who know of the town's golden past shrug, "It's part of the town's character." □

—Margie Webber photo

—Photo courtesy of Medford *Mail Tribune*.

Compact gold sponge after retorting of amalgam in crucibles at the Greenback mine, Josephine County, about 1910. Each gold "biscuit" is about 2½ inches in diameter. The total weight of the four biscuits is estimated to be 250 - 300 ounces. On March 2, 1995, this would have been worth between $93,687 and $1,12425 per New York quotation.

FOREWORD

"Gold mining was originally the mainstay of the economy of southern (and northeastern) Oregon. It started settlements, built roads and schools, promoted local government, and established law and order. It was about the only source of new wealth and was a common means of earning a livelihood. It is now at best only a token of its past. Not only is gold mining as an industry dead, but its history and the knowledge of its individual mines, which formerly represented a large part of the area's payrolls, are fading into the hazy past. The critical point in its downfall was World War II's Administrative Order L-208, which was designed to stop the mining of gold, thus forcing gold miners to seek employment in base-metal mines, especially copper, in which there was supposed to be a shortage of miners. The order failed essentially to accomplish its objective, but the final result was to deal a crushing blow to gold mining. Shutdowns, always a serious operating matter in an underground mine because of the maintenance problem, compounded the gold miners' difficulties. After the war and the termination of L-208, costs of labor and supplies had multiplied but the price of gold remained the same. Thus gold mining was effectively killed."

Thus wrote Mr. F.W. Libbey, formerly Director of the Department, in his article "Lest We Forget," that appeared in The ORE BIN in June 1963. What Mr. Libbey said in 1963 is just as true today (April 1968). However, the Department was of the opinion that the need and demand for gold would increase and therefore that a bulletin collecting the salient facts on gold and silver in Oregon as known today would aid in the future development of Oregon's mineral industry.

The authors of this publication have spent the past five years researching the literature and compiling the data found herein. They have been greatly aided by Margaret Steere, staff geologist of the Department, in coordinating and arranging the information, and by other members of the staff in preparation of the report for publication — namely, Miriam Roberts in editing and typing the text, James Powell in drafting the numerous charts and maps, and Raymond Corcoran in reviewing the stratigraphic data.

But this publication was not turned out merely for its historic interest. Events of recent months have indicated that a rise in the price of gold is imminent. This will come as no surprise to those who have attended the Gold and Money Sessions of the Pacific Northwest Metals and Minerals Conference, the Proceedings of which were distributed by the Department. Not only is great concern expressed by many monetary authorities, especially those in countries outside the United States, over the need for more gold to back money, but also the use of gold in the arts and industries is increasing at such a rate that its value will likely have to be raised if only to meet these demands. We consider the outlook for a more realistic gold price as good. Consequently, the information contained in this volume should prove of great value to those who wish to evaluate Oregon's gold mining regions.

It is hoped that study of this publication will suggest properties or areas that should be investigated more thoroughly even with the price of gold at $35 an ounce. We feel that attention should be paid especially to those districts where there have been considerable prospecting and tunneling. Such districts may still contain unrecognized areas of gold mineralization of sufficient grade and tonnage to warrant mining by the large-scale methods used in copper mining today. The U.S. Department of Interior, through its Heavy Metals Program, is looking at some of these areas and results of its work should be followed closely.

In addition to this broad coverage of gold and silver in Oregon, the Department has in the past published on two regions in greater detail than is found in this volume. These publications are: Bulletin 34, "Mines and Prospects of the Mount Reuben Mining District, Josephine County, Oregon" (1947); and Bulletin 49, "Lode Mines of the Central Part of the Granite Mining District, Grant County, Oregon" (1959). Reports on specific properties are: GMI Short Paper 23, "Oregon King Mine, Jefferson County, Oregon" (1962); and GMI Short Paper 24, "The Almeda Mine, Josephine County, Oregon" (1967).

Gold mining was once the backbone of Oregon's mining industry. Its early importance to the State is demonstrated by the fact that a pick occurs with other symbols representing her industry on the Great Seal of the State of Oregon. It is our hope that this publication will aid in re-establishing the mining for gold once again as a significant industry in our State.

Hollis M. Dole
State Geologist

April 10, 1968

Editor's note: On March 2, 1995, gold was $375 an ounce per New York quotation.

GENERAL CONTENTS

*Enlarged table of contents accompanies each section.

ILLUSTRATIONS

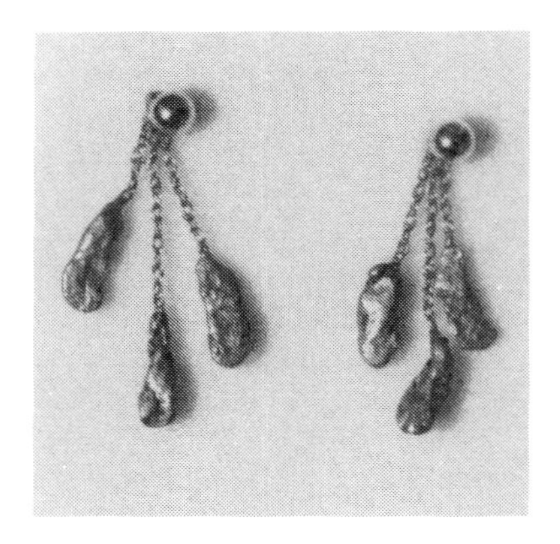

Figures

(Continued on page vi)

Figures, continued

Tables

Part I General Discussion

Cornucopia Mine at Cornucopia, Oregon. The gold concentrator line (lower) **in the Cornucopia mill.**
—Archives: Baker County Library

PART I. GENERAL DISCUSSION

Table of Contents

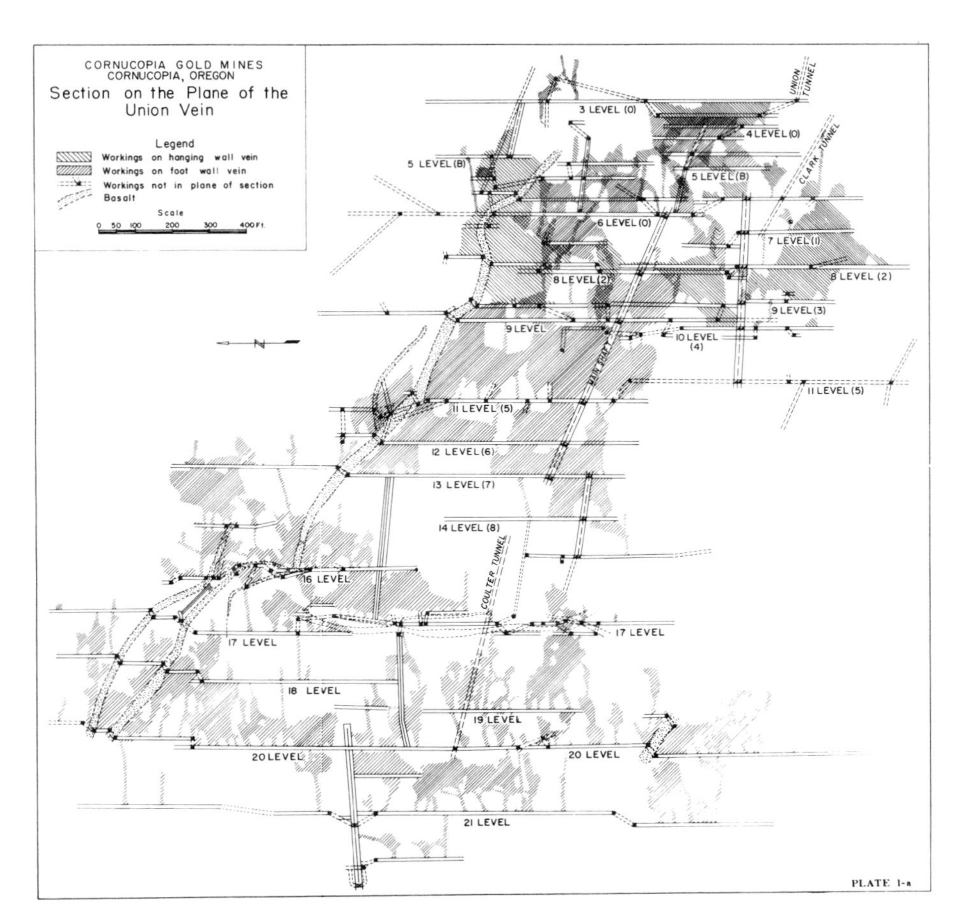

PLATE 1-a

GOLD AND SILVER IN OREGON

BY HOWARD C. BROOKS and LEN RAMP

PART I. GENERAL DISCUSSION

INTRODUCTION

Status of the industry

Gold mining was the primary reason for the settlement and early growth of northeastern and southwestern Oregon--the two regions where the principal deposits are found. Between 1850 and 1965, Oregon produced roughly 5.8 million fine ounces of gold and 5.4 million fine ounces of silver worth a total of about $210,000,000 at today's prices, and probably 60 percent of the gold was mined before 1900.

Prior to World War II and for a few years afterward the annual dollar value of gold and silver far surpassed that of any other metallic product of Oregon mines. During the late 1930's and early 1940's the industry flourished and showed every indication of continuing strength. But in 1942, gold mining was stopped by government order for the duration of World War II. Owing to subsequent inflation of costs without compensating increase in the price of gold, the industry never recovered. Oregon's gold mining has since been only a token of what it was in the past -- or what it could be in the future under a more realistic price structure.

Purpose and scope of report

This bulletin represents a compendium of information on the known gold and silver occurrences in the state. It gathers together significant facts which otherwise are scattered throughout a great number of published and unpublished records. No new field work was done in the preparation of this report.

In order to facilitate the handling of the material, the bulletin is divided into three parts, as follows: Part I discusses the economics of gold and silver mining and the production, geologic occurrence, and distribution of these metals in Oregon. Parts II and III describe the principal gold-mining areas of the state and give information on individual mines. For convenience in presentation, the state is divided into eastern and western Oregon along the 121st meridian. Part II is concerned with (a) the "gold belt" of the eastern Blue Mountains and (b) isolated districts of eastern Oregon. Part III treats (a) the Klamath Mountains of southwestern Oregon and (b) the Western Cascades in western Oregon.

Method of presentation

Oregon's gold deposits tend to occur in clusters in certain areas. This natural segregation of mineralization is caused by fundamental differences in the geology and is the basis for dividing and describing the deposits. Geologically distinct segments of mineralized regions are referred to here as "areas," such as the Gold Hill-Applegate-Waldo area in southwestern Oregon and the Wallowa Mountains area in northeastern Oregon.

In some of the mineralized areas, particularly those in northeastern Oregon, the deposits have been further separated into groups known as mining districts. In the early days of mining, before county lines and other legal subdivisions were made, it was the custom to refer to territory containing a group of deposits as a "mining district." Each district was given a name, commonly that of some geographic feature within it such as Canyon, Connor Creek, and Greenhorn districts in northeastern Oregon, to name only a few. The early inhabitants often set up and administered local mining regulations. Boundaries of most of the districts were never formally defined, and eventually many of the smaller districts were combined to form larger ones. District names in use today are those that have become firmly established in published reports.

The description of each mineralized area is accompanied by an index map showing mine and prospect locations. Available data concerning the geology, development, and production of each mine is summarized, and souces of this information are listed. A number of prospects and small mines have been omitted because of lack of information; some of these, however, may have greater economic potential than many that are described.

Production records are incomplete for most mines and many of the available figures are based in part on estimates. Unless otherwise stated, production figures given in dollars reflect metal prices current at the time output was made.

Most of the information given in the bulletin has been compiled from published works listed in the bibliography. The remainder is from unpublished reports and records supplied by the Department staff and others. Nearly all of the published material is now out of print, but copies of these can be consulted at some libraries and at the three offices of the Department. Additional information, not printed in this or previous reports, is available for a number of the mines and is kept in Department files for public use.

ECONOMICS OF GOLD AND SILVER

Gold: uses, prices, and controls

The gold production of the world is used mainly for monetary purposes, either as money or, more generally, as reserves in the form of bullion to back currencies which in themselves have no intrinsic value. Large quantities of gold are used in the decorative arts, particularly for jewelry and watches and for gilding and plating. Smaller amounts are used in industry, dentistry, and medicine.

In the United States the Federal Government has long controlled the price of gold -- first by establishing its coinage value and then, in 1934, by withdrawing it from circulation as a medium of exchange and eliminating the right of citizens to buy or sell refined gold without a license.

In 1792 the U.S. Treasury price for gold was set at $19.393939 per fine troy ounce. The price was raised to $20.689658 on June 28, 1834, then lowered to $20.671835 on January 18, 1837. The latter price remaining in effect until 1933, when various Government decrees and legislation led to the establishment on January 31, 1934, of the present price of $35.00 per fine troy ounce. The average weighted price for gold in 1933 was $25.56; for 1934 it was $34.95. Since 1934, the price paid for gold purchased by the mints has been $35.00 per fine troy ounce, less a quarter of one percent for handling.

As a result of the 1933-34 legislation, gold was withdrawn from circulation in the United States and it has since been unlawful for any unlicensed person or group to buy, sell, or transport gold except in its natural state. Natural gold is defined by the Treasury as gold recovered from a natural occurrence and not melted, smelted, refined, or otherwise treated by a chemical or electrical process. All newly mined gold which does not fall into the category of "natural gold" must be sold to the U.S. mint or to a licensed buyer. Gold amalgam produced from domestic sources may be dealt with in the same manner as gold in its natural state. Retort sponge resulting from heating the amalgam may be held and transported without a license by the person who mined or panned the gold, except that he may not hold at any one time an amount which exceeds 200 troy ounces in fine gold content. Unlicensed persons may purchase retort sponge containing up to 200 ounces gold, but can then resell it to no one except the United States or a licensed buyer.

Silver: uses, prices, and controls

Silver, like gold, has long been monetized, but unlike gold, silver is still in circulation in the form of coins in the United States and the general public can legally buy and sell silver bullion. The principal non-monetary consumers of silver are tableware, photographic, and electroplating industries.

The coinage value of silver was set at $1.2929+ in 1837 by Government legislation establishing the amount of contained silver in the silver dollar as 412½ grains 900 fine or 0.7734 troy ounce of pure silver. Except for a brief period during 1873-1878, this has remained the standard weight and fineness of the silver dollar. Silver values have fluctuated widely over the years (figure 1), reaching a record annual low of 28 cents an ounce in 1932. In 1933 the Treasury buying price for newly mined domestic silver began to be set by legislative action under the coinage law. On December 21, 1933, the Treasury buying price for newly mined domestic silver was set at $0.6464 per fine troy ounce. Changes in Treasury price have been as follows: It decreased to 0.5001 in August 1934; increased to $0.7111 on April 10, 1935; increased to $0.7757 on April 24, 1935; decreased to $0.6464 on January 1, 1938; increased to $0.7111 on July 1, 1939; and increased to $0.90505 on July 1, 1946. The last price cited held until June 4, 1963, when Public Law 88-36 went into effect. This law repealed all existing silver-purchase laws and for the first time in nearly 30 years freed silver from special legislative restrictions. Prices in the open market rose rapidly and on September 9, 1963 reached 129.29 cents per ounce, at which price the silver dollar contains exactly $1.00 worth of silver.

Effect of price on the industry

In most other segments of the metal-mining industry, increases in mining costs are generally, although often belatedly, offset by increases in metal prices. This is not true of gold mining. Because gold prices remain fixed by law, the health of the industry fluctuates with the national economy. Periods of inflation are unfavorable for gold mining for, while the costs of labor and materials for mining increase, the value of the product remains the same and is thus less able to cover rising expenses. Increasing costs must, therefore, be offset by decreasing expenditures for maintenance of equipment and for the development of future reserves, and generally the minimum grade of ore mined must be raised, with the result that the amount of ore available for mining is proportionately lowered. Eventually, the mine must close. Often considerable ore remains that could be mined under more favorable economic conditions.

The great depression of the 1930's and the increase in the price of gold in 1934 resulted in a peak output for the United States industry in 1940. At the beginning of World War II, production began to decrease because of high wages and scarcity of materials. Then in 1942 War Production Board Order L-208 was promulgated and effectively stopped gold mining for the avowed purpose of providing additional manpower and equipment for mines producing metals and minerals more directly essential to the war effort. The gold-mining industry contended generally that the order was arbitrary and foredoomed to failure and that it violated constitutional rights. In 1951 and 1952 litigation was started to obtain compensation from the Government for damages caused by the order. More than 200 claims for damages were eventually filed with the United States Court of Claims. On February 20, 1956 the Court ruled in favor of the claimants, but in 1958 this decision was reversed by the U.S. Supreme Court. Subsequent industry appeals to Congress died in committee.

Order L-208 was rescinded in mid-1945 but the death knell of the industry had already rung in the form of inflation. Most of the mines that were productive in pre-war years were not reopened. The immediate reasons were that materials and labor costs had risen greatly while the price of gold remained fixed, and the years of idleness had resulted in such deterioration of plants and workings that prohibitively large expense would have been required for rehabilitation. These conditions have steadily deteriorated.

The manner in which costs of mining have risen over the years is reflected by a comparison of the equivalent prices of gold for certain years in terms of 1940 and 1965 dollars.

	1935	1940	1950	1960	1965
Quoted price	35	35	35	35	35
Equivalent price in 1940 dollars	34	35	17	15	14
Equivalent price in 1965 dollars	81	83	41	36	35

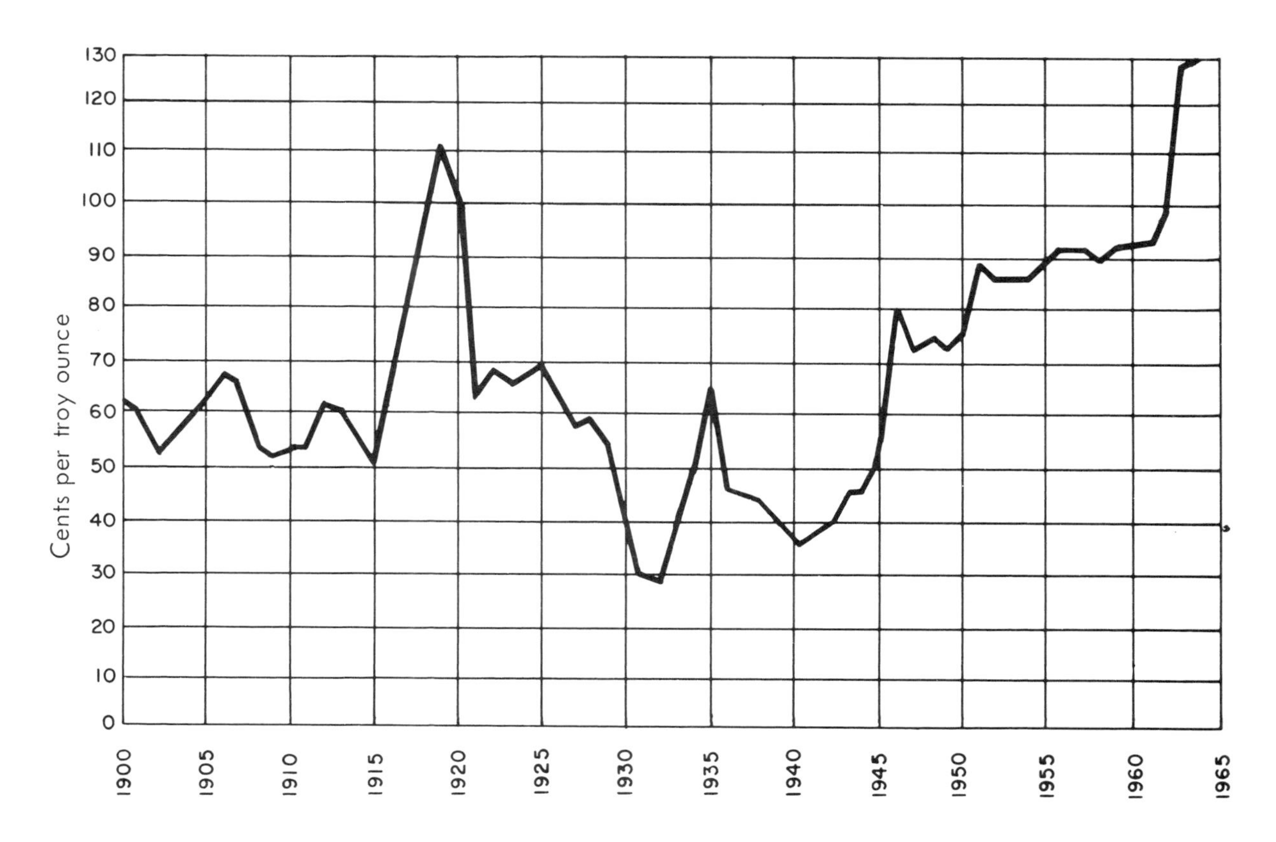

Figure 1. Trends in annual average price of silver, 1900 - 1965.
(From Engineering and Mining Journal Annual Survey and Outlook, February 1965.)
U.S. Treasury buying prices for newly mined domestic silver, 1933 - 1963, are given in the text.

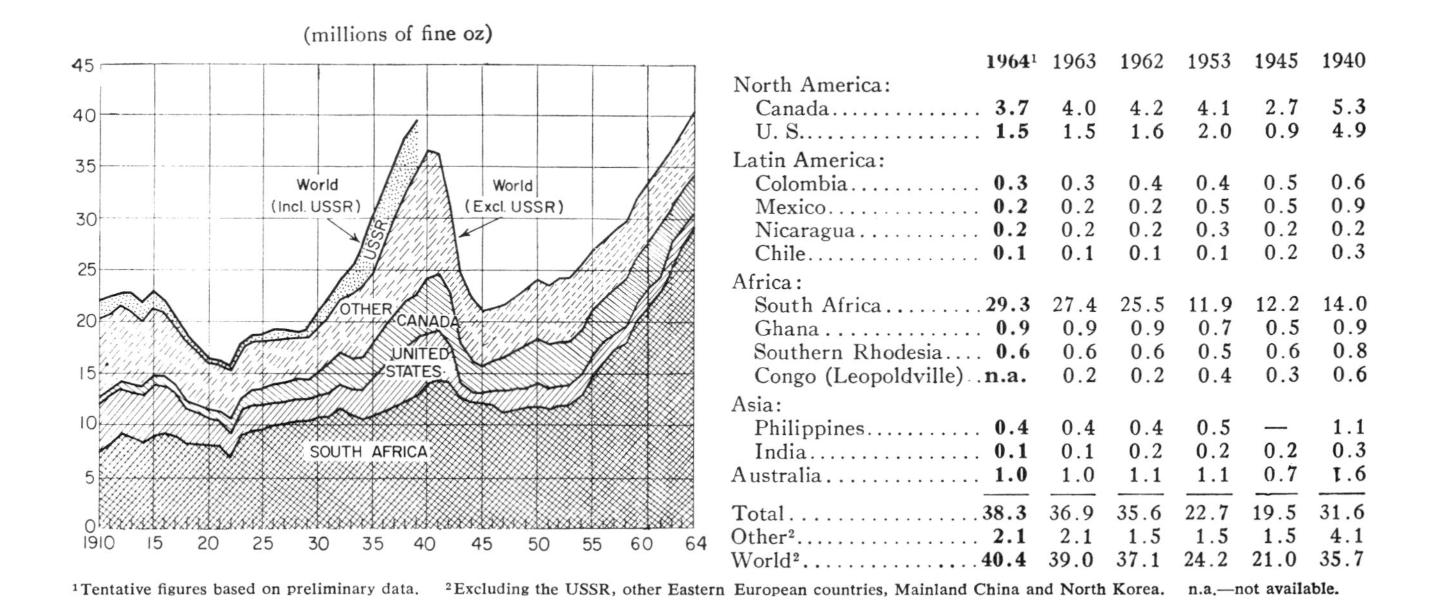

	1964[1]	1963	1962	1953	1945	1940
North America:						
Canada	**3.7**	4.0	4.2	4.1	2.7	5.3
U. S.	**1.5**	1.5	1.6	2.0	0.9	4.9
Latin America:						
Colombia	**0.3**	0.3	0.4	0.4	0.5	0.6
Mexico	**0.2**	0.2	0.2	0.5	0.5	0.9
Nicaragua	**0.2**	0.2	0.2	0.3	0.2	0.2
Chile	**0.1**	0.1	0.1	0.1	0.2	0.3
Africa:						
South Africa	**29.3**	27.4	25.5	11.9	12.2	14.0
Ghana	**0.9**	0.9	0.9	0.7	0.5	0.9
Southern Rhodesia	**0.6**	0.6	0.6	0.5	0.6	0.8
Congo (Leopoldville)	**n.a.**	0.2	0.2	0.4	0.3	0.6
Asia:						
Philippines	**0.4**	0.4	0.4	0.5	—	1.1
India	**0.1**	0.1	0.2	0.2	0.2	0.3
Australia	**1.0**	1.0	1.1	1.1	0.7	1.6
Total	**38.3**	36.9	35.6	22.7	19.5	31.6
Other[2]	**2.1**	2.1	1.5	1.5	1.5	4.1
World[2]	**40.4**	39.0	37.1	24.2	21.0	35.7

[1] Tentative figures based on preliminary data. [2] Excluding the USSR, other Eastern European countries, Mainland China and North Korea. n.a.—not available.

Figure 2. World gold production, 1910 - 1964.
(Engineering and Mining Journal, Feb. 1965.)

Thus, during the decade 1940-1950, the purchasing power of an ounce of gold shrank from $35 to $17. In 1965 an ounce of gold was worth only $14 in terms of purchasing power compared to 1940. Stated another way, if gold prices had been allowed to keep pace with average commodity price increases since 1940, in 1965 they would have reached $83.00 per ounce. By continuing to buy gold from the Treasury at $35 per ounce, U.S. consumers are getting a real bargain.

Looking at the problem another way, labor costs, which constitute about 60 percent of the cost of mining gold, have risen greatly since 1934, when gold was revalued at $35 per ounce. In 1934 underground miners received about $4 to $5 per shift. Today, with all fringe benefits, the wage is between $30 and $40 per shift. The cost of mining is many times what it was in 1934 and may be as much as 10 times the 1934 figure. Greater productivity per man, as a result of mechanization, offsets only part of this increase.

Domestic v. world production

United States mine output of gold dwindled from 4.9 million ounces from 5,393 lode mines and 4,176 placer mines in 1940 to 1.5 million ounces from 355 mines in 1964. In 1940 output from 30 of the largest producers accounted for only 50 percent of the total; comprising these 30 mines, none of which produced less than 19,000 ounces, were 18 lode gold mines, 5 placer gold mines, 6 copper mines, and 1 lead-zinc mine. Of the 1964 production, more than one third came from the Homestake mine at Lead, South Dakota and 93 percent of the total was contributed by this and 24 other mines. Of the latter, 9 were large-scale open-pit copper mines and 6 were copper-lead-zinc and lead-zinc mines in which gold was a relatively unimportant by-product. While United States output is dwindling, world production has been rising, due almost entirely to escalation of output from the Rand in South Africa. Trends in world output and the major sources are shown in figure 2.

The United States which, during the latter half of the 1800's, led the world in gold output and for many years thereafter ran second only to the Union of South Africa, now ranks a poor fourth in world production. On the other hand the United States is, and long has been, the world's leading industrial consumer of gold and this consumption is increasing rapidly. During the late 1950's consumption averaged about 1.7 million ounces per year. In 1962, industrial consumption was 3.6 million ounces and in 1964 it was 4.8 million ounces. In contrast to these figures, the 1955-1964 average output was only 1.65 million ounces. This growing supply-demand deficit is balanced by withdrawals from gold stocks of the United States Treasury.

OREGON'S GOLD AND SILVER INDUSTRY

Production

The U.S. Bureau of Mines' Minerals Yearbooks report the total production for Oregon from 1852 through 1964 as 5,795,987 ounces of gold (valued at $130,822,000) and 5,454,106 ounces of silver (valued at $5,030,000). Of the total ounces of gold, about 2.3 million came from lode deposits and 3.5 million from placers. The production figures given above are but crude approximations because records were poorly kept prior to 1880 when organized Government canvassing of mine output began*. Earlier statistics are based on records of shipping agencies, banks, and the Director of the Mint.

Most of Oregon's early gold output eventually reached the San Francisco Mint, but the source of much of it became obscured because it passed through several hands en route, or it was mixed with

---

* F. W. Libbey (1963) reports: "An organized canvass of mineral production in Western States by the Government began about 1880, although U.S. Mineral Commissioners J. Ross Browne (in the 1860's) and Rossiter W. Raymond (in the 1870's) reported on the mineral industry in Western States and included incomplete production statistics. These pioneer efforts grew into the reliable annual Mineral Resources volumes of the U.S. Geological Survey, the statistical duties of which were, in 1933, assigned to the U. S. Bureau of Mines. Since then mineral industry statistics have been assembled and published annually in the Bureau's comprehensive Minerals Yearbook. Figures for annual production of gold in Oregon, beginning in 1881, are believed to be reliable."

shipments from other states, or it was circulated as a medium of exchange. Some of the gold produced never reached the mint. As was common in most early-day gold camps, Chinese labor was often employed in working the placers; when the cream was skimmed off the better diggings, the white miners moved on leaving the semi-depleted ground to the Chinese, who hoarded much of what they produced and even carried some of it to their homeland.

Production figures for Oregon for the years 1852-1876 can only be estimated. One of the lowest figures comes from the Report of the Director of the Mint for 1882 (p. 44) which indicates that the gold and silver from Oregon deposited at the United States Mints and assay offices from the time of their organization to June 30, 1882, amounted to $16,816,275.39 in gold.

A higher value is obtained by subtracting the recorded 1877-1964 totals from the totals for the entire period 1852-1964, as given by the U.S. Bureau of Mines. This gives 2,095,838 ounces ($43,320-971) gold and 8,805 ounces ($11,358) silver for the 1852-1876 period, the dollar values being calculated at $20.67 per ounce for gold and $1.29 for silver. Even these figures are low as compared to those of Lindgren (1901, p. 569), who suggested that Oregon's output during the four years 1862-1865 alone might approach $50,000,000. Browne (1867, p. 9) states: "In 1865 the generally accepted estimate for Oregon was $19,000,000 though that was probably above the actual product."

Table 1 shows the amount of gold and silver carried to Portland and sold to banks or transmitted directly to the San Francisco Mint by Wells-Fargo between 1864 and 1870. However, much Oregon gold probably never reached Portland but was shipped to San Francisco by other routes. Some of the gold sold in Portland was carried in from mines in Idaho, Washington, Montana, and British Columbia. Taking these factors into account, Browne and Raymond estimated Oregon and Washington production for 1866-1875 to be as shown in table 2. No official statistics are available for 1876, but Lindgren (1901, p. 570) estimated output at roughly $1,100,000. Oregon production for 1877-1899 as reported by the Director of the Mint is shown in table 3.

Relative amounts of lode and placer gold and silver produced annually from 1900 through 1965 are given in table 4. These statistics were taken from the Mineral Resource volumes of the U.S. Geological Survey and from Mineral Yearbooks of the U.S. Bureau of Mines. Figure 3-A illustrates the trends in gold production in Oregon from 1877 through 1964 and compares the annual output of lode versus placer mines for the period 1897-1964. Figure 3-B shows the placer production by various methods of mining for 1897-1962. Of the total 1902-1965 placer output of 1,076,821 fine ounces, 542,850 fine ounces came from the bucketline dredges operating mainly between 1913 and 1954. Figure 4 compares graphically the production by bucketline and dragline dredges between 1933 and 1954.

Early gold and silver production by counties (Baker and Grant, 1880-1899; Jackson and Josephine, 1852-1901) is given in tables 5 and 6. Figure 5 illustrates graphically the production from these counties during 1902-1965. Of the 2,185,778 fine ounces of gold recorded for this period, about 94 percent came from four counties: 57 percent from Baker County; 16 percent from Grant County; 12 percent from Josephine County; and 9 percent from Jackson County. The remaining 6 percent came mainly from Malheur and Lane Counties. Output from other counties has been very small. During the 1902-1965 period, Baker County furnished 75 percent of the total lode gold output for the state.

Figure 6 compares the value of gold in Oregon with the total value of gold, silver, copper, lead, and zinc during the period 1905-1964. Production of gold or silver as a by-product of copper, lead, or zinc mining has been small in comparison to that of some other western states. Table 7 gives the annual output of gold, silver, copper, lead, and zinc from Baker, Grant, Josephine, Jackson, Lane, and Malheur Counties from 1902 through 1965.

Total mine output of copper from Oregon, according to United States Bureau of Mines statistics, has been about 12,500 tons. Lead and zinc output is much less.

Deposits in which copper is the most valuable metal occur in several areas, but only a few of them have been productive on an important scale. The more productive copper deposits are in the Homestead area of Baker County in the northeastern part of the state and the Waldo area of Josephine County in the southwestern part.

Table 1. Treasure shipments of gold and silver from Portland, Oregon, 1864-1870 (as recorded by R.W. Raymond, 1872)

Year	Gold and Silver
1864	$ 6,200,000[1]
1865	5,800,000[1]
1866	5,400,000[1]
1867	4,000,000[1]
1868	3,677,850[1,2]
1869	2,979,137[1,3]
1870	1,797,800[1,4]
	$ 29,854,787

1. Shipments from the Portland office of Wells-Fargo & Co.; mostly from Oregon mines but also includes bullion from Washington, Idaho, Montana, and British Columbia.
2. Includes $640,850 shipped by Ladd and Tilton, Portland bankers.
3. Includes $419,657 shipped by Ladd and Tilton, and $480 by private hands.
4. Includes an estimated $250,000 shipped by private hands.

Table 2. Production of gold and silver in Oregon and Washington from 1866 to 1875, inclusive. (Compiled from the official reports on the production of the precious metals by Browne and Raymond.)

Year	Gold and Silver
1866	$ 8,000,000[1]
1867	3,000,000
1868	4,000,000
1869	3,000,000
1870	3,000,000
1871	2,500,000
1872	2,000,000
1873	1,376,400[2]
1874	609,070[3]
1875	1,246,978[3]
	$ 28,732,448

1. Estimated by some as high as $20,000,000.
2. Estimate of total by Wells Fargo Express Company; Oregon only.
3. Oregon only.

Table 3. Production of gold and silver in Oregon from 1877 to 1899.

(Values compiled from reports of the Director of the Mint on the production of the precious metals. Ounces of gold calculated.)

Year	Gold: Fine ounces	Gold: Value at $20.67 per ounce	Silver: Coinage value $1.29+ per ounce	Total value
1877	48,379	$ 1,000,000	$ 100,000	$ 1,100,000
1878	48,379	1,000,000	100,000	1,100,000
1879	55,636	1,150,000	20,000	1,170,000
1880	52,734	1,090,000	15,000	1,105,000
1881	53,217	1,100,000	50,000	1,150,000
1882	40,155	830,000	35,000	865,000
1883	31,930	660,000	3,000	663,000
1884	31,930	660,000	20,000	680,000
1885	38,704	800,000	10,000	810,000
1886	47,896	990,000	5,000	995,000
1887	43,541	900,000	10,000	910,000
1888	39,913	825,000	15,000	840,000
1889	58,055	1,200,000	38,787	1,238,787
1890	52,588	1,087,000	129,199	1,216,199
1891	96,498	1,994,622	296,280	2,290,902
1892	72,171	1,491,781	64,080	1,555,861
1893	81,807	1,690,951	13,557	1,704,508
1894	102,243	2,113,356	10,315	2,123,671
1895	88,906	1,837,682	15,192	1,852,874
1896	62,456	1,290,964	71,811	1,362,775
1897	65,534	1,354,593	109,643	1,464,236
1898	58,862	1,216,669	165,916	1,382,585
1899	70,991	1,467,379	187,932	1,655,311
	1,342,525	$27,749,997	$1,485,712	$29,235,709

1. Census reports: Gold, $964,000; silver, $23,383; total, $987,383.

TABLE 4. GOLD AND SILVER PRODUCTION IN OREGON, 1900 - 1965. *

Year	Gold				Silver				No. of mines		Tons of ore & old tailings treated
	Lode (fine oz.)	Placer (fine oz.)	Total (fine oz.)	Total value	Lode (fine oz.)	Placer (fine oz.)	Total (fine oz.)	Total value	Lode	Placer	
1900[1]			81,980	$1,694,700			115,400	$ 71,548			
1901[1]			87,950	1,818,100			160,100	96,060			
1902[1]			87,881	1,816,700			93,300	49,449			
1903[1]			62,411	1,290,000			118,000	62,720			
1900[2]	68,319	15,268	83,587	1,727,894			132,042	81,866			
1901[2]	68,790	19,969	88,759	1,834,808			163,873	98,324			
1902[2]	77,086	11,798	88,884	1,837,392			109,463	58,015			
1903[2]	55,447	10,000	65,447	1,352,907			125,599	67,823			
1904[3]	51,426	16,895	68,321	1,412,186	131,981	96	132,077	75,284	84	211	121,189
1905	55,806.17	12,172.06	67,978.23	1,405,235	88,691	1,945	90,636	54,744			150,268
1906	48,633.32	17,490.46	66,123.78	1,366,900	75,992	3,354	79,346	53,162			138,274
1907	38,596.23	16,031.77	54,628.00	1,129,261	83,927	2,791	86,718	57,234			110,698
1908	28,661.37	13,186.68	41,848.05	865,076	37,751	5,851	43,602	23,109	66	173	50,684
1909	27,121.25	10,706.26	37,827.51	781,964	25,773	2,054	27,827	14,470	66	96	59,281
1910	24,601.73	8,268.50	32,870.23	679,488	34,739	1,239	35,978	19,428	64	116	82,132
1911	22,500.81	8,140.25	30,641.06	633,407	43,964	1,257	45,221	23,967	40	136	98,558
1912	28,103.21	9,147.52	37,250.73	770,041	55,140	1,941	57,081	35,105	54	156	90,945
1913	56,941.34	21,799.13	78,740.47	1,627,710	173,543	5,493	179,036	108,139	79	94	155,901
1914	50,462.09	26,524.83	76,986.92	1,591,461	137,040	5,512	142,552	78,831	28	77	124,331
1915	66,739.41	23,324.97	90,064.38	1,861,796	112,420	5,527	117,947	59,799	30	65	155,791
1916	49,809.90	42,208.01	92,017.91	1,902,179	222,093	9,249	231,342	152,223	33	75	159,071
1917	36,979.40	35,186.33	72,165.73	1,491,798	118,496	7,160	125,656	103,541	23	53	142,438
1918	37,355.95	24,102.79	61,458.74	1,270,465	103,221	4,102	107,323	107,323	42	78	126,681
1919	28,889.26	18,413.99	47,303.25	977,845	107,451	3,670	111,121	124,455	23	55	96,173
1920	27,398.29	21,822.79	49,221.08	1,017,490	77,890	4,853	82,743	90,190	20	47	82,156
1921	19,509.68	23,158.71	42,668.39	882,034	37,829	4,291	42,120	42,120	23	84	40,333
1922	8,970.24	16,744.36	25,714.60	531,568	148,801	3,011	151,812	151,812	44	126	27,522
1923	10,702.39	13,388.75	24,091.14	498,008	94,260	2,062	96,322	78,984	55	151	50,385
1924	10,945.32	15,750.04	26,695.36	551,842	35,824	2,279	38,103	25,529	43	71	44,842
1925	9,945.42	9,037.38	18,982.80	392,409	31,662	1,131	32,793	22,758	53	94	27,483
1926	7,304.68	5,938.42	13,243.10	273,759	28,890	843	29,733	18,553	49	100	21,727
1927	5,789.80	8,886.36	14,676.16	303,383	44,675	1,155	45,830	25,986	36	150	26,839
1928	5,100.79	5,830.40	10,931.19	225,968	30,145	779	30,924	18,091	41	90	12,676
1929	5,144.87	11,947.13	17,092.00	353,323	28,231	1,778	30,009	15,995	45	111	10,509
1930	4,028.80	10,372.54	14,401.34	297,702	7,421	1,579	9,000	3,465	47	143	8,994
1931	4,231.07	11,119.03	15,350.10	317,315	5,690	1,564	7,254	2,104	57	139	7,092
1932	3,659.31	16,201.90	19,861.21	410,568	6,371	2,245	8,616	2,430	99	169	5,195
1933	5,456.90	14,782.76	20,239.66	517,326	18,594	2,166	20,760	7,266	111	292	11,557
1934	11,471.68	22,239.91	33,711.59	1,178,220	42,983	3,577	46,560	30,099	95	332	62,145
1935	21,456.08	32,704.03	54,160.11	1,895,604	106,410	3,975	110,385	79,339	115	268	184,543

TABLE 4 (CONTINUED)

Year	Gold				Silver				No. of mines		Tons of ore & old tailings treated
	Lode (fine oz.)	Placer (fine oz.)	Total (fine oz.)	Total value	Lode (fine oz.)	Placer (fine oz.)	Total (fine oz.)	Total value	Lode	Placer	
1936	21,332	39,421	60,753	$ 2,126,335	79,411	5,650	85,061	$ 65,880	93	166	136,338
1937	18,443	34,219	52,662	1,843,170	55,540	5,024	60,564	46,846	104	150	77,230
1938	27,398	54,331	81,729	2,860,515	92,206	8,301	100,507	64,974	84	157	74,936
1939	32,593	60,779	93,372	3,268,020	94,794	10,594	105,388	71,536	116	201	69,025
1940	41,825	71,577	113,402	3,969,070	206,317	12,795	219,112	155,813	112	192	113,402
1941	36,135	60,430	96,565	3,379,775	264,953	11,205	276,158	196,379	91	153	96,565
1942	7,897	38,336	46,233	1,618,155	80,235	7,141	87,376	62,134	48	83	31,728
1943	889	208	1,097	38,395	10,495	28	10,523	7,483	16	16	2,680
1944	1,076	293	1,369	47,915	20,205	38	20,243	14,395	13	10	4,217
1945	481	3,986	4,467	156,345	9,672	789	10,461	7,439	9	10	1,378
1946	1,096	16,502	17,598	615,930	3,698	3,229	6,927	5,597	23	37	3,264
1947	1,331	17,648	18,979	664,265	26,852	3,527	30,379	27,493	20	49	3,277
1948	2,089	12,522	14,611	511,385	10,939	2,657	13,596	12,305	23	38	3,119
1949	1,761	14,465	16,226	567,910	9,488	2,707	12,195	11,037	28	29	6,215
1950	2,036	9,022	11,058	387,030	11,706	1,859	13,565	12,227	32	42	4,257
1951	763	7,164	7,927	277,445	4,455	1,763	6,218	5,628	14	21	1,495
1952	613	4,896	5,509	192,815	2,952	1,085	4,037	3,654	13	25	931
1953	1,199	7,289	8,488	297,080	10,333	1,926	12,259	11,095	8	21	1,215
1954	1,528	4,992	6,520	228,200	13,049	1,286	14,335	12,974	20	26	2,916
1955	1,505	203	1,708	59,780	8,795	20	8,815	7,978	19	21	3,835
1956	2,384	354	2,738	95,830	13,491	51	13,542	12,256	15	15	1,991
1957	3,202	179	3,381	118,335	15,897	27	15,924	14,412	25	17	2,594
1958	878	545	1,423	49,805	2,663	65	2,728	2,469	17	33	1,947
1959	236	450	686	24,010	180	62	242	219	10	27	356
1960	167	668	835	29,225	183	101	284	257	13	34	1,231
1961	349	705	1,054	36,890	1,918	104	2,022	1,830	15	27	782
1962	411	411	822	28,770	5,989	58	6,047	5,473	14	21	2,117
1963	1,281	528	1,809	63,315	58,172	62	58,234	52,702	16	14	4,599
1964	339	322	661	23,135	14,327	45	14,372	18,583	5	14	1,782
1965	265	234	499	17,465	8,771	30	8,801	11,353	2	2	500
	1,246,065.76	1,112,058.06	2,358,123.82	$66,382,147			4,147,319	$ 3,213,761			

Compiled from annual reports of the U.S. Geological Survey and U.S. Bureau of Mines.

(1) As estimated by the Director of the Mint.
(2) As estimated by officers and agents of the Mint.
(3) 1908-1952 figures.

* Silver production 1904 - 1965 inclusive totals 3,616,312 ounces: 3,435,584 ounces from lode mines and 180,758 ounces from placers. Lode and placer silver production figures for 1900-1903 are combined.

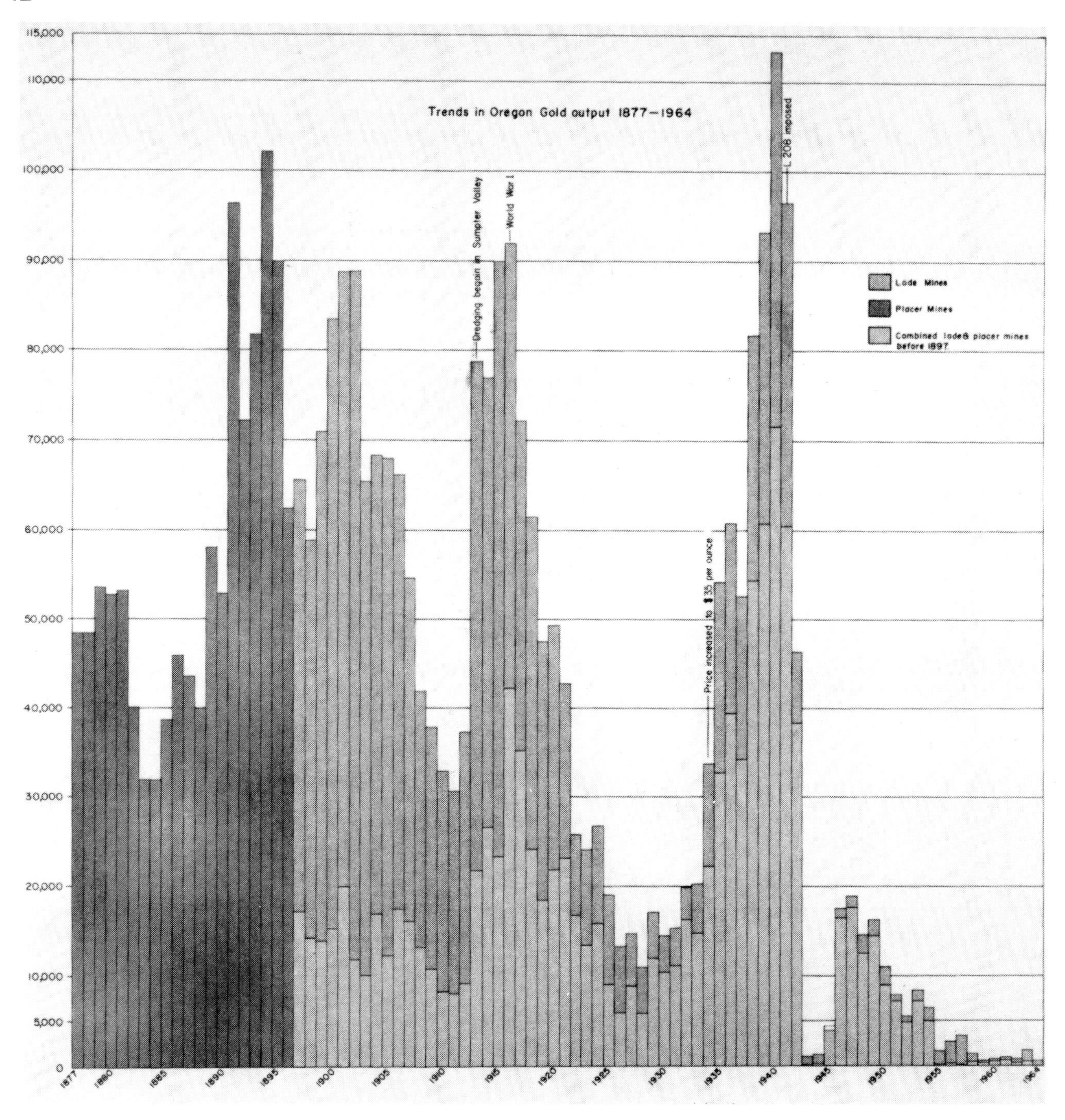

Figure 3-A. Trends in Oregon gold output, 1877 - 1964

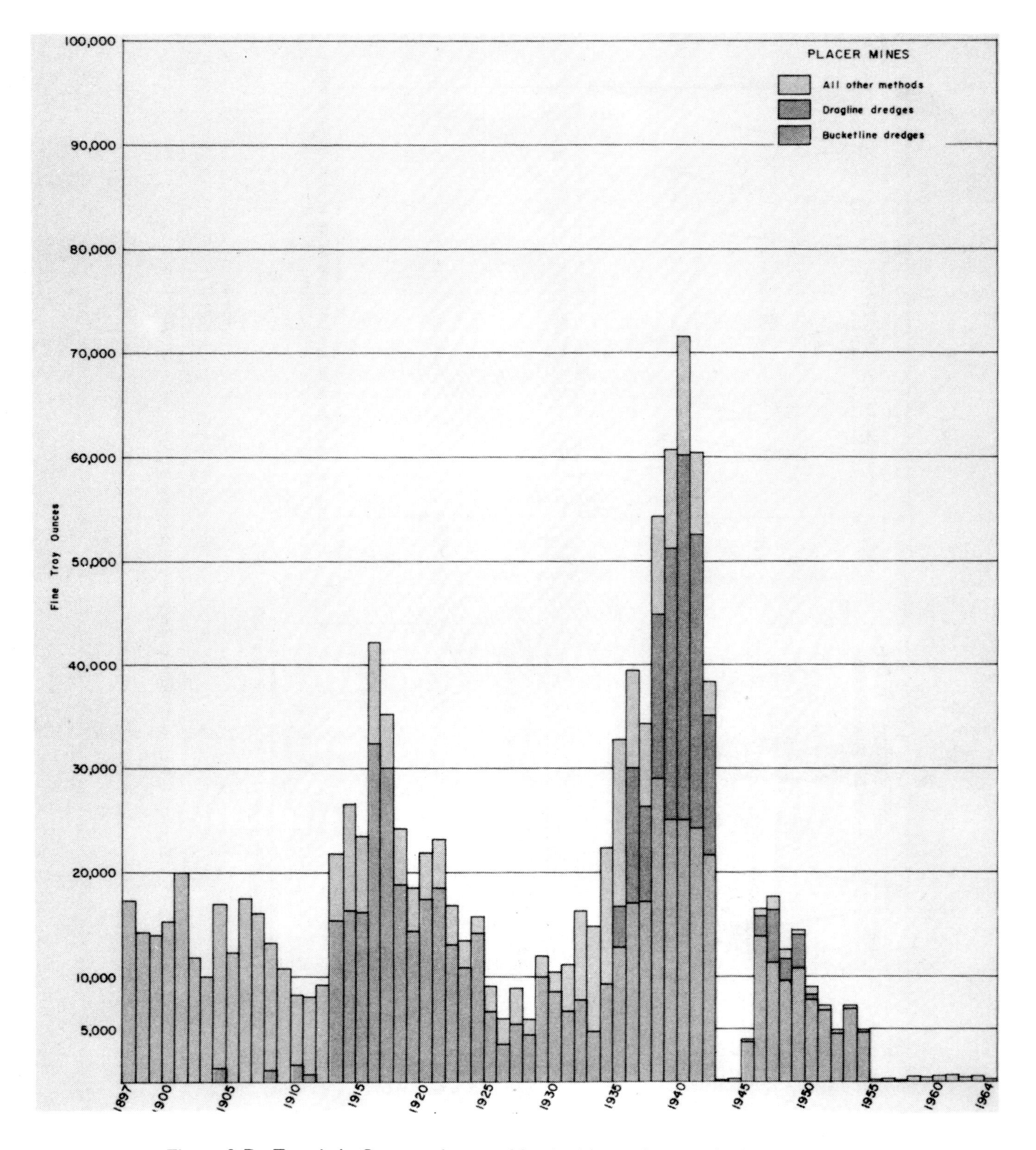

Figure 3-B. Trends in Oregon placer gold output by various methods, 1897 - 1964

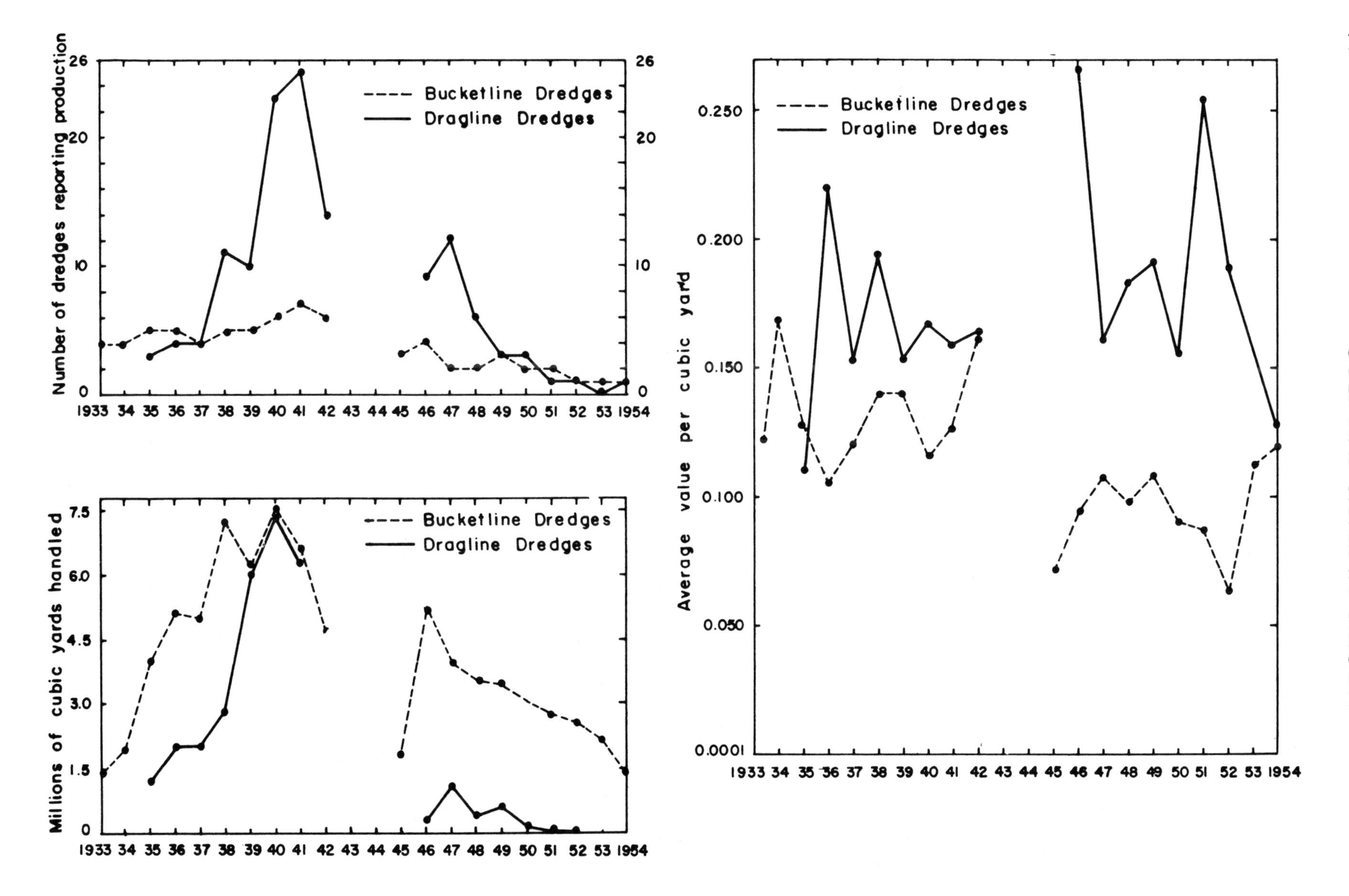

Figure 4. Production from bucketline and dragline dredges, 1933 - 1954.

Table 5. Production of gold and silver in Baker, Grant, and Union [1] Counties, Oregon, from 1880 to 1899. [Compiled from the reports of the Director of the Mint.] *

Year	Baker			Grant			Union			Total
	Gold	Silver	Total	Gold	Silver	Total	Gold	Silver	Total	
1880	$ 226,647	$ 400	$ 227,047	$ 85,400	$ 543	$ 85,943	$ 60,347	$	$ 60,347	$ 373,337
1881	250,000	10,000	260,000	280,000	20,000	300,000	40,000		40,000	600,000
1882	190,000	5,000	195,000	240,000	25,000	265,000	60,000	800	60,800	520,800
1883	160,000	2,500	162,500	200,000	15,000	215,000	45,000	300	45,300	422,800
1884	160,000	2,500	162,500	200,000	15,000	215,000	45,000	300	45,300	422,800
1885	348,044		348,044	194,600		194,600	7,322		7,322	549,966
1886	396,115	9,005	405,120	198,580		198,580	20,650		20,650	624,350
1887	173,558	5,153	178,711	163,896	11,797	175,693	15,000		15,000	369,404
1888	190,000	5,000	195,000	140,000	10,000	150,000	15,000		15,000	360,000
1889	463,604	7,500	471,104	73,989	9,550	83,539	574,989	1,028	576,017	1,130,660
1890	335,000	127,540	462,540	90,000	129	90,129	400,000		400,000	952,669
1891	873,058	217,833	1,090,891	124,487	4,297	128,784	625,956	3,500	629,456	1,849,131
1892	367,587	3,257	370,844	53,780	40	53,820	753,715	1,900	755,615	1,180,279
1893	728,947	10,454	739,401	198,650		198,650	420,237	3,046	423,283	1,361,334
1894	447,996	2,251	450,247	129,853		129,853	1,059,070	8,100	1,067,170	1,647,270
1895	942,483	7,963	950,446	101,853		101,853	144,800	3,000	147,800	1,200,099
1896 [2]	800,000	20,000	820,000	100,000		100,000	300,000		300,000	1,220,000
1897	796,741	50,088	846,829	86,969	4,880	91,841	211,699	36,071	247,770	1,186,440
1898	525,945	42,690	568,635	143,463	32,769	176,232	292,324	67,816	360,140	1,105,007
1899	582,348	55,418	637,766	217,054	86,626	303,680	114,212	19,466	133,678	1,075,124
Total	$8,958,073	$ 584,552	$9,542,625	$3,022,574	$ 235,631	$3,258,197	$5,205,321	$ 145,327	$5,350,648	$18,151,470

(1) Output from Union County came mostly from areas that have since been ceeded to Baker County.
(2) No product by counties given in mint reports. Figures for this year are only rough estimates.
* From Lindgren, 1901, p. 573, totals adjusted.

Table 6. Production of gold and silver in Jackson and Josephine Counties, Oregon from 1852 to 1901 *

Years	Jackson County		Josephine County		Total value
	Gold	Silver	Gold	Silver	
1852-1863	$1,500,000	$75,000	$1,000,000	$50,000	$2,625,000
1864-1869	1,200,000	60,000	1,000,000	50,000	2,310,000
1870-1879	800,000	40,000	800,000	40,000	1,680,000
1880	180,000	8,800	195,000	6,200	390,000
1881	240,000	10,000	195,000	5,000	450,000
1882	135,000	2,000	175,000	2,000	314,000
1883	100,000	1,000	110,000	1,000	212,000
1884	100,000	1,000	110,000	1,000	212,000
1885	57,000	1,000	70,000	1,000	129,000
1886	105,000	840	150,000	2,500	258,340
1887	238,000	900	140,000	5,400	384,300
1888	195,000	1,000	130,000	5,000	331,000
1889	59,000	5,250	38,000	1,000	103,250
1890	85,000	1,275	85,000	255	171,530
1891	140,000	2,000	47,000	1,000	190,000
1892	41,000	1,000	72,000	300	114,300
1893	107,000	500	113,000	400	220,900
1894	107,000	650	123,000	700	231,350
1895	142,800	2,200	282,000	500	427,500
1896	100,000	1,000	200,000	1,000	302,000
1897	28,000	1,500	132,000	4,700	166,200
1898	29,300	1,800	147,000	3,900	182,000
1899	103,400	3,100	236,000	7,000	349,500
1900	100,000	3,000	250,000	7,000	360,000
1901	100,000	3,000	300,000	9,000	413,000
Total	$6,441,000	$227,815	$6,100,000	$205,855	$12,527,170

* From Diller (1914, p. 29).

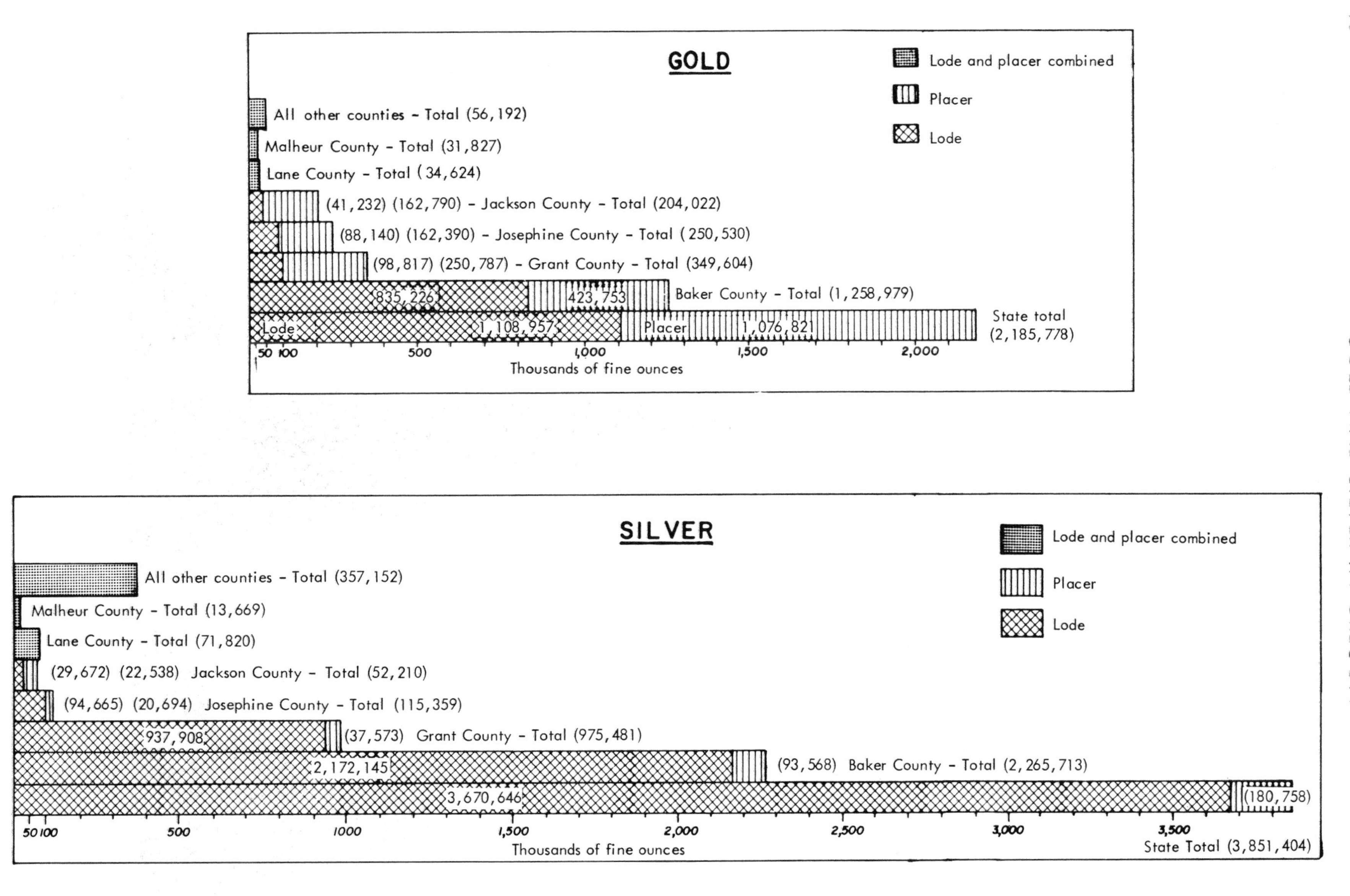

Figure 5. Ounces of gold and silver produced; total and by counties, 1902 - 1965.

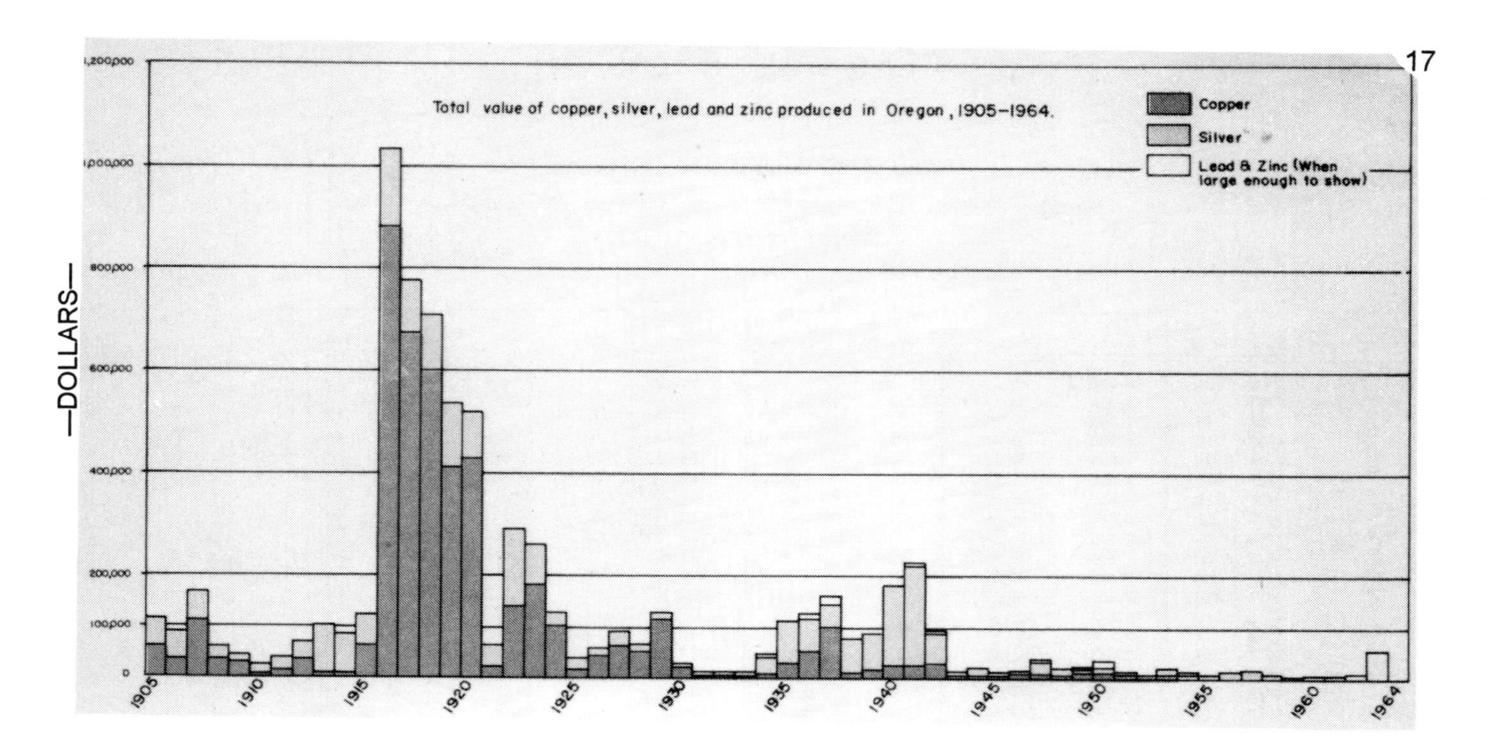
Total value of copper, silver, lead and zinc produced in Oregon, 1905-1964.
Copper
Silver
Lead & Zinc (When large enough to show)
—DOLLARS—
1,200,000
1,000,000
800,000
600,000
400,000
200,000
100,000
0
1905
1910
1915
1920
1925
1930
1935
1940
1945
1950
1955
1960
1964

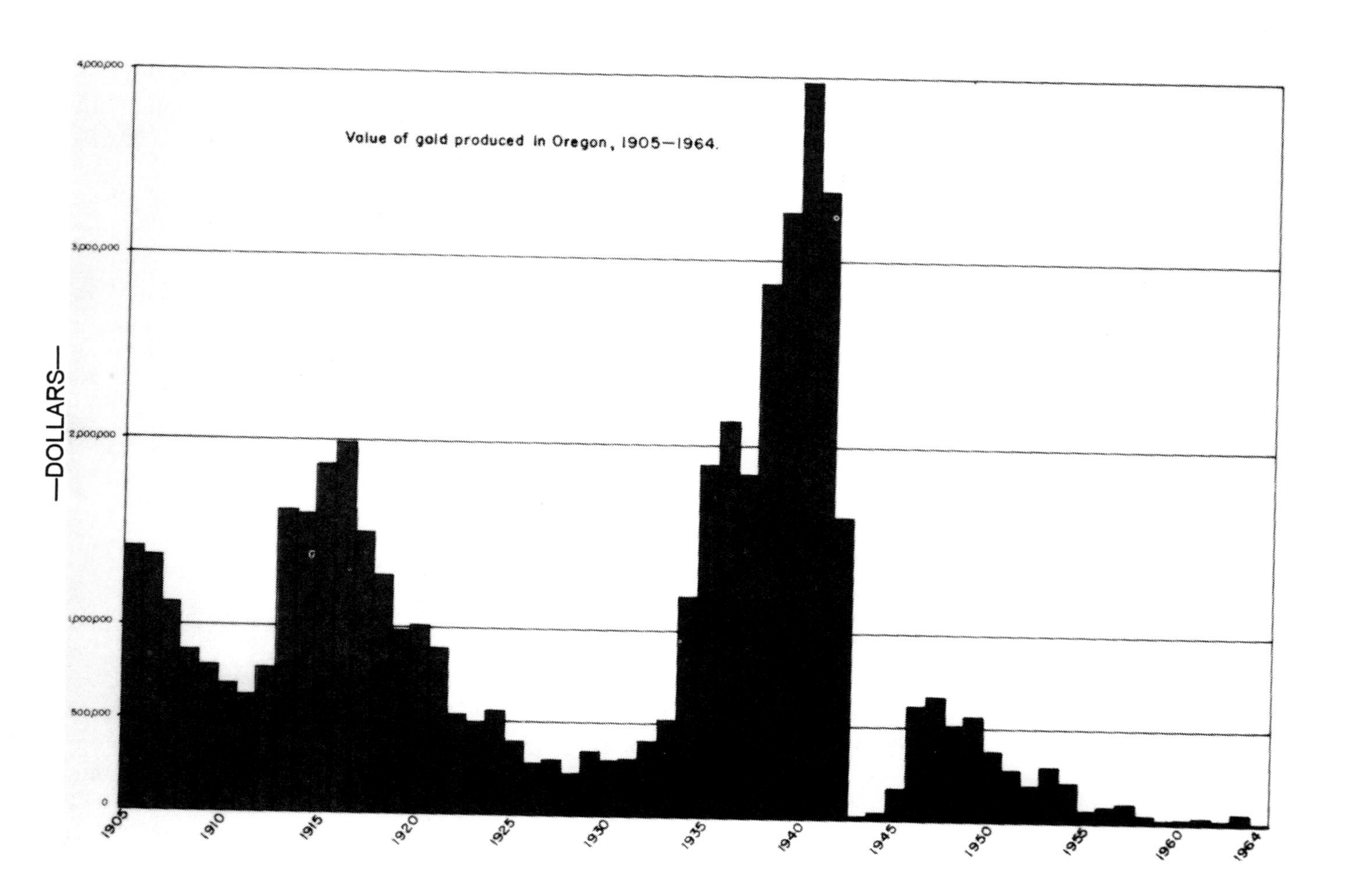
Value of gold produced in Oregon, 1905—1964.
—DOLLARS—
4,000,000
3,000,000
2,000,000
1,000,000
500,000
0
1905
1910
1915
1920
1925
1930
1935
1940
1945
1950
1955
1960
1964

GOLD AND SILVER IN OREGON

TABLE 7A. PRODUCTION OF GOLD, SILVER, COPPER, AND LEAD, BAKER COUNTY, OREGON, 1902 - 1965.

From U.S.B.M. statistics.

	PLACER MINES						LODE MINES			
		Gold		Silver		Total		Gold		
Year	No. of mines	Troy ounces	Value $	Troy ounces	Value $	placer value $	No. of mines	Short tons	Troy ounces	Value $
1902	24	5,814	$ 98,758	-	-	98,758	17	43,885	53,055	1,096,659
1903	36	4,036	83,440	20	$ 11	83,451	19	47,010	29,958	619,297
1904	23	2,508	51,855	7	4	51,859	31	69,973	35,748	738,973
1905	19	1,232	25,476	297	179	25,655	18	75,053	36,094	746,131
1906	29	1,638	33,859	376	252	34,111	22	58,889	32,014	661,794
1907	25	2,987	61,754	561	370	62,124	21	69,881	27,539	569,291
1908	29	1,857	38,397	409	217	38,614	21	38,413	22,713	469,532
1909	22	797	16,470	183	95	16,565	20	32,521	19,402	401,068
1910	13	545	11,276	117	63	11,339	25	56,263	18,853	389,726
1911	23	841	17,381	178	94	17,475	11	74,558	18,015	372,405
1912	26	1,830	37,834	574	353	38,187	23	70,631	21,585	446,207
1913	27	17,401	359,701	4,275	2,582	362,283	14	117,040	49,042	1,013,779
1914	19	17,874	369,482	4,450	2,461	371,943	12	97,722	45,689	944,485
1915	14	17,604	363,908	4,540	2,302	366,210	12	143,727	63,926	1,321,474
1916	14	32,028	662,075	8,076	5,314	667,389	12	144,129	47,337	978,551
1917	10	25,011	517,016	6,263	5,160	522,176	11	131,208	34,403	711,183
1918	22	15,073	311,589	3,125	3,125	314,714	20	118,960	33,769	698,067
1919	10	11,975	247,539	2,797	3,132	250,671	12	95,062	28,557	590,324
1920	8	16,432	339,684	3,849	4,196	343,880	9	80,243	26,640	550,696
1921	12	14,541	300,593	3,235	3,235	303,828	10	38,569	14,748	304,872
1922	36	9,195	190,089	2,148	2,148	192,237	18	24,350	6,414	132,599
1923	44	5,476	113,204	1,177	965	114,169	22	47,608	8,572	177,199
1924	17	8,817	182,255	1,511	1,012	183,267	12	37,239	6,869	141,997
1925	19	3,542	73,210	509	353	73,563	-	24,874	5,226	108,041
1926	24	1,341	27,714	320	200	27,914	17	12,950	2,315	47,850
1927	30	2,962	61,234	483	274	61,508	13	16,518	1,019	21,071
1928	17	206	4,255	39	23	4,278	9	3,111	346	7,153
1929	15	1,054	21,796	150	80	21,876	11	566	184	3,805
1930	12	395	8,171	79	30	8,201	10	4,907	592	12,231
1931	20	1,578	32,622	231	67	32,689	15	701	689	14,233
1932	32	2,098	43,377	384	108	43,485	15	767	711	14,691
1933	53	1,519	31,408	318	111	31,519	24	5,842	2,948	60,949
1934	65	4,570	159,730	797	515	160,245	25	37,376	6,895	240,968
1935	51	7,288	255,094	1,296	932	256,026	28	144,338	12,796	447,873
1936	28	10,110	353,850	2,239	1,734	355,584	26	88,494	12,700	444,500
1937	24	8,999	314,965	2,072	1,603	316,568	30	47,409	11,800	413,000
1938	29	14,301	500,535	3,247	2,099	502,634	23	51,339	22,478	786,730
1939	39	22,002	770,070	5,193	3,525	773,595	30	38,632	21,994	769,790
1940	36	21,206	742,210	4,848	3,447	745,657	26	55,941	26,446	925,610
1941	31	23,025	805,875	4,881	3,471	809,346	22	38,448	18,503	647,605
1942	11	16,455	575,925	3,811	2,710	578,635	11	6,623	2,731	95,585
1943	1	34	1,190	4	3	1,193	8	1,283	378	13,230
1944	-	27	945	4	3	948	5	2,488	468	16,380
1945	2	2,961	103,635	686	488	104,123	3	1,015	179	6,265
1946	6	9,540	333,900	2,133	1,723	335,623	5	2,579	176	6,160
1947	10	10,800	378,000	2,288	2,071	380,071	6	969	207	7,245
1948	5	8,019	280,665	1,717	1,554	282,219	8	1,187	345	12,075
1949	6	5,774	202,090	1,019	922	203,012	9	900	185	6,475
1950	9	5,981	209,335	1,190	1,077	210,412	10	1,029	397	13,895
1951	4	5,536	193,760	1,193	1,080	194,840	3	302	140	4,900
1952	6	4,630	162,050	1,058	957	163,007	2	420	61	2,135
1953	6	7,040	246,400	1,884	1,705	248,105	3	101	86	3,010
1954	5	4,790	167,650	1,262	1,142	168,792	7	710	374	13,090
1955	7	38	1,330	- -	- -	1,330	6	1,921	261	9,135
1956	2	108	3,780	21	19	3,799	8	545	201	7,035
1957	1	34	1,190	7	6	1,196	7	118	144	5,040
1958	3	W	W	W	W	W	W	W	W	W
1959	1	W	W	W	W	W	W	W	W	W
1960	8	W	W	W	W	W	W	W	W	W
1961	3	W	W	W	W	W	W	W	W	W
1962	4	63	2,205	2	2	2,207	2	60	101	3,535
1963	2	10	350	3	4	354	6	132	26	910
1964	2	W	W	W	W	W	W	W	W	W
1965	-	-	-	-	-	-	2	W	W	W
Total		423,753	$11,511,326	93,568	$ 71,339	$11,582,665			835,226	$19,274,879

(W) Indicates information withheld because of individual company data.

TABLE 7 A (CONTINUED)

LODE MINES

Silver		Copper		Lead		Total lode value $	Total value lode and placer	Year
Troy ounces	Value $	Pounds	Value $	Pounds	Value $			
37,387	18,721	-	-	-	-	1,115,380	1,214,138	1902
27,053	14,609	7,000	910	-	-	634,816	718,267	1903
76,467	44,351	1,800	232	8,121	290	783,846	835,705	1904
54,864	33,138	3,200	499	-	-	779,768	805,423	1905
53,332	35,732	185	36	-	-	697,562	731,673	1906
69,029	45,559	-	-	-	-	614,850	676,974	1907
17,086	9,056	-	-	-	-	478,588	517,202	1908
19,331	10,052	-	-	-	-	411,120	427,685	1909
29,718	16,048	13,861	1,760	-	-	407,534	418,873	1910
30,766	16,306	-	-	-	-	388,711	406,186	1911
38,566	23,718	951	157	-	-	470,082	508,269	1912
74,907	45,244	2,783	431	19,796	871	1,060,325	1,422,608	1913
53,817	29,761	692	92	11,179	436	974,774	1,346,717	1914
105,584	53,531	397,103	69,493	1,872	88	1,444,586	1,810,796	1915
205,407	135,158	2,631,727	647,405	-	-	1,761,114	2,428,503	1916
94,405	77,790	1,386,271	378,452	-	-	1,167,425	1,689,601	1917
87,260	87,260	1,655,319	408,864	7,663	544	1,194,735	1,509,449	1918
87,666	98,186	2,095,491	389,761	-	-	1,078,271	1,328,942	1919
72,580	79,112	2,355,276	433,371	-	-	1,063,179	1,407,059	1920
36,274	36,274	174,300	22,485	-	-	363,631	667,459	1921
140,434	140,434	1,041,196	140,561	75,694	4,163	417,757	609,994	1922
83,084	68,129	1,277,494	187,792	26,666	1,867	434,987	549,156	1923
25,070	16,797	761,141	99,709	-	-	258,503	441,770	1924
19,787	13,732	106,325	15,098	2,079	181	137,052	210,615	1925
8,741	5,454	234,578	32,841	-	-	86,145	114,059	1926
8,465	4,800	486,480	63,729	-	-	89,600	151,108	1927
1,373	803	71,841	10,345	-	-	18,301	22,579	1928
317	169	126	22	-	-	3,996	25,872	1929
781	301	-	-	-	-	12,532	20,733	1930
144	42	-	-	-	-	14,275	46,964	1931
628	177	-	-	-	-	14,868	58,353	1932
16,713	5,850	9,071	581	2,480	92	67,472	98,991	1933
37,040	23,945	26,150	2,092	517	19	267,024	427,269	1934
83,054	59,695	352,945	29,294	2,610	104	536,966	792,992	1935
49,136	38,056	442,000	38,824	14,000	644	522,024	877,608	1936
32,578	25,199	556,000	67,276	32,000	1,888	507,363	823,931	1937
83,750	54,141	74,000	7,252	42,000	1,932	850, 055	1,352,689	1938
84,043	57,047	92,000	9,568	18,000	846	837,251	1,610,846	1939
128,129	91,114	132,000	14,916	16,000	800	1,032,440	1,778,097	1940
79,165	56,295	88,000	10,384	14,000	798	715,082	1,524,428	1941
7,148	5,083	6,000	726	-	-	101,394	680,029	1942
672	478	6,300	819	-	-	14,527	15,720	1943
1,502	1,068	-	-	-	-	17,448	18,396	1944
927	659	-	-	-	-	6,924	111,047	1945
981	793	-	-	-	-	6,953	342,576	1946
1,590	1,439	-	-	-	-	8,684	388,755	1947
560	507	-	-	-	-	12,582	294,801	1948
1,861	1,684	400	79	-	-	8,238	211,250	1949
595	539	1,100	228	-	-	14,662	225,074	1950
524	474	400	97	-	-	5,471	200,311	1951
14	13	-	-	-	-	2,148	165,155	1952
12	11	-	-	-	-	3,021	251,126	1953
136	123	-	-	-	-	13,213	182,005	1954
171	155	700	261	-	-	9,551	10,881	1955
922	834	900	383	-	-	8,252	12,051	1956
317	287	-	-	-	-	5,327	6,523	1957
W	W	W	W	W	W	W	W	1958
W	W	W	W	W	W	W	W	1959
W	W	W	W	W	W	W	W	1960
W	W	W	W	W	W	W	2,861	1961
9	10	-	-	-	-	3,545	5,752	1962
7	9					919	1,273	1963
W	W	W	W	W	W	W	4,384	1964
W	W	-	-	-	-	W	W	1965
2,172,145	$1,586,196	16,318,806	$3,086,825	294,677	$15,563	$23,963,463	$35,546,128	

TABLE 7B. PRODUCTION OF GOLD, SILVER, COPPER, LEAD, AND ZINC IN GRANT COUNTY, OREGON, 1902-1965.

From U.S.B.M. statistics.

	PLACER MINES						LODE MINES			
		Gold		Silver		Total	Gold			
Year	No. of mines	Troy ounces	Value $	Troy ounces	Value $	Total placer value $	No. of mines	Short tons	Troy ounces	Value $
1902	10	1,989.70	33,570	-	-	33,570	8	7,403	2,690.99	40,871
1903	8	819.52	16,941	-	-	16,941	12	11,576	4,129.92	85,372
1904	16	1,081.42	22,355	-	-	22,355	20	11,099	2,898.87	59,925
1905	13	1,163.76	24,057	262	158	24,215	11	12,662	3,105.14	64,189
1906	13	1,562.42	32,298	415	278	32,576	12	2,346	1,102.66	22,794
1907	12	741.93	15,337	164	108	15,445	10	9,058	3,159.27	65,308
1908	16	1,106.87	22,881	623	330	23,211	11	5,251	3,180.90	65,755
1909	11	531.45	10,986	131	68	11,054	16	7,058	1,467.79	30,342
1910	13	871.77	18,021	148	80	18,101	14	3,215	879.46	18,180
1911	13	573.20	11,849	94	50	11,899	8	4,382	1,233.19	25,492
1912	17	561.73	11,612	138	85	11,697	5	1,400	486.70	10,061
1913	17	399.19	8,252	68	41	8,293	3	23,711	2,679.98	55,400
1914	3	998.65	20,644	118	65	20,709	3	13,147	1,898.91	39,254
1915	3	66.23	1,369	14	7	1,376	2	1,864	235.97	4,878
1916	7	2,903.27	60,016	315	207	60,223	3	874	173.04	3,577
1917	5	5,685.22	117,524	518	427	117,951	1	4,600	976.98	20,196
1918	8	4,687.39	96,897	408	408	97,305	5	378	245.94	5,084
1919	6	2,929.68	60,562	306	343	60,905	5	360	190.26	3,933
1920	7	2,816.88	58,230	716	781	59,011	5	1,766	542.91	11,223
1921	10	4,771.52	98,636	525	525	99,161	2	900	463.09	9,573
1922	7	4,737.95	97,942	507	507	98,449	9	1,978	1,080.03	22,326
1923	14	6,283.92	129,900	657	539	130,439	16	1,581	1,348.22	27,870
1924	9	5,680.70	117,430	584	391	117,821	8	6,137	2,600.19	53,751
1925	10	3,753.41	77,590	401	278	77,868	10	1,210	1,645.58	34,017
1926	11	3,044.88	62,943	324	202	63,145	6	717	2,495.62	51,589
1927	24	3,777.80	78,094	400	227	78,321	5	9,836	3,621.11	74,855
1928	13	2,166.12	44,674	224	131	44,805	5	7,170	1,798.95	37,188
1929	14	1,855.46	38,356	245	131	38,487	5	3,410	1,688.09	34,896
1930	8	4,425.96	91,493	685	264	91,757	4	991	645.78	13,349
1931	8	2,571.94	53,167	332	96	53,263	6	549	588.11	12,157
1932	17	3,334.97	68,940	435	123	69,063	14	429	607.64	12,561
1933	25	1,446.98	29,912	215	75	29,987	16	887	442.07	9,138
1934	33	5,450.02	190,478	897	580	191,058	18	1,828	740.80	25,891
1935	41	9,308.61	325,801	1,134	815	326,616	17	4,302	1,466.56	51,330
1936	24	17,803.00	623,105	2,082	1,613	624,718	17	15,345	2,434.00	85,190
1937	23	10,187.00	356,545	1,255	971	357,516	15	8,928	1,663.00	58,205
1938	28	22,501.00	787,535	2,844	1,839	789,374	19	2,926	1,146.00	40,110
1939	32	26,648.00	932,680	3,955	2,685	935,365	17	11,222	6,008.00	210,280
1940	24	24,541.00	858,935	4,421	3,144	862,079	17	23,889	8,112.00	283,920
1941	21	17,383.00	608,405	3,610	2,567	610,972	14	31,788	11,522.00	403,270
1942	15	11,541.00	403,935	2,011	1,430	405,365	7	4,721	1,502.00	52,570
1943	4	32.00	1,120	4	3	1,123	-	-	-	-
1944	1	17	595	-	-	595	1	23	110	3,850
1945	1	830	29,050	75	53	29,103	-	-	-	-
1946	6	5,735	200,725	911	736	201,461	1	31	259	9,065
1947	11	3,834	134,190	834	755	134,945	2	623	W	W
1948	7	3,553	124,355	803	727	125,082	3	1,652	W	W
1949	7	8,369	292,915	1,648	1,492	294,407	2	1,049	W	W
1950	6	2,227	77,945	566	512	78,457	6	687	631	22,085
1951	1	1,262	44,170	514	465	44,635	2	422	W	W
1952	4	20	700	3	3	703	2	386	438	15,330
1953	3	64	2,240	12	11	2,251	3	1,113	1,069	37,415
1954	2	W	W	W	W	W	3	2,108	1,032	36,120
1955	-	-	-	-	-	-	5	1,869	1,132	39,620
1956	2	10	350	1	1	351	2	1,417	2,148	75,180
1957	1	1	35	1	1	36	3	2,079	2,566	89,810
1958	4	27	945	2	2	947	2	1,650	564	19,740
1959	2	40	1,400	7	6	1,406	-	-	-	-
1960	2	W	W	-	-	W	3	741	W	W
1961	3	14	490	1	1	491	3	402	W	W
1962	1	3	105	2	2	107	4	1,098	176	6,160
1963	2	W	W	W	W	W	2	1,430	W	W
1964	1	4	140	1	1	141	2	657	W	W
1965	1	W	W	-	-	W	2	1,159	W	W
TOTAL		250,787	7,632,767	37,573	27,348	7,660,115		283,490	98,817	2,723,035

W = Withheld because of company confidential data.

TABLE 7 B (CONTINUED)

LODE MINES

Silver		Copper		Lead		Zinc		Total lode	Total value (Lode and placer)	
Troy ounces	Value $	Pounds	Value $	Pounds	Value $	Pounds	Value $	Value $		Year
76,837	40,076	-	-	-	-			80,947	114,517	1902
63,775	34,439	4,000	376	-	-			120,187	137,128	1903
50,718	29,417	4,710	590	500	20			89,952	112,307	1904
24,493	14,794	1,000	156	-	-			79,139	103,354	1905
3,304	2,214	42,886	8,277	-	-			33,285	65,861	1906
10,603	6,998	54,440	10,888	-	-			83,194	98,639	1907
17,364	9,203	583	67	-	-			75,025	98,236	1908
1,563	813	-	-	-	-			31,155	42,209	1909
2,174	1,174	-	-	-	-			19,354	37,455	1910
2,226	1,180	-	-	-	-			26,672	38,571	1911
446	274	-	-	-	-			10,335	22,032	1912
91,589	55,320	424	66	162	7			110,793	119,086	1913
68,438	37,846	406	54	154	6			77,160	97,869	1914
1,428	724	-	-	-	-			5,602	6,978	1915
1,284	845	1,204	296	-	-			4,718	64,941	1916
22,534	18,568	-	-	-	-			38,764	156,715	1917
12,618	12,618	-	-	-	-			17,702	115,007	1918
19,356	21,678	-	-	-	-			25,611	86,516	1919
5,270	5,745	-	-	-	-			16,968	75,979	1920
1,088	1,088	-	-	-	-			10,661	109,822	1921
6,164	6,164	4,724	638	-	-			29,128	127,577	1922
10,178	8,346	5,976	878	-	-			37,094	167,533	1923
10,274	6,884	947	124	513	41			60,800	178,621	1924
11,138	7,730	-	-	-	-			41,747	119,615	1925
15,415	9,619	2,566	359	11,273	902			62,469	125,614	1926
35,780	20,287	1,720	225	5,300	334			95,701	174,022	1927
25,306	14,804	1,422	205	13,246	768			52,965	97,770	1928
20,573	10,965	2,040	359	20,180	1,271			47,491	85,978	1929
4,835	1,861	502	65	6,053	303			15,578	107,335	1930
3,944	1,144	308	28	1,711	63			13,392	66,655	1931
2,120	598	111	7	1,211	36			13,202	82,265	1932
137	48	100	6	-	-			9,192	39,179	1933
1,206	780	2,838	227	689	25			26,923	217,981	1934
5,998	4,311	6,807	565	1,079	43			56,249	382,865	1935
17,732	13,733	-	-	-	-			98,923	723,641	1936
11,385	8,806	-	-	-	-			67,011	424,527	1937
7,090	4,583	2,000	196	2,000	92			44,981	834,355	1938
9,090	6,170	2,000	208	4,000	188			216,846	1,152,211	1939
53,197	37,829	10,000	1,130	28,000	1,400			324,279	1,186,358	1940
93,340	66,375	16,000	1,888	64,000	3,648			475,181	1,086,153	1941
2,925	2,080	4,000	484	-	-			55,134	460,499	1942
-	-	-	-	-	-			-	1,123	1943
900	640	-	-	-	-			4,490	5,085	1944
-	-	-	-	-	-			-	29,103	1945
821	663	-	-	-	-	-	-	9,728	211,189	1946
W	W	-	-	W	W	-	-	W	W	1947
W	W	W	W	W	W	-	-	W	W	1948
W	W	W	W	W	W	-	-	W	W	1949
4,391	3,974	700	146	4,300	581	1,000	142	26,928	105,385	1950
2,565	2,321	300	73	1,600	277	1,800	328	12,519	57,154	1951
2,570	2,326	400	97	1,900	306	2,000	332	18,391	19,094	1952
10,310	9,331	18,000	5,166	10,000	1,310	-	-	53,222	55,473	1953
12,879	11,656	10,000	2,950	10,000	1,370	-	-	52,096	52,764	1954
8,604	7,787	7,300	2,723	6,000	894	-	-	51,024	51,024	1955
12,554	11,362	3,500	1,487	10,000	1,570	-	-	89,599	89,950	1956
15,449	13,982	40,000	12,040	10,000	1,430	-	-	117,262	117,298	1957
2,473	2,238	20,000	5,260	2,000	234	-	-	27,472	28,419	1958
-	-	-	-	-	-	-	-	-	1,406	1959
W	W	W	W	-	-	-	-	W	W	1960
W	W	-	-	W	W	-	-	W	W	1961
2,478	2,689	600	185	1,600	147	-	-	9,181	9,288	1962
W	W	W	W	W	W	W	W	W	W	1963
W	W	W	W	W	W	W	W	W	W	1964
W	W	W	W	W	W	W	W	W	W	1965
937,908	637,933	328,614	75,715	260,071	23,659	17,300	2,461	3,462,803	11,122,918	

TABLE 7 C. PRODUCTION OF GOLD, SILVER, COPPER, AND LEAD IN JOSEPHINE COUNTY, OREGON, 1902 - 1965.

From U.S.B.M. statistics.

	PLACER MINES				LODE MINES								
	Gold		Silver		Gold		Silver		Copper		Lead		
Year	Troy ounces	Value $	Troy ounces	Value $	Troy ounces	Value $	Troy ounces	Value $	Pounds	Value $	Pounds	Value $	Total value $
1902	1,659.58	19,806	-	-	1,494.70	30,339	-	-	28,154	2,800	-	-	52,945
1903	8,837.78	182,758	46	22	7,768.35	160,506	23,894	12,903	14,000	1,563	-	-	357,752
1904	6,969.00	144,062	17	10	8,688.54	179,608	647	375	263,000	34,190	-	-	358,245
1905	5,872.23	121,390	793	479	9,179.25	189,752	3,494	2,110	842,615	131,447	-	-	445,178
1906	6,248.16	129,161	1,103	739	7,211.84	149,082	1,752	1,174	372,732	71,937	-	-	352,093
1907	5,866.10	121,263	894	590	1,906.22	39,405	806	532	499,664	99,933	-	-	261,723
1908	5,639.94	116,588	3,664	1,942	1,747.99	36,134	2,343	1,242	289,645	38,233	-	-	194,139
1909	5,158.37	106,633	663	345	2,049.36	42,364	981	510	235,000	30,550	-	-	180,402
1910	4,197.74	86,775	517	279	3,060.83	63,273	448	242	-	-	-	-	150,569
1911	4,187.20	86,557	441	234	619.49	12,806	9,995	5,297	82,808	10,351	-	-	115,245
1912	2,816.35	58,219	389	239	863.10	17,842	10,385	6,387	254,380	41,973	-	-	124,660
1913	1,992.86	41,196	273	165	365.13	7,548	187	113	32,558	5,046	6,928	305	54,373
1914	5,006.43	103,492	579	320	383.90	7,936	12,499	6,912	38,150	5,074	-	-	123,734
1915	3,400.71	70,299	578	293	715.13	14,783	3,470	1,759	53,663	9,391	52,617	2,473	98,998
1916	3,128.12	64,664	384	253	1,442.40	29,817	14,582	9,595	939,122	231,024	11,652	804	336,157
1917	2,381.12	49,222	208	171	1,195.97	24,723	67	55	1,087,216	296,810	-	-	370,981
1918	2,802.36	57,930	339	339	732.59	15,144	242	242	795,697	196,537	-	-	270,192
1919	2,037.41	42,117	305	341	50.07	1,035	151	169	118,501	22,041	-	-	65,703
1920	1,953.82	40,389	194	211	155.72	3,219	26	28	-	-	-	-	43,847
1921	2,301.78	47,582	301	301	3,789.51	78,336	169	169	-	-	-	-	126,388
1922	1,368.51	28,290	179	179	369.89	7,646	309	309	435	59	-	-	36,483
1923	609.91	12,608	88	72	197.04	4,073	18	15	-	-	-	-	16,768
1924	647.09	13,377	81	54	447.57	9,252	71	48	-	-	-	-	22,731
1925	851.42	17,600	89	62	2,041.71	42,206	245	170	-	-	-	-	60,038
1926	826.78	17,091	96	60	1,920.68	39,704	157	98	12,118	1,697	-	-	58,650
1927	1,460.99	30,201	184	104	1,070.34	22,126	328	186	-	-	-	-	52,617
1928	672.57	13,903	89	52	970.65	20,065	376	220	148,250	21,348	-	-	55,588
1929	1,271.13	26,277	139	74	1,211.99	25,054	1,018	543	464,393	81,733	-	-	133,681
1930	1,201.87	24,845	139	54	1,408.10	29,108	332	128	149,274	19,406	-	-	73,541
1931	2,241.10	46,328	289	84	1,298.35	26,839	336	97	-	-	-	-	73,348
1932	3,438.51	71,080	437	123	750.10	15,506	165	47	-	-	-	-	86,756
1933	3,602.78	74,476	462	162	793.01	16,393	188	66	100	6	1,048	39	91,142
1934	4,345.45	151,873	596	385	896.59	31,336	933	603	932	75	-	-	184,272
1935	9,283.59	324,926	880	633	732.04	25,621	141	101	167	14	50	2	351,297
1936	8,156.00	285,460	869	673	1,516.00	53,060	351	272	-	-	-	-	339,465
1937	9,759.00	341,565	971	751	1,591.00	55,685	565	437	4,000	484	-	-	398,922
1938	10,371.00	362,985	1,126	728	2,671.00	93,485	648	419	-	-	-	-	457,617
1939	3,499.00	122,465	379	257	3,098.00	108,430	424	288	-	-	-	-	231,440
1940	5,813.00	203,455	697	496	5,256.00	183,960	744	529	26,000	2,938	-	-	391,378

1941	3,435.00	120,225	389	277	4,215.00	147,525	502	357	16,000	1,888	-	-	270,273
1942	1,521.00	53,235	149	106	1,859.00	65,065	377	268	16,000	1,936	-	-	120,610
1943	81.00	2,835	10	7	-	-	-	-	-	-	-	-	2,842
1944	198.00	6,930	24	17	-	-	-	-	-	-	-	-	6,947
1945	119.00	4,165	14	10	14	490	5	4	-	-	-	-	4,669
1946	484.00	16,940	64	52	38	1,330	10	8	-	-	-	-	18,330
1947	369.00	12,915	48	43	16	560	4	4	-	-	-	-	13,522
1948	290.00	10,150	35	32	35	1,225	16	14	W	W	-	-	W
1949	205.00	7,175	22	20	36	1,260	30	27	5,800	1,142	-	-	9,624
1950	556.00	19,460	56	51	120	4,200	135	122	-	-	-	-	23,833
1951	244.00	8,540	31	28	28	980	35	32	8,700	2,105	-	-	11,685
1952	152.00	5,320	15	13	15	525	2	2	-	-	-	-	5,860
1953	74.00	2,590	8	7	-	-	-	-	-	-	-	-	2,597
1954	39.00	1,365	-	-	4	140	-	-	-	-	-	-	1,505
1955	67.00	2,345	2	2	2	70	-	-	-	-	-	-	2,417
1956	96.00	3,360	10	9	18	630	11	10	9,200	3,910	-	-	7,919
1957	71.00	2,485	9	8	7	245	20	18	6,000	1,806	-	-	4,562
1958	W	W	W	W	W	W	W	W	-	-	-	-	W
1959	W	W	W	W	W	W	W	W	-	-	-	-	W
1960	W	W	W	W	W	W	W	W	6,000	1,926	-	-	W
1961	W	W	W	W	W	W	-	-	-	-	-	-	W
1962	W	W	W	W	W	W	W	W	-	-	-	-	W
1963	W	W	W	W	W	W	-	-	-	-	-	-	W
1964	W	W	W	W	-	-	-	-	-	-	-	-	W
1965	W	W	W	W	W	W	W	W	-	-	-	-	W
Total	162,389.76	4,226,863	20,694	13,294	88,140.15	2,140,351	94,665	55,286	6,822,274	1,371,807	72,295	3,623	7,811,225

W = Withheld because of company confidential data.

TABLE 7 D. PRODUCTION OF GOLD, SILVER, COPPER, AND LEAD IN JACKSON COUNTY, OREGON, 1902 - 1965.

From U.S.B.M. statistics.

	Placer Mines				Lode Mines									
	Gold		Silver		Tons	Gold		Silver		Copper		Lead		Total
Year	Ounces	Value $	Ounces	Value $	ore	Ounces	Value $	Ounces	Value $	Pounds	Value $	Pounds	Value $	value
1902	6,565	111,631	-	-	2,200	2,327	48,103	3	2	-	-	-	-	$159,736
1903	4,961	102,558	-	-	1,719	988	20,421	20	11	-	-	-	-	122,990
1904	4,121	85,208	36	21	1,310	1,129	23,345	120	70	-	-	-	-	108,644
1905	2,148	44,403	356	215	5,919	2,210	45,697	541	327	-	-	-	-	90,642
1906	6,342	131,094	1,118	749	9,644	2,296	47,467	2,051	1,374	-	-	-	-	180,684
1907	5,181	107,099	944	623	14,923	3,502	72,391	2,324	1,534	-	-	-	-	181,647
1908	2,517	52,043	769	408	3,293	720	14,890	585	310	-	-	-	-	67,651
1909	3,416	70,612	927	482	5,329	1,432	29,606	1,011	526	-	-	400	17	101,243
1910	1,835	37,296	298	161	63	220	4,540	780	421	-	-	-	-	42,418
1911	1,245	25,740	293	155	8,980	1,006	20,787	464	246	-	-	-	-	46,928
1912	2,094	43,284	361	222	3,375	953	19,701	1,088	669	-	-	3,532	159	64,035
1913	1,065	22,031	182	110	922	721	14,905	975	589	-	-	433	19	37,654
1914	1,633	33,752	231	128	265	278	5,754	40	22	-	-	5,103	199	39,855
1915	1,216	25,150	221	112	315	244	5,051	546	277	-	-	2,489	117	30,707
1916	1,765	36,491	220	145	2,306	313	6,478	21	14	-	-	-	-	43,128
1917	839	17,353	87	72	20	17	351	2	2	-	-	-	-	17,778
1918	1,093	22,594	176	176	1,810	1,296	26,790	419	419	-	-	1,576	112	50,091
1919	1,292	26,715	239	267	86	41	848	163	183	823	153	-	-	28,166
1920	405	8,374	67	73	13	5	105	1	1	-	-	-	-	8,553
1921	1,152	23,813	143	143	451	411	8,496	43	43	-	-	-	-	32,495
1922	1,049	21,679	133	133	43	187	3,862	44	44	-	-	-	-	25,718
1923	673	13,906	94	77	718	322	6,656	302	248	-	-	155	11	20,898
1924	404	8,344	73	49	98	163	3,371	28	19	-	-	-	-	11,783
1925	401	8,286	58	40	779	815	16,859	350	243	-	-	3,533	307	25,735
1926	425	8,792	64	40	160	105	2,178	19	12	-	-	-	-	11,022
1927	543	11,229	71	40	60	39	800	8	5	-	-	-	-	12,074
1928	2,689	55,585	412	241	88	51	1,051	203	119	-	-	-	-	56,996
1929	7,596	157,025	1,210	645	48	76	1,572	19	10	-	-	-	-	159,252
1930	4,250	87,848	662	255	271	1,111	22,976	257	99	-	-	-	-	111,178
1931	3,687	76,224	516	150	356	897	18,535	386	112	-	-	1,786	66	95,087
1932	6,419	132,685	870	245	1,352	719	14,857	410	116	-	-	692	21	147,924
1933	6,941	143,481	1,000	350	2,581	702	14,518	952	333	-	-	926	34	158,716
1934	6,627	231,613	1,107	716	12,789	1,527	53,367	2,172	1,404	2,213	177	535	20	287,297
1935	5,766	201,819	586	421	25,358	3,746	131,103	5,091	3,659	4,918	408	1,920	77	337,487
1936	2,690	94,150	374	290	17,222	2,597	90,895	3,312	2,565	-	-	4,000	184	188,084
1937	4,799	167,965	668	517	2,301	861	30,135	533	412	-	-	2,000	118	199,147
1938	6,434	225,190	991	641	2,587	940	32,900	687	444	-	-	2,000	92	259,267
1939	7,651	267,785	946	642	2,743	1,119	39,165	859	583	-	-	2,000	94	308,269
1940	15,729	550,515	2,354	1,674	3,159	1,530	53,550	1,152	819	-	-	8,000	400	606,958
1941	14,789	517,615	2,014	1,432	2,227	909	31,815	900	640	-	-	6,000	342	551,844
1942	8,423	294,805	1,104	785	524	354	12,390	232	165	-	-	2,000	134	308,279
1943 *	W	W	W	W	W	W	W	W	W	-	-	W	W	W
1944	48	1,680	10	7	30	117	4,095	31	22	-	-	-	-	5,804
1945	76	2,660	14	10	1	59	2,065	10	7	-	-	-	-	4,742
1946	518	18,130	74	60	61	76	2,660	15	12	-	-	-	-	20,862
1947	1,781	62,335	227	205	59	12	420	1	1	-	-	-	-	62,961
1948	290	10,150	43	39	107	121	4,235	23	21	-	-	-	-	14,445
1949	96	3,360	16	14	55	76	2,660	14	13	-	-	-	-	6,047
1950	221	7,735	35	31	28	21	735	4	4	-	-	-	-	8,505
1951	76	2,660	12	11	106	16	560	6	5	-	-	-	-	3,236
1952	40	1,400	4	4	11	12	420	3	3	-	-	-	-	1,827
1953	108	3,780	22	20	*	W	W	W	W	-	-	-	-	W
1954	104	3,640	18	16	31	90	3,150	26	24	-	-	-	-	6,830
1955	56	1,960	9	8	43	110	3,850	20	18	-	-	-	-	5,836
1956	104	3,640	15	14	3	17	595	3	3	-	-	-	-	4,252
1957	60	2,100	8	7	200	473	16,555	108	98	-	-	-	-	18,760
1958	69	2,415	8	7	120	239	8,365	54	49	-	-	-	-	10,836
1959	51	1,785	9	8	255	150	5,250	35	32	-	-	-	-	7,075
1960	W	W	8	7	320	W	W	-	-	-	-	-	-	987
1961	63	W	W	W	113	W	W	55	51	W	W	-	-	11,981
1962	W	W	W	W	406	W	W	W	W	18,700	5,760	-	-	7,931
1963	53	1,855	4	5	141	27	945	13	17	-	-	W	W	2,833
1964	W	W	W	W	4	W	W	W	W	-	-	-	-	2,113
1965	-	-	-	-	21	34	1,190	8	10	-	-	-	-	1,200
Total	162,790	4,440,982	22,538	14,144	140,581	41,232	1,080,636	29,672	19,589	34,654	8,898	49,380	2,549	5,566,798

W = Withheld because of company confidential data.

* Less than one-half ton.

TABLE 7 E. PRODUCTION OF GOLD, SILVER, COPPER, LEAD, AND ZINC IN LANE COUNTY, OREGON, 1902 - 1965. *

Year	Gold		Silver		Copper		Lead		Zinc		Total value $
	Ounces	Value $	Ounces	Value $	Pounds	Value $	Pounds	Value $	Pounds	Value $	
1902	2,419	50,004	1,007	503	-	-	-	-	-	-	50,507
1903	1,105	31,115	1,019	535	-	-	-	-	-	-	31,650
1904	2,419	50,000	-	-	-	-	-	-	-	-	50,000
1905	4,395	90,844	2,814	1,319	-	-	-	-	-	-	92,163
1906	4,066	84,046	13,833	9,268	-	-	-	-	-	-	93,314
1907	1,846	38,158	727	480	-	-	-	-	-	-	38,638
1908	42	876	143	76	629	83	2,138	90	-	-	1,125
1909	686	14,172	1,335	694	-	-	-	-	-	-	14,866
1910	196	4,054	67	36	-	-	-	-	-	-	4,090
1911	1,459	30,154	466	247	-	-	-	-	-	-	30,401
1912	2,411	49,834	1,655	1,018	5,098	841	35,785	1,610	-	-	53,303
1913	343	7,098	1,820	1,099	7,565	1,173	59,204	2,605	-	-	11,975
1914	127	2,627	9	5	-	-	-	-	-	-	2,632
1915	470	9,718	57	29	406	71	5,979	281	-	-	10,099
1916	160	3,307	128	84	1,390	342	16,348	1,128	-	-	4,861
1917	96	1,984	328	270	-	-	-	-	-	-	2,254
1918	51	1,054	93	93	-	-	1,362	97	-	-	1,244
1919	-	-	-	-	-	-	-	-	-	-	-
1920	-	-	-	-	-	-	-	-	-	-	-
1921	48	991	232	232	-	-	-	-	-	-	1,223
1922	-	-	-	-	-	-	-	-	-	-	-
1923	W	W	W	W	W	W	-	-	-	-	W
1924	97	2,006	46	31	-	-	-	-	-	-	2,037
1925	63	1,312	29	20	-	-	-	-	-	-	1,332
1926	82	1,689	28	17	-	-	-	-	-	-	1,706
1927	-	-	-	-	-	-	-	-	-	-	-
1928	64	1,318	29	17	-	-	-	-	-	-	1,335
1929	15	303	9	5	-	-	-	-	-	-	308
1930	20	405	8	3	-	-	-	-	-	-	408
1931	271	5,594	120	35	-	-	-	-	-	-	5,629
1932	280	5,797	182	51	125	8	1,361	41	-	-	5,897
1933	W	W	W	W	W	W	-	-	-	-	W
1934	910	31,792	655	423	3,850	308	11,800	437	-	-	32,960
1935	1,857	65,007	4,440	3,191	28,541	2,369	48,106	1,924	-	-	72,491
1936	1,807	63,245	4,476	3,467	34,000	3,128	140,000	6,440	122,000	6,100	82,380
1937	2,292	80,220	4,853	3,754	28,000	3,388	184,000	10,856	48,000	3,120	101,338
1938	15	525	9	6	-	-	-	-	-	-	531
1939	259	9,065	337	229	2,000	208	6,000	282	-	-	9,784
1940	13	455	2,638	1,876	2,000	226	-	-	-	-	2,557
1941	23	805	1,620	1,152	2,000	236	-	-	-	-	2,193
1942	W	W	W	W	W	W	W	W	-	-	W
1943	-	-	-	-	-	-	-	-	-	-	-
1944	-	-	-	-	-	-	-	-	-	-	-
1945	W	W	W	W	W	W	W	W	W	W	W
1946	454	15,890	838	677	4,000	648	4,000	436	-	-	17,651
1947	361	12,635	1,802	1,631	20,000	4,200	6,000	864	2,000	242	19,572
1948	-	-	-	-	-	-	-	-	-	-	-
1949	717	25,095	3,853	3,487	33,000	6,501	17,400	2,749	12,000	1,488	39,320
1950	810	28,350	4,688	4,243	35,500	7,384	28,500	3,847	41,000	5,822	49,646
1951	272	9,520	1,320	1,195	12,600	3,049	2,400	415	4,200	764	14,943
1952	83	2,905	362	327	1,600	387	100	16	-	-	3,635
1953-57	-	-	-	-	-	-	-	-	-	-	-
1958	W	W	W	W	-	-	-	-	-	-	W
1959	-	-	-	-	-	-	-	-	-	-	-
1960	W	W	W	W	-	-	-	-	-	-	W
1961	W	W	W	W	-	-	W	W	W	W	W
1962	W	W	W	W	W	W	W	W	-	-	W
1963-65	-	-	-	-	-	-	-	-	-	-	-
Total	34,624	884,326	71,820	51,681	393,059	55,424	605,883	36,807	237,200	18,456	1,046,694

W = Withheld because of company confidential data.

* From U.S.B.M. statistics.

TABLE 7F. PRODUCTION OF GOLD, SILVER, AND COPPER IN MALHEUR COUNTY, OREGON, 1902 - 1965.*

From U.S.B.M. statistics

Year	Gold		Silver		Total
	Ounces	Value $	Ounces	Value $	value $
1902	1,715	31,255	-	-	31,255
1903	1,414	29,227	-	-	29,227
1904	1,116	23,069	17	10	23,079
1905	895	18,495	133	80	18,575
1906	687	14,207	121	81	14,288
1907	925	19,127	132	87	19,214
1908	711	14,699	96	51	14,750
1909	2,049	42,353	1,540	801	43,154
1910	1,513	31,274	1,574	850	32,124
1911	684	14,133	126	67	15,491
1912	2,101	43,430	407	250	43,680
1913	3,969	82,041	3,355	2,026	84,067
1914	2,421	50,052	2,279	1,260	51,312
1915	1,627	33,639	1,377	698	34,337
1916	691	14,276	71	47	14,323
1917	92	1,892	13	11	1,903
1918	46	947	6	6	953
1919	62	1,285	12	13	1,298
1920	10	200	2	2	202
1921	165	3,410	40	40	3,450
1922	461	9,536	96	96	9,632
1923	250	5,159	65	53	5,212
1924	398	8,224	112	75	8,299
1925	161	3,336	84	58	3,394
1926	401	8,288	343	214	8,502
1927	63	1,306	99	56	1,362
1928	52	1,076	8	5	1,081
1929	109	2,262	21	11	2,273
1930	-	-	-	-	-
1931	768	15,884	166	48	15,932
1932	384	7,930	71	20	7,950
1933	189	3,898	29	10	3,908
1934	268	9,353	174	112	9,465
1935	117	4,105	18	13	4,118
1936	61	2,135	11	9	2,144
1937	67	2,345	12	9	2,354
1938	146	5,110	27	17	5,127
1939	291	10,185	38	26	10,211
1940	1,658	58,030	311	221	58,251
1941	1,130	39,550	177	126	39,676
1942	247	8,645	38	27	8,672
1943	-	-	-	-	-
1944	-	-	-	-	-
1945	-	-	-	-	-
1946	W	W	6	W	W
1947	W	W	W	W	W
1948	357	12,495	59	53	12,548
1949	5	175	1	1	176
1950	8	280	2	2	282
1951	8	280	2	2	282
1952	-	-	-	-	-
1953	W	W	-	-	W
1954-55	-	-	-	-	-
1956	9	315	2	2	317
1957	10	350	2	2	352
1958	W	W	W	W	W
1959	W	W	W	W	W
1960	W	W	-	-	W
1961	W	W	W	W	W
1962	W	W	W	W	W
1963	W	W	W	W	W
1964	-	-	-	-	-
1965	W	W	W	W	W
Total	31,827	735,323	13,669	8,025	744,639

* Only copper production was in 1911 - 10,328 pounds valued at $1,291.
W = Withheld because of company confidential data.

History

Mining of placer gold deposits began in southwestern Oregon in 1850 and in northeastern Oregon in 1862. These principal gold-mining regions of Oregon owe their discovery to the wave of prospectors that invaded California in 1849 and from there spread through all the mountain areas of the West, following the constantly shifting centers of excitement. Many of the virgin placers were very rich and the first few years after discovery mark the high point of Oregon gold production.

The thoroughness of the early search for gold in Oregon is attested to by the fact that by 1865 placer deposits were being worked in nearly all the districts of prominence known today. No doubt many of the lode deposits were discovered during these early days, but the comparative difficulty of exploitation made the miners reluctant to abandon the easy pickings provided by the placers, and so the development of lode mining was much slower. Despite this, the initial development of most of the important lode mines in the state dates back at least to the 1880's.

Because of the great influx of miners, the better placer diggings were depleted within a few years, and a marked decline in output began with production dropping from possibly as much as $19,000,000 in 1865 to less than a million annually during 1882-1888. Lode mining then began to develop rapidly, pushing output to above $2 million in 1891 and 1894. From then until 1921 lode mines were the chief source of Oregon gold. Combined lode and placer output for 1891-1907 averaged $1.6 million annually. In 1904 production was made by 84 lode mines and 211 placers. Between 1908 and 1912 inclusive, the combined annual value of gold and silver for the state again fell below the million-dollar mark, mainly as a result of a decline in lode-mine production.

Placer mining in Oregon was done largely with hand-operated equipment until early in the present century, when dredges came into use. The first floating bucketline dredge known to have been employed in the state was operating on Burnt River near Durkee in northeastern Oregon at the time of Lindgren's visit in 1900. In southwestern Oregon a steam-powered dredge began operations on Foots Creek in 1903. It was converted to electric power in 1905. In 1913 the first of several large-scale dredges began work in Sumpter Valley, the largest dredge field in the state. Placer output for the state that year more than doubled the 1912 production, the increase being due almost entirely to the introduction of the dredge. Placer output nearly doubled again in 1916, when dredging began at John Day. The apparent success and scale of these operations stimulated use of mechanized equipment for placer mining elsewhere and by 1921 placer-mine output again began to exceed that of lode mines and continued to do so until 1954.

The general prosperity of the 1920's provided an economic climate unfavorable for gold mining and Oregon production declined sharply. The great depression which began in 1929 brought labor and materials costs back into line with gold prices; then when gold was revalued at $35 per ounce in 1934 the industry flourished. Output for the seven years, 1935-1941, averaged about $3,000,000 per year. The 1940 gold and silver production was valued at $4,124,883, the highest figure since the heyday of placer mining in the 1860's.

In 1940 output was made by 112 lode mines and 192 placer mines. Of the latter, 29 were operated by floating dredges and 29 used non-floating mechanized washing plants. Sixty-two percent of the total output for 1940 was produced by 10 different properties. In the order of their productive rank these were: Cornucopia Gold Mines, Inc. (lode); Sumpter Valley Dredging Co. (connected bucket dredge); Northwest Development Co. (dragline dredge); Porter & Co. (connected bucket dredge); Cougar-Independence Lessees (lode); Murphy-Murray Dredging Co. (connected bucket dredge); Ferris Mining Co. (dragline dredge); Lewis Investment Co. (lode); Timms Gold Dredging Co. (connected bucket dredge); and the B-H Co. (dragline dredge).

With the curtailment of gold mining brought about by War Production Board Order L-208 in 1942, Oregon gold production virtually ceased. Output in 1943 totaled 1,097 ounces gold and 10,523 ounces silver having a combined value of $55,400.

Oregon gold production reached its post-war high in 1947 of 18,979 ounces, only 17 percent of the 1940 output. Ninety-three percent of this came from placer mines. Since 1947, production has followed a generally diminishing trend and in 1965 totaled a mere 499 ounces.

Total gold output for the 21 years since the end of World War II (1945-1965) is only 126,999 ounces, 78 percent of which was produced from placer-mine operations and was mainly from dredges in northeastern Oregon during 1945-1954. The largest part of this came from the Sumpter Valley dredge, which resumed operations in 1945 and continued almost without interruption through September 1954. Next in rank was the Porter & Co. dredge working in Clear, Olive, and Crane Creeks west of Granite from 1946 through early 1951. Since the shut-down of the Sumpter Valley dredge in 1954, output from placer mines has averaged little more than 400 ounces per year from small, periodic operations.

Very few lode mines have been operated since World War II and only one has any consistent record of production. Of the 23,800 ounces of gold produced from lode mines during 1945-1965, more than 60 percent came from the Buffalo mine in the Granite district of Grant County. Output from this mine has been small since 1958.

In 1940, lode mines in Oregon produced 41,825 ounces of gold. The greatest output of lode gold for a single year since World War II was 3,202 ounces in 1957, of which 91 percent was from the Buffalo. Lode mines produced 264,953 ounces of silver in 1941. The largest output since was 58,172 ounces in 1963. Most of the latter came from the then newly reopened Oregon King mine near Ashwood in Jefferson County. Activity at this mine dwindled in 1964 and ceased the following year.

Future

Oregon's gold-mining industry has been nearly defunct for several years, mainly because the price of gold has remained fixed at a point set in 1934 while operating costs have soared. Government legislation to raise the price of gold or to subsidize gold mining could revitalize this very important basic industry. Such legislation may eventually become inevitable, because of our continual loss of monetary gold in international trade and our need for gold to back the ever-increasing amount of currency required to serve our expanding population.

It seems probable that the larger placer deposits of the state, particularly dredgeable areas, are for the most part worked out or so seriously depleted that under any price structure that can reasonably be anticipated for the future, placer mining cannot be expected to resume its pre-war importance. On the other hand, there are many lode deposits in the state that have not been fully investigated. Few productive deposits have been developed to depths of 1,000 feet below surface croppings and the workings of many mines open only small portions of potentially ore-bearing zones. Of particular significance are the facts that there are a great number of old mines and little-developed prospects in the northeastern and southwestern parts of the state and that most of these deposits are grouped together in zones of interrelated fracturing and widespread mineralization. Explorations for gold within these zones have, in the past, been largely confined to the investigation of narrow, well-defined veins found initially by the prospector with a gold pan. The recent improvements in geochemical and geophysical equipment for prospecting make possible the investigation of intervening areas for buried deposits or for zones of low-grade mineralization that would be amenable to large-scale open-pit mining.

Unfortunately, the important workings of nearly all of Oregon's lode gold mines are now caved or inaccessible. Consequently, the search for new ore will require costly rehabilitation or driving of new accessways into deeper or lateral portions of previously worked ore bodies. At some mines important information can be acquired by drilling from the surface or from accessible undergound stations, but this is at best only a preliminary step in the development of ore bodies and must, where warranted, be followed by underground work.

Whatever the potential for future developments might be, the fact remains that until something occurs to narrow the gap greatly between the fixed price of gold on the one hand and the increasing cost of labor and materials on the other, or until large, low-grade deposits amenable to open-pit mining are found, Oregon gold output must be expected to remain very small.

OCCURRENCE OF GOLD AND SILVER IN OREGON

Geology and distribution

Gold and silver deposits occur in many parts of Oregon (figure 7), but most of the production has come from mines in the Blue Mountains in the northeastern part of the state and from the Klamath Mountains in the southwestern part (figure 8). In these widely separated areas folded, faulted, and

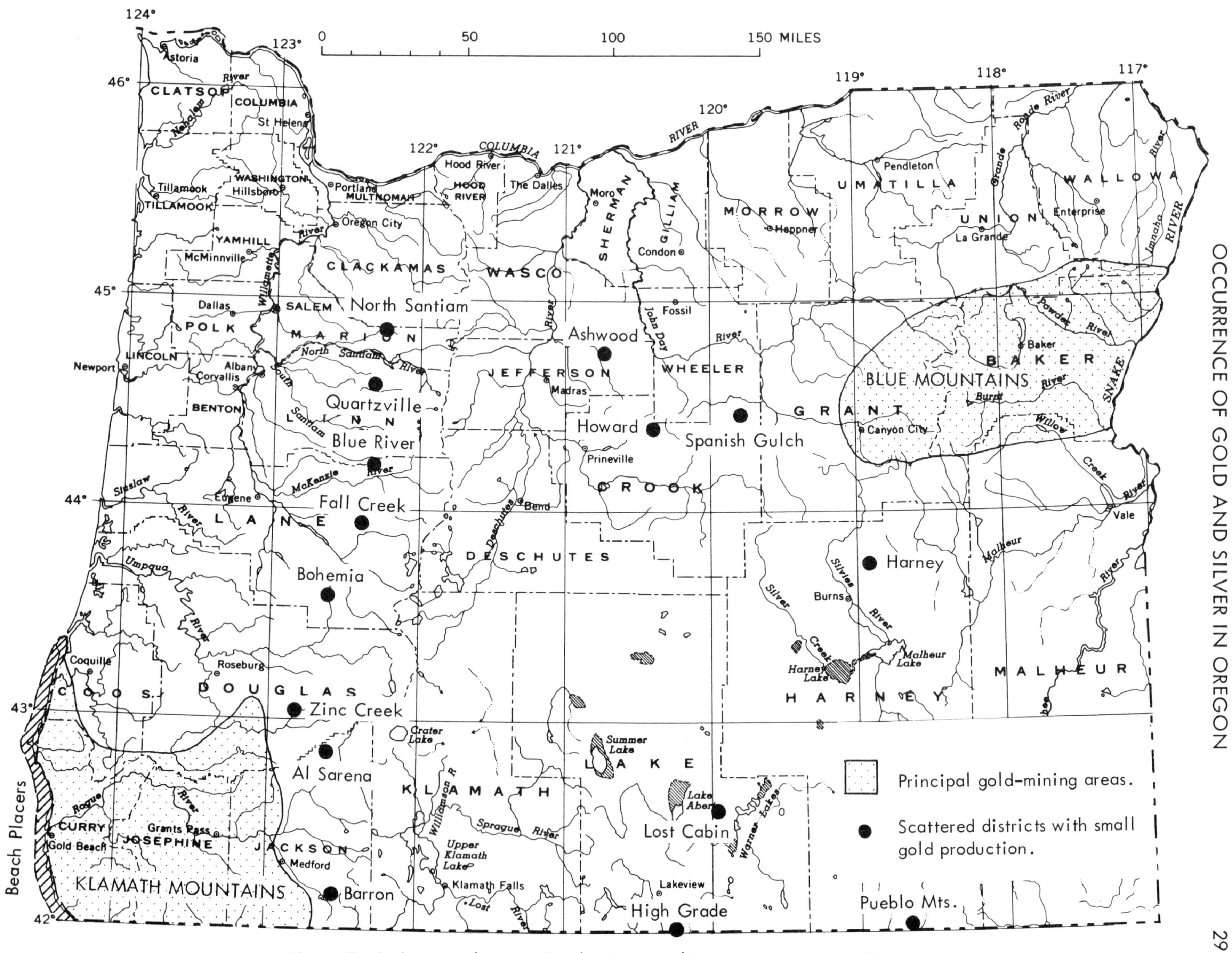

Figure 7. Index map showing distribution of gold-producing areas in Oregon.

metamorphosed sedimentary, volcanic, and intrusive rocks of pre-Late Jurassic age have been invaded by Late Jurassic-Early Cretaceous granitic rocks. Lode deposits are clustered in certain areas along the edges of the granitic intrusives and occur both in the intrusive bodies and in the older invaded rocks. Placer gravels in and near the lode-mining areas have yielded more than half of the gold produced in Oregon.

A few gold prospects occur in pre-Tertiary rocks in the Pueblo Mountains in southern Harney County and in the Spanish Gulch area of Wheeler County (figures 7 and 8).

Other gold deposits shown on figure 7 are contained in Cenozoic volcanic and sedimentary rocks, and most appear to be related to small intrusive bodies of Tertiary age. Deposits in eastern Oregon include those of the Ashwood district in Jefferson County, the Harney district in Harney County, the High Grade and Lost Cabin districts in Lake County, and the Howard district in Crook County. Deposits in western Oregon occur in volcanic rocks of the Western Cascades. Mineralized areas in this province are, from north to south, the North Santiam, Quartzville, Blue River, Fall Creek and Bohemia districts in Marion, Linn, Lane, and Douglas Counties, and scattered occurrences in Douglas and Jackson Counties.

Mineralogy

Gold occurs chiefly as the native metal. In ores it is often in particles too small to be seen with the unaided eye. Silver is always alloyed with the gold and many primary deposits also contain silver sulfides, the most common of which are freibergite, pyrargyrite, stephanite, argentite, and argentiferous galena. Tellurides are rare. Petzite and hessite occur in the ores of the Cornucopia district and at a few other places in northeastern Oregon. Petzite is also one of the ore minerals of the Bunker Hill mine in

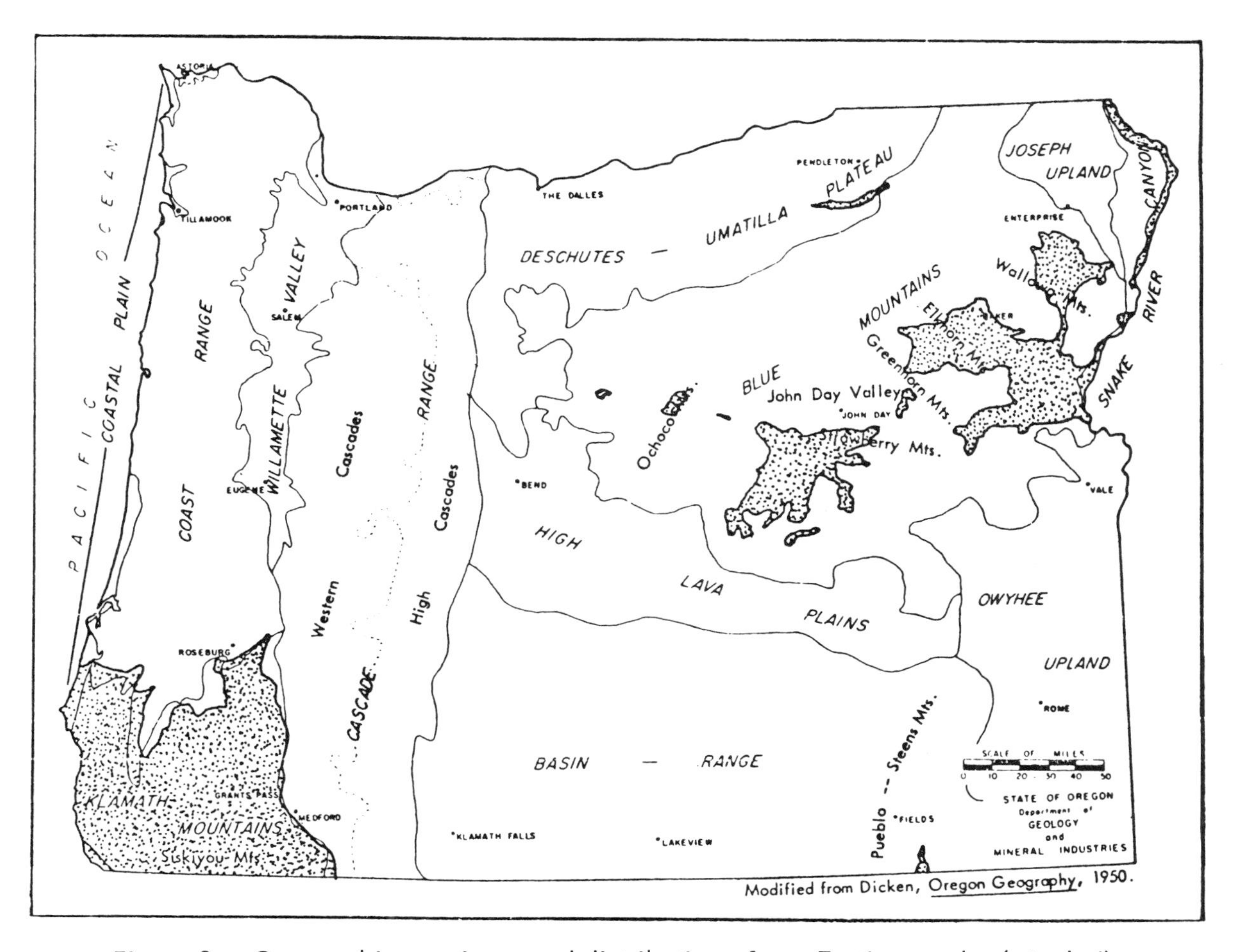

Figure 8. Geomorphic provinces and distribution of pre-Tertiary rocks (stippled).

the Galice district of southwestern Oregon.

The chief gangue mineral is quartz, which is nearly always accompanied by a little calcite. More rarely calcite predominates. Except in near-surface oxidized portions of gold veins, sulfide minerals are usually present. Pyrite generally predominates, although arsenopyrite is common and is the more abundant sulfide mineral in some deposits. The sulfides of copper, lead, and zinc are contained in many gold-silver ores and antimony sulfides are occasionally present. Cobalt minerals occur in several of the gold deposits in the Quartzburg district in Grant County.

In its ores, gold may occur as discrete particles or it may occur in a finely divided state intermixed with sulfide minerals. If the gold particles can be separated from associated gangue by simple grinding and amalgamation, it is said to be "free milling." When intimately mixed with sulfides which must be roasted or smelted before separation of the gold is accomplished, the ore is generally referred to as "sulfide" or "base" ore.

The ratio of gold to silver varies widely from one deposit to the other, but few deposits contain one of these metals to the total exclusion of the other. Silver often exceeds gold in quantity, but rarely in sufficient amount to supersede it in value because of the wide difference in the price of the two metals. Notable exceptions include the Oregon King mine in Jefferson County, where the ratio of silver to gold was as much as 100 to 1 in some portions of the ore body, and the Bay Horse silver mine in Baker County, where the ores contain almost no gold. The purest bullion from Oregon mines of significant size came from the Virtue mine in the northeastern part of the state, where the gold averaged more than 920 fine. (Fineness defines the proportion of gold or silver in bullion or coin expressed in parts per thousand.) Bullion shipments from the Columbia mine, several miles west of the Virtue, averaged 518 fine gold. Placer gold usually contains 10 to 20 percent silver.

Nature and origin of gold deposits

Most of Oregon's gold and silver output has been from narrow vein (lode) deposits valued mainly for these two metals and from placers derived from them by erosion. Deposits amenable to large-scale, open-pit mining, such as recently developed in Nevada, have not yet been found.

Gold-bearing veins are rarely uniform in value and many are too small or contain insufficient values to sustain a profitable mining operation. Portions of veins that contain valuable minerals in minable quantity are termed "ore shoots" or "ore bodies." Although some ore bodies have been mined continuously along the strike for many hundreds of feet, most are interrupted at intervals by zones too lean to pay for mining. In Oregon's gold deposits, ore shoots vary considerably in width as a result of pinching and swelling. Few attain minable widths of more than 10 feet and most average less than 4 feet.

Lode deposits: The majority of the primary gold-silver deposits were formed in and along fissures in the earth's crust which provided conduits for mineral-rich hot waters or vapors migrating upward from deep-seated igneous sources. Because the gold-silver deposits are generally clustered along the edges of acid to intermediate intrusive rocks, it appears likely that the metallizing fluids originated during late stages of the igneous activity and that the deposits were formed shortly after emplacement of the intrusives. Because the intrusive rocks solidified at considerable depths, deposits that formed from accompanying mineralizing fluids have since been exposed at the surface only as the result of long periods of uplift and erosion. Some ore bodies may have been partly or wholly destroyed by erosion; others may still exist along the edges of granitic masses hidden far below the earth's surface.

In most, if not all, important lode-gold districts the workable deposits are closely related to zones fractured by folding, faulting, or shearing. Such zones of broken rock provide the channelways or conduits for the circulation of mineralizing solutions and the openings in which deposition takes place. Ore deposits occur as fissure veins and replacement veins. In the former, deposition of the ore and gangue minerals was confined mainly within the fissure openings. Replacement veins were developed by substitution of ore and gangue minerals for minerals of the rocks along the fissures. Many deposits show evidence of both processes of mineralization.

Certain wall rocks are more susceptible to replacement by mineralizing solutions than others, and as the vein passes from one formation to another the width and grade of ore may change, or the ore may spread out along a favorable contact. Ore bodies are often richer and wider in brittle, easily shattered rocks.

Where the ore-forming solutions followed clear-cut fissures, the resulting veins are likely to be massive, well defined, tabular bodies with fairly constant strike and dip. In more intensely broken zones through which the solutions spread widely by way of a broad, interconnected series of openings, the resulting deposits commonly comprise a network of small veins and stringers interspersed through shattered and often highly silicified country rock. Gouge, which develops from the grinding together of rocks along a fault, is commonly associated with gold veins.

Pockets: Many of the rich accumulations of gold found in oxidized portions of deposits are believed to be the result of supergene or near-surface enrichment. Miners commonly refer to such accumulations as "pockets." Supergene enrichment involves the action of circulating ground water at or near the surface. When attacked by these waters for sufficient time, certain gangue minerals, particularly the sulfides, oxidize to other compounds, some of which may be carried away in solution. Regarding the role of gold in the process, Emmons (1933) states: "In certain gold deposits the gold seems to be dissolved sparingly if at all. By subtraction of other material, the gold in the oxidized zone is enriched. It accumulates at the outcrop from which it may be washed away to form placer deposits. Since more material is removed from the upper than the lower part of the oxidized zone, more enrichment is likely to take place near the surface. Where gold is dissolved by acid waters containing chlorides in the presence of an oxidizing agent, such as manganese oxide, the outcrop may be impoverished and the gold carried downward so that the lower part of the oxidized zone is enriched."

In his discussion of pockets in the Kerby quadrangle, Wells (1949, p. 21) writes: "Owing to the nature of pockets and pockethunters, record of all but the largest pockets does not exist, traces of the holes are soon obliterated, and memory of them is gone with the passage of the pockethunter."

Because supergene pockets formed near the surface, erosional processes eventually removed them and transferred them to placer deposits, thus preserving occasional large nuggets such as those described below.

Placer deposits: Because gold has a very high specific gravity (19.33 when pure as compared to 2.5 to 3.5 for the quartz and other major rock components with which it is usually associated) and great resistance to weathering, it becomes concentrated in alluvial deposits as the parent rocks are disintegrated and carried away by the processes of erosion. The principal concentrations occur in the beds of streams and gulches below the veins.

Occasionally placer gold accumulates at or very near its bedrock source during disintegration and removal of the matrix. "Residual placers" are thus formed. More often, running water carries the loosened material and much of the gold away from its place of origin. Because of its greater weight, the gold works toward the bottom of the moving debris, and with time eventually reaches bedrock or some other impervious layer known to miners as "false bedrock." The coarser gold, tending to work downward more rapidly, is often found mixed with sand and gravel in crevices in the bedrock.

Generally, only parts of an auriferous gravel deposit are rich enough to be worked profitably. The "pay streaks" are commonly very irregular and have to be located by hit-or-miss prospecting. Gold may be transported long distances, but as a rule the greater the distance from the source the finer and more scattered it becomes and the more rounded or flattened are the particles. The general run of gold recovered from placers ranges in size from that of a mustard seed to a wheat grain.

Gold may become so finely divided that several thousand "colors" may be required to equal the value of one cent. In such condition it is extremely difficult to "save" by any known method of placer mining that is of a sufficiently large scale to be profitable.

At the other extreme, many nuggets the size of chicken eggs or larger have been won from Oregon placer deposits. Lindgren (1901, p. 636) states that a nugget worth $14,000 (at $20.67 per ounce) was reportedly discovered in McNamee Gulch in the Greenhorn district of western Baker County. Such a nugget would be equivalent to 677.31 ounces of pure gold. The Armstrong nugget, weighing 80½ ounces troy, was found at Susanville in Grant County; it is on display at the United States National Bank of Oregon branch in Baker. Spreen (1939), in covering the 1850 to 1870 history of placer mining in Oregon, reported sources of information on several remarkably large nuggets found in southwestern Oregon. The largest was discovered in 1859 on the east fork of Althouse Creek in Josephine County. It weighed

204 ounces (17 pounds troy) and was then valued at $3,500. A single piece weighing more than 15 pounds was reportedly found near Sailors Diggings at Waldo in Josephine County some time between 1860 and 1864. It was valued then at more than $3,100. It should be pointed out that values of large nuggets as collector's items or museum pieces are often more than twice the value of their metal content.

Workable gold placers usually lie in or near districts where gold veins occur. Thus, the tracing of placer gold up hill to its source has led to the discovery of important lode mines. Most lode mines were located in this manner. On the other hand, the source of the gold in some placers has never been determined and for others the source is thought to have been numerous small veins and stringers, none of which alone are large enough to be mined profitably.

Streams continually change course or disappear entirely as erosion and orogenic movements alter the configuration of the earth's surface. Thus, gold-bearing gravels may be found on the sides or even the tops of hills far from present drainages. Graveled terraces or "high bars" along valley walls may mark the former courses of streams. Old gravels may be buried by lava, by glacial or landslide debris, or by gravels from rejuvenated streams. Gravels long buried may become firmly cemented and thus prove difficult to work with ordinary placer-mining methods.

MINING METHODS AND TREATMENT

Lode mining

Most of Oregon's lode-gold deposits occur in steeply dipping veins and mineralized fault zones and rarely exceed 10 feet in economic width. Therefore, underground methods of mining are required except within a few feet of the surface. The mines are developed by shafts or adits with haulage levels at convenient intervals. Raises are run in the ore or just beneath it from level to level. The ore is then drilled and blasted from the vein, loaded into cars, and taken to the surface.

Gold is recovered from its ores by several processes. The treatment utilized at any particular mine depends primarily on the mineralogical character of the ore. Each process requires crushing and fine grinding as initial steps. Free gold and gold-bearing sulfide minerals may then be removed from the finely ground ore by amalgamation, flotation, cyanidation, jigging, table concentration, or a combination of these processes. Free gold recovered by amalgamation or cyanidation is sold as bullion. Sulfide concentrates are usually shipped to a smelter.

Some of the earliest operators in Oregon utilized the arrastra (figure 9), a device that ground ores by dragging a heavy stone around a circular bed. For many years, however, the mainstay of ore reduction was the stamp mill, which employed heavy iron pestles (stamps) working mechanically in a huge iron mortar. Stamps ranged up to 2,000 pounds in weight, rising and dropping 6 or 8 inches a hundred or more times a minute. The pulverized ore was then brought into contact with mercury, which combined with the free gold to form amalgam. The gold was then recovered by distilling off the mercury to make gold sponge (see frontispiece).

Gold-bearing sulfides from which the gold could not be liberated economically by simple crushing were separated from waste material with concentrating devices such as the vanner or concentrating table. Compared with today's standards, these methods of concentration were inefficient, often resulting in losses of 25 percent or more of the values contained in the sulfides. This factor is in part responsible for the lack of deep development of a great many mines in Oregon, for as depth was gained in mining, the amount of sulfides in the ore generally increased until the bulk of the gold values was contained in the sulfides. Photographs (figure 10) taken about 1905 at the Granite Hill mine, Josephine County, show a typical early-day mining camp and concentration mill.

Eventually flotation and cyanidation began to supplant gravity concentration in the treatment of sulfide ores, thus making possible much greater recovery of the contained values.

In the flotation process, finely ground ore mixed with water is agitated and aerated with small amounts of certain compounds which adhere to the desirable minerals and float them to the surface, where they are skimmed off as a concentrate. Waste materials remain submerged and are discarded. In the cyanidation process, finely crushed ores or concentrates are placed in vats containing a dilute solution of sodium cyanide. The gold dissolves to form sodium gold cyanide. The solution is then brought into contact with zinc or aluminum, which causes the gold to precipitate.

Figure 9. An arrastra was a primitive device used to grind gold and silver ores in the early days of mining in Oregon. The ore was spread on a floor, usually made of rock, and ground beneath heavy stones suspended from arms attached to a vertical shaft. At some arrastras, horses or oxen were utilized to turn the shaft instead of a water wheel as shown in the above photograph.

Figure 10. Granite Hill mining camp, Josephine County, between 1902 and 1907. The 20-stamp mill housed boiler, stamps, amalgamation plates, Frue vanners, jaw crusher, and 150-HP electric motor. Hoist was steam operated. Views on opposite page show the mill building, equipment, and portal of mine adit. (Photographs courtesy of W. R. Graham)

a) Exterior of Granite Hill mill building

(See Figure 10 on opposite page)

b) Frue vanners at Granite Hill mill

c) Two 5-stamp batteries and amalgamation plates

d) Collar of shaft and head fram at portal of adit

Figure 11. Sterling hydraulic mine near Jacksonville about 1880.

Placer mining

The methods of extracting gold from placer gravels are numerous, but all take advantage of the fact that gold is much heavier than the accompanying rock debris and thus works downward during agitation. Running water is nearly always used to wash away the rock debris, leaving the gold behind. For this reason, the amount of water available is a prime factor in determining methods of mining and the length of the annual operating season. In the early days ditches, some of great length, were constructed to carry water to placer ground.

Among the simplest placer-mining devices are the common miner's pan, the sluice box, and the rocker or cradle. Miner's pans are round, shallow containers for shaking the heavier material to the bottom and washing lighter particles off the top. Sluice boxes are slightly inclined wooden or iron troughs with transverse ridges or "riffles" fixed across the bottom. As the auriferous gravel is washed down through the box, gold and heavy sands are caught by the riffles. In "ground sluicing," gravels are washed across bedrock and the gold trapped by natural riffles in the rock surface. The rocker and cradle are, in effect, short sluice boxes with superimposed detachable sieves or riddles for screening out coarse material ahead of the apron and riffles. The apron is a sagging canvas on a removable sloping frame that effectively pans the sieved material. A rocking motion, generally supplied by hand, is employed to keep the sand and gravel moving.

Where water is abundant, "hydraulicking" greatly increases the amount of gravel that can be worked in a given time. Hydraulic mining employs a large, pipe-fed nozzle called a giant or monitor to direct a powerful jet of water against banks of auriferous gravel (figure 11). The excavated gravels flow through a long sluice where the gold is recovered. Significant amounts of gold may also be recovered by cleaning bedrock after the gravel is stripped away.

Dredging is by far the fastest method of working placer gravels and, after 1913, produced more gold in Oregon than all other processes combined. Although cost of installation is high, operating costs amount to only a few cents per cubic yard of gravel handled. The capacity of dredges ranges from 1,000 yards to 15,000 yards of gravel per day.

The bucketline dredge consists of a wood- or steel-hulled barge upon which is mounted a continuous chain of buckets for excavating, a screening and washing plant, and one or more conveyor belts for stacking tailings (figure 12a). Some dredges, instead of utilizing connected bucket excavators, are fed by dragline (figure 12b). In Oregon much placer gravel has been worked by washing plants or "doodlebugs" which are mounted on skids, or sometimes on wheels so they can be operated on dry land. The dry-land rigs must be dragged from place to place and are fed by dragline or power shovel.

Figure 12-a. Murphy-Murray dredge on Foots Creek, Jackson County, in January 1941. Capacity 4000 cubic yards daily, electric powered, 67 buckets, of $3\frac{1}{2}$-cubic-foot capacity; dug 20 feet below water line. Steel hull 81 by 37 by 6 feet, gantries, and superstructure. Refer to book *Dredging For Gold.*

Figure 12-b. Dragline dredge (doodlebug) at the Johnson placer on upper Pleasant Creek, Jackson County, in 1960.

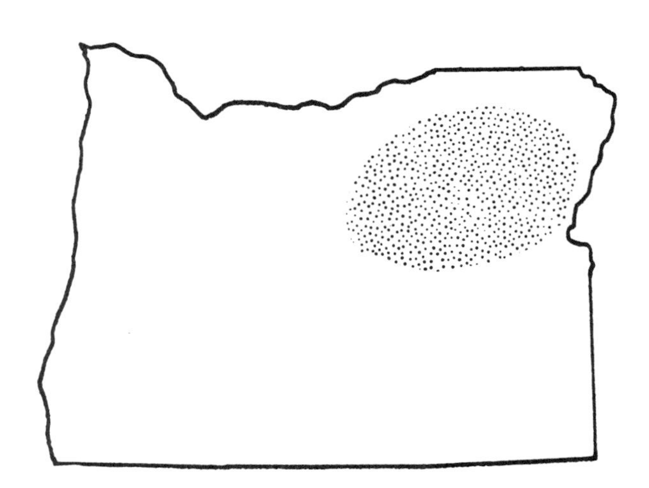

Part II-A Deposits in Eastern Oregon Gold Belt of the Blue Mountains

Currently active mine (top) **near Big Boy between Granite and Olive Lake. Photographed October 9, 1993. The condition of Independence Mine** (right) **on Granite Creek on October 9, 1993.**

—Nancy Wilson photos

Editor's Note: In an earlier printing, Page No. 38 was blank. Blank pages have been eliminated in the present book thus that page number does not appear.

GOLD BELT OF THE BLUE MOUNTAINS

Table of Contents

GOLD BELT OF THE BLUE MOUNTAINS

INTRODUCTION

Approximately three-fourths of the gold produced in Oregon has come from lode and placer deposits in the Blue Mountains geomorphic province, which occupies much of the northeastern part of the state. The deposits lie in a region named by Lindgren (1901) "the gold belt of the Blue Mountains." The belt is about 50 miles wide and 100 miles long, extending from John Day on the west to Snake River on the east. The principal mining areas are in Baker and Grant Counties, although some mining has been done in adjacent parts of Malheur and Union Counties. All of the lode deposits are in pre-Tertiary rocks and are believed to be associated with Jurassic-Cretaceous dioritic intrusions (figure 13).

GEOGRAPHY

The Blue Mountains comprise a complex system of mountain ranges, high plateaus, and deep canyons interrupted in places by broad, fertile valleys. In the eastern part of the province, elevations range from less than 1600 feet in the Hells Canyon of the Snake River to 9845 feet on Matterhorn Peak in the Wallowa Mountains. The larger valleys lie between 2000 and 4000 feet above sea level. The elevation at Baker, the largest city (9500 population) in the gold belt, is 3450 feet. The main drainage systems are the Burnt, Powder, Grande Ronde, and Imnaha Rivers which flow eastward to the Snake River, and the vast John Day River system which flows westward, then north to the Columbia. The Blue Mountains complex continues eastward into Idaho and is there known as the Seven Devils Mountains. Between the Blue Mountains and the Seven Devils Mountains the Snake River has cut a canyon even deeper than the Grand Canyon of the Colorado.

The average annual precipitation varies from about 10 inches in the lower valleys to about 45 inches in the high mountains. Areas of high elevation are commonly snowbound between November and May. Much of the region between 4000 and 7000 feet elevation is heavily forested, chiefly with pine. Below 4000 feet the hills and valleys support desert grasses and brush. The larger valleys are farmed. Lumber and cattle long ago supplanted gold as the most valuable products of the region.

PREVIOUS WORKERS

The earliest systematic geologic report on the gold belt of the Blue Mountains was by Lindgren (1901) who studied the gold deposits in detail and prepared a large-scale reconnaissance geologic map of the region. His report remains the best source of information on many of the old mines. Pardee discussed faulting and vein structure in the Cracker Creek district (1909) and the placer gravels of the Sumpter and Granite districts (1910). A geologic map and description of the mines in the Sumpter quadrangle was published by Pardee and Hewett (1914). The map and text were later revised (Pardee, Hewett, and others, 1941). The earlier report was accompanied by a description of several mines near Baker (Grant and Cady, 1914). Lindgren's study of the mines of the Blue Mountains was complemented by a comprehensive re-examination

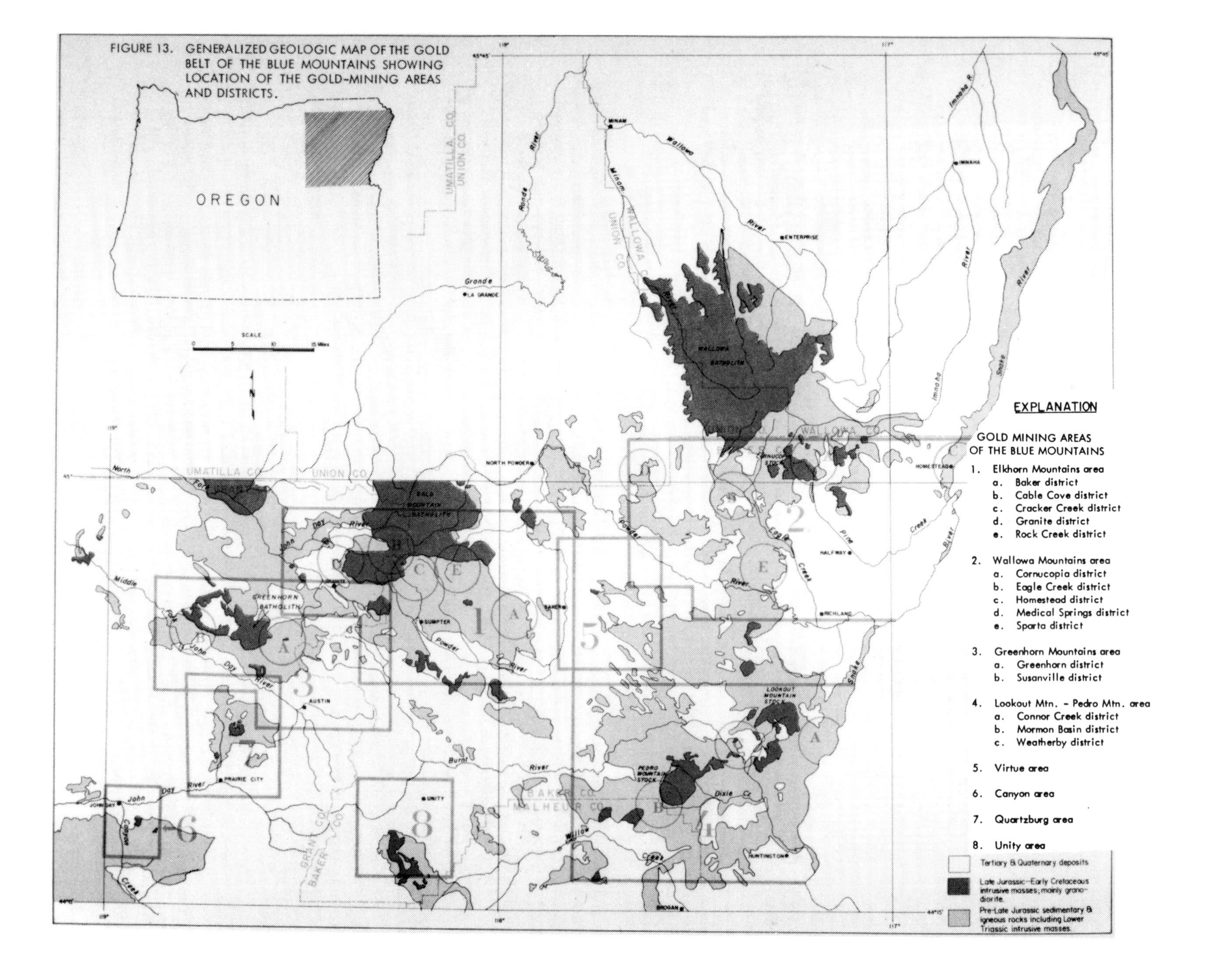

FIGURE 13. GENERALIZED GEOLOGIC MAP OF THE GOLD BELT OF THE BLUE MOUNTAINS SHOWING LOCATION OF THE GOLD-MINING AREAS AND DISTRICTS.

by Swartley (1914). Parks and Swartley (1916) prepared descriptions in alphabetical order of a great many mines throughout the state. Much of the data came from previous publications. A report by Livingston (1925) contains a geologic map of the Snake River canyon below Huntington and a description of the Bay Horse mine. Hewett (1931) discussed mineralogic variations observed in the lodes of the Sumpter quadrangle. Reconnaissance investigations were made of many of the mining districts of the region during 1929 and 1930 by geologists of the U.S. Geological Survey. This work led to publications by Gilluly (1931), Gilluly, Reed, and Park (1933), and Gilluly (1937). The latter report describes the geology of the Baker quadrangle. Lorain (1938) briefly summarized the mining geology of the region and described some of the mines, mostly those that were then active, with emphasis on mining and milling methods. Goodspeed (1939) discussed the gold quartz veins of Cornucopia. Oregon Department of Geology and Mineral Industries (1939 and 1941) summarized most of the previously accumulated data on the mines and mining districts of the region. The information is supplemented by interim reports and notes supplied by the Department staff and others. G.S. Koch (1959) described the lode mines of the central part of the Granite district including the Buffalo, the only consistently active lode-gold mine in the region at the time.

In addition to coverage given in some of the reports cited above, geologic maps and other stratigraphic, structural, and petrologic data have been compiled for certain areas by geologists of the Oregon Department of Geology and Mineral Industries and the U.S. Geological Survey and by geology professors and candidates for master's and doctoral degrees at several universities. Published reports which should be mentioned are those by Ross (1938) and Smith and Allen (1941) on the Wallowa Mountains; Prostka (1962 and 1967) on the Sparta and Durkee quadrangles; and the various papers listed in the bibliography by Taubeneck on the Bald Mountain batholith, Wallowa batholith, and Cornucopia stock; and by Thayer and Brown on the John Day - Canyon City area.

HISTORY AND PRODUCTION

Placer Mining

According to legend, gold was discovered somewhere in eastern Oregon in 1845 by the Meek immigrant party bound for the Willamette Valley. During the 1850's small groups of prospectors reportedly found placer gold in the Burnt and John Day River areas. In the fall of 1861 Henry Griffin, with a party of prospectors from Portland, discovered gold in what is now Griffin Gulch, a tributary of Powder River a few miles south of Baker. The following spring another group of prospectors found gold in Canyon Creek near the present site of John Day. In early 1862 the first white settlement in the region was established at Auburn about 5 miles south of Griffin Gulch. So great was the influx of miners and adventurers that by the end of that year the population of the camp was estimated at between 5000 and 6000 persons. A camp of similar proportions sprang up near what is now Canyon City.

From Auburn and Canyon City prospectors spread in all directions, and by 1864 almost every placer-mining district in eastern Oregon was being exploited. Supplies were hauled from The Dalles. Because of the difficulty of access and transportation costs, gravels which did not yield $8 per day per man were not considered workable.

Among the most productive of the early-day placer-mining districts were the Canyon, Dixie Creek, Granite, and Susanville in Grant County, and the Sumpter, Auburn, Pocahontas, Sanger, Sparta, Malheur, Mormon Basin, Rye Valley, Eldorado, and Connor Creek in Baker County (figure 14).

Water for working the gravels in several areas was scarce and for that reason ditches, some of great length, were constructed. The Auburn Ditch was completed in 1863, the Rye Valley Ditch in 1864, and the Sparta and Eldorado Ditches in 1873. The Eldorado Ditch, which took 10 years to complete, carried water more than 100 miles from near the head of Burnt River to the placers near Malheur. Use of these ditches for mining ceased long ago. The Auburn Ditch is now part of the Baker city water system and the Sparta Ditch is used for irrigation.

Sketchy records indicate that many of the early diggings were rich, but there are no reliable statistics to show the total amount of gold produced in eastern Oregon during the period of 1861-1880. Production probably was at its peak during 1863-1866, then began to decline gradually as the richest placers were worked out. Raymond (1870) estimated that production from the placers of Canyon Creek averaged

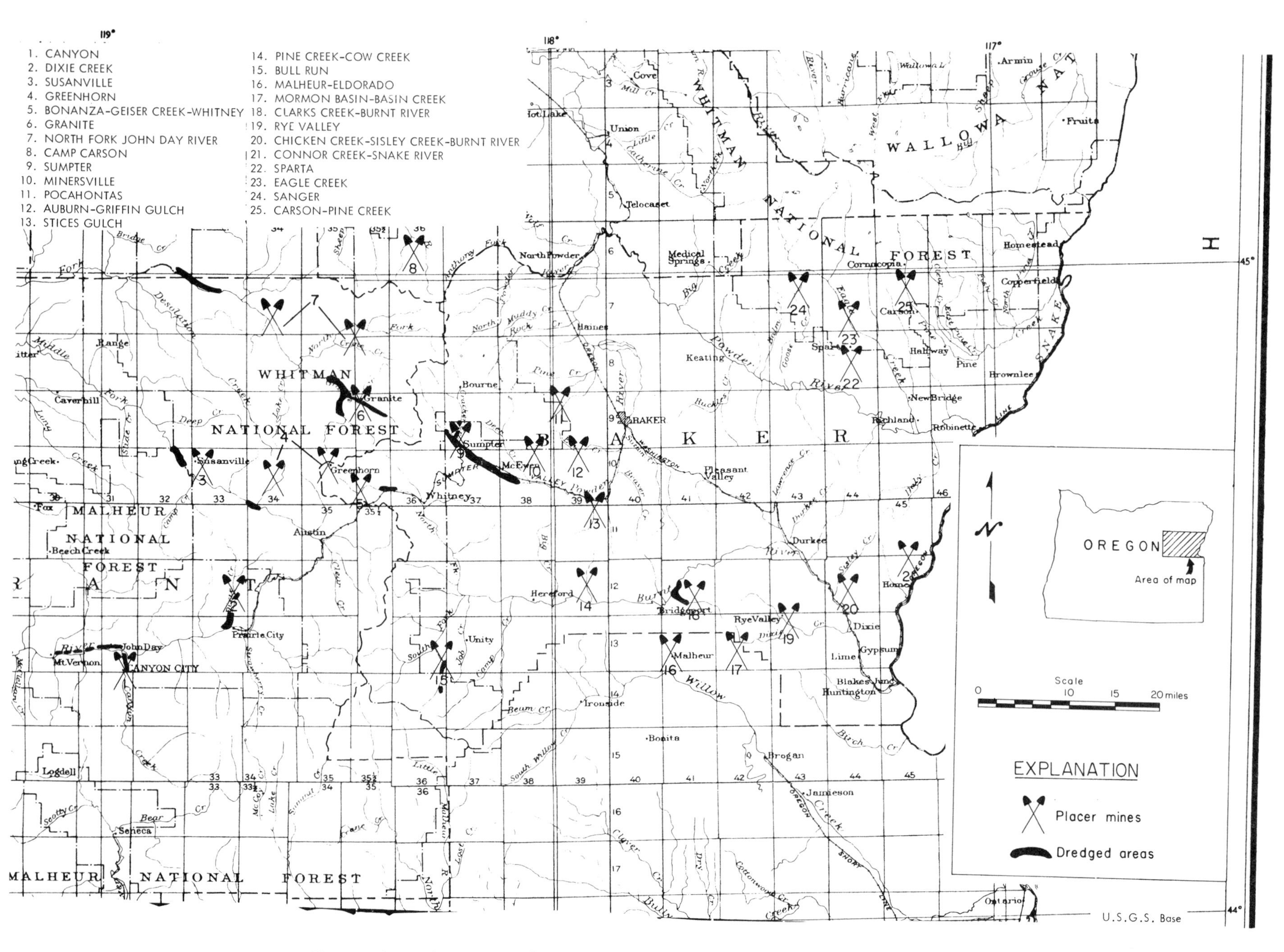

Figure 14. Index map of the principal placer mining areas in the Blue Mountains.

about $22,000 per week during the mining season of 1865. In 1869, output averaged about $8,000 per week from April to October and perhaps $20,000 per month the rest of the year. For the year 1870, Raymond (1872) estimated "...the total shipments of gold from eastern Oregon exclusive of Cañon City and other districts west of the Blue Range" at $600,000. W. H. Packwood (in Raymond, 1872, p. 184) stated that the yearly gold output of Baker and Union Counties "...cannot have been less than from one to one and one-half million dollars from 1863 to 1870. The gold has been, we may say, the sole product of labor. The number of miners has been from one to three thousand averaging for several years about fifteen hundred."

By 1914, according to Pardee and Hewett (1914, p. 10-11), the placer mines of the Sumpter quadrangle had produced a minimum of $5,231,000 in gold and silver. They state:

"During the most active period of placer mining in the Sumpter quadrangle, records of production were seldom kept. Therefore the exact yield from this source is not known. It is possible, however, to estimate approximately the amount that must have been produced to have made operations profitable.... In a very few instances authentic records of production were obtained and were used for the mines they represent in arriving at the totals. The cubic contents of the principal placer excavations in the quadrangle were estimated to aggregate 21,500,000 cubic yards, the minimum yield of which was computed to be $5,231,000, or an average of about 25 cents a cubic yard. The minimum yields computed for the various districts are as follows:

North Fork drainage basin	$ 893,000
Granite Creek	1,033,000
Sumpter district	1,691,000
Greenhorn district	1,140,000
Bonanza	396,000
Minersville district	36,000
Whitney Valley	32,000
Scattering	10,000
Total	$5,231,000

"The actual placer production of the quadrangle is doubtless much larger than the total arrived at. Miners and others familiar with the region invariably make higher estimates than those given, in some instances estimating more for a single district than the total for the whole as given above. As nearly as can be ascertained the portion of silver alloyed with placer gold ranged from 5 to 30 percent and probably averaged 17 percent, equivalent to a gold fineness of 830 or value of $17.15 an ounce. Assuming the gold and silver values combined in the total given, it is calculated that of the total amount about $30,000 was silver."

Estimates of production are available for few placers outside the Sumpter quadrangle. In his account on the Canyon area (1901, p. 712-720), Lindgren stated "...it is scarcely probable that the total production [of that area, Ed.] much exceeds $15,000,000." Swartley (1914, p. 169) credits the placers of Elk Creek near Susanville with total output of approximately $600,000 and suggests the same figure for the Dixie Creek placers (p. 198). Swartley also states (p. 228) that production from the Rye Valley placers amounted to more than $1,000,000. The placers near Sanger reportedly produced about $500,000 in gold up to 1901 (Lindgren, 1901, p. 738).

The first successful large-scale bucketline dredge employed in eastern Oregon began work in Sumpter Valley in 1913 (one of the later dredges is shown in figure 15). Bucketline dredges were used on Canyon Creek and on the John Day River near John Day; along the Middle Fork of the John Day below Bates and at Susanville; on Granite, Clear, and Bull Run Creeks in the Granite area; on Burnt River near Whitney; and on Clarks Creek and Burnt River near Bridgeport. Numerous dragline dredges and dry-land washing plants worked gravels in the Sumpter, Granite, Prairie City, and John Day areas.

Recorded output of gold from the bucketline dredges alone is nearly 480,000 ounces. Calculated at the present price of gold, the value of recorded gold production from both bucketline and dragline dredges operating in the Sumpter area during 1913-1954 amounts to about $12,250,000. Value of output from the John Day - Prairie City area, 1916-1949, amounts to a little more than $5,000,000.

Figure 15. Sumpter Valley dredge working the gravels in Sumpter Valley a few miles below Sumpter about 1941. Elkhorn Ridge in background. (Photograph courtesy of Brooks Hawley.) Refer to book *Dredging For Gold.*

Figure 16. Main Street of Sumpter, Baker County, Oregon about 1914. Note board construction of street. (Photograph courtesy of Brooks Hawley.)

Lode Mining

Vein deposits were discovered soon after the advent of placer mining. Development of the Virtue mine about 8 miles northeast of Baker began in 1862, and a 10-stamp mill used to treat the ore was erected on the outskirts of Baker in 1864. An arrastra was in use near the present Sanger mine in the Eagle Creek district before 1865. Quartz mines were worked as early as 1865 and 1868 in the Susanville and Mormon Basin districts and in the Connor Creek, Granite, and Cable Cove districts in the early 1870's. The first claims on the great North Pole - Columbia Lode in the Cracker Creek district near Sumpter were located in 1877. Lode-gold deposits were discovered near the present site of Cornucopia in about 1880.

Early production from lode mines was dominated by the Virtue, Sanger, and Connor Creek mines, whose ores were amenable to the crude treatment methods of the time. The Virtue mine was almost continuously active during 1864-1884 and 1893-1899. Production totaled about $2,200,000. At the Sanger mine the principal vein was discovered in 1870 and worked more or less actively through 1897 with a total output of about $1,500,000. The Connor Creek mine on Connor Creek about 2½ miles from Snake River was placed in operation in 1872, but the period of greatest productivity was between 1880 and 1890. Production from this mine totals about $1,250,000. There has been little production from the Virtue, Sanger, or Connor Creek mines since 1900.

With the exception of the early production described above, the development of lode mining in the Blue Mountains region was slow and sporadic until after extension of the transcontinental railroad to Baker in 1886. Completion of the Sumpter Valley Railroad to Sumpter in 1896 caused a mining boom in that area that lasted until about 1908. Sumpter grew from a small hamlet of a few hundred population to a town of 3000 or more (figure 16). Many of the most productive mines of the surrounding area were developed during this period. There was also considerable speculation. Much money was unwisely invested in the development of worthless prospects.

From 1904 to 1908 a 100-ton-per-day smelter erected at Sumpter was operated on ores and concentrates hauled in from the mines (figure 17). The plant was well built and efficiently run, but the amount of ores and concentrates supplied by the mines was inadequate to sustain operations. According to an article in the Blue Mountain American newspaper of July 30, 1910 only 19,068 tons of ore and concentrates were treated between November 15, 1904 and November 15, 1907. In regard to the mines in the Sumpter quadrangle, Pardee and Hewett (1914, p. 10-11) stated:

"The greater part of the deep mine production of the quadrangle has been made since 1900 and for this period authentic records show an output of $8,943,486 from 53 of the mines, only eight of which exceeded $100,000 production. Some of the records are incomplete, however, and for some of the mines no reliable records are available. This additional production, as based on estimates believed to be fairly reliable, amounts to $800,000. Prior to 1900 few records are available but estimates which seem to be reasonable place the production for that period at $1,600,000. Thus it appears that the total deep mine production of the Sumpter quadrangle is not far from $11,350,000 in gold and silver. Of this amount probably about 5 per cent, or $565,000, has been derived from silver."

Of the total production cited, about $8,000,000 came from the North Pole, E. & E., Taber Fraction, Columbia, and Golconda mines which worked contiguous parts of the North Pole - Columbia Lode near Bourne in the Cracker Creek district. Other mines whose total output at the time was near or above the $1,000,000 mark were the Bonanza and Red Boy mines in the Greenhorn district and the Baisley-Elkhorn mine in the Rock Creek district. The principal period of activity of each of these mines occurred between 1892 and 1916. Although there have been attempts to reactivate some of these mines, production has been relatively small.

Lode-mine output for Baker and Grant Counties reached a high of 55,746 ounces from 27 mines in 1902 when the Sumpter "boom" was at its peak. Output in 1911 after most of the large mines had closed was only 19,248 ounces from 19 mines. In 1913 production of the two counties bounded back to 51,721 ounces from 17 mines. The increase was largely the result of increased productivity of the Union-Companion and Last Chance mines in the Cornucopia district and the Rainbow mine in the Mormon Basin district. Production from the Rainbow from the year of its discovery in 1901 through 1919 was more than $2,323,000. During 1913-1915 it was the largest producer in the state.

Figure 17. Oregon Smelting & Refining Company smelter at Sumpter in 1903. The plant was erected to service the numerous gold and silver mines in the area. The railroad tracks are those of the Sumpter Valley Railroad, a narrow-gauge line that ran between Baker and Prairie City.

Sumpter Valley Railroad hauled logs from the forests to the mills, hauled building materials from Baker into the mountains, hauled gold from the smelters to Baker for transhipment on the mainline trains. The genesis of the idea to build the line came in 1890. By 1896 the rails reached Sumpter. Eventually, after pushing through the pristine but tortuous mountains, the line terminated, 80 miles later, at Prairie City. By 1937, having lived out its usefulness, passenger service ceased, then freight service terminated ten years later. Abandoned and almost forgotten, some cars and the property deteriorated in the weather. In 1971, there was a rebirth with the formation of the Sumpter Valley Restoration Railroad (Inc.), a non-profit group of railroaders, historians and well-wishers. They placed all 80 miles of right-of-way on the *National Register of Historic Places,* then set about re-laying some track to carry summer tourists. The depot and shops were built within a stone's through of the skeleton of old Sumpter Dredge No. I, at McEwan, once a major stop on the line. Now the line runs west to the edge of Sumpter, but soon will be extended to encircle Sumpter Dredge No. 3, in town, that is the focal point of the new Sumpter Valley Dredge State Park. For history of the railroad see Chapter 10 in *Flagstaff Hill on the National Historic Oregon Trail.* For colorful details and pictures of all three Sumpter Dredges, and some others, refer to *Dredging For Gold - Documentary,* in special bibliography on page *xviii-A.* —Photo from Jim Eccles collection

Sumpter Valley Railroad

From the point of view of productivity and long life the Cornucopia mines, 12 miles north of Halfway, rank first among the lode mines of Oregon. Total output of these mines has been estimated at more than $10,000,000. Except for short periods of inactivity during 1921-1922 and 1926-1933, Cornucopia dominated lode-mine output from 1915 through 1941 and for several of those years accounted for more than half of the total lode-gold production of the entire state. The mine closed in October 1941 as a result of escalated costs of production and shortage of men and materials brought about by the threat of World War II.

Gold production since 1945 has been largely from the Buffalo mine in the Granite district.

Most of the mines mentioned by name in this discussion have recorded output of more than $1,000-000; however, the fact should not be overlooked that a large percentage of the gold produced from lode mines in the Blue Mountains has come from intermittent operation of a great number of smaller mines and prospects. During the 41-year period from 1902 to 1942, the number of lode mines reporting production each year in Baker County ranged from 9 to 30, averaging about 18. For Grant County the average was about 9. There are at least 20 small mines in the region whose production is known or reliably estimated to be in excess of $100,000 but less than $1,000,000.

GEOLOGY

The rocks of the gold belt of the Blue Mountains fall into two main groups, those of pre-Tertiary age and those of Tertiary and Quaternary age. The two groups are separated by a major unconformity.

Pre-Tertiary Rocks

The pre-Tertiary rocks comprise thick sequences of eugeosynclinal sedimentary and volcanic strata of late Paleozoic and Late Triassic-Jurassic age and at least two independent and dissimilar suites of plutonic intrusive rocks, one of Late Permian-Early Triassic age and the other of Late Jurassic-Early Cretaceous age. The bedded rocks and older intrusives have been tightly folded and regionally metamorphosed. The dominant structural trends are east to northeast and dips of bedding and foliation planes are generally steep. Mineral transformations tend toward the greenschist facies. The Paleozoic sediments and volcanics are in general more intensely deformed than the younger strata and they are locally foliated.

Rocks older than Permian are unknown in the eastern part of the Blue Mountains, but it is possible that earlier periods are represented by some of the more highly metamorphosed and deformed rocks. Older Paleozoic beds are exposed in the Suplee area to the west (Merriam and Berthiaume, 1943; Kleweno and Jeffords, 1961). Bedded rocks of Early Triassic age have not been recognized.

Sedimentary and volcanic strata

Elkhorn Ridge Argillite: Typifying the major part of the upper Paleozoic sequence are thick sections of argillite and chert, with subordinate metavolcanic rocks and lenticular limestones that occupy the middle third of the Sumpter quadrangle (Pardee, Hewett, and others, 1941).

In general the rocks are fine grained and gray to black in color owing to incorporated carbonaceous matter. Gradations of argillite to chert or tuff or conglomerate are common, as is repetitious interlayering of the various components. No well-marked type of beds characterizes any particular portion of the series. The argillite layers commonly range from several inches to several feet in thickness, while the cherty rocks more often occur as contorted bands a fraction of an inch to 3 inches thick separated by thin layers of argillite.

The limestone occurs as small detached bodies with angular boundaries. It is light gray to dark blue in color. Most of the masses are less than 300 feet in longest dimension, although some exceed 1500 feet in length and 500 feet in width. The sequence is particularly well exposed on Elkhorn Ridge and along Cracker Creek above Sumpter. Gilluly (1937) traced the argillite series from Elkhorn Ridge eastward across the middle third of the Baker quadrangle and named it the Elkhorn Ridge Argillite. Contiguous rocks extend easterly to the Snake River and beyond into Idaho (Prostka, 1967; Brooks and Williams, in progress). The thickness of the series is unknown, but Gilluly's provisional estimate of 5000 feet appears conservative. Limited fossil evidence indicates that most of the formation is Permian in age but according to Bostwick and Koch (1962) younger rocks may be included.

Burnt River Schist: A severely deformed sequence of thinly layered phyllite and chert, massive to schistose greenstone and tuff, and marble that occurs along Burnt River in the southern part of the Baker quadrangle was named the Burnt River Schist by Gilluly (1937). These rocks extend westward into the southern part of the Sumpter quadrangle, southward through Mormon Basin, and eastward into Idaho. The age and stratigraphic relationships of the series is unknown. However, Prostka (1967) considers it to be partly Permian and partly Late Triassic in age.

Clover Creek Greenstone: A group of rocks consisting of altered lavas, pyroclastics, and volcaniclastic sandstones, breccias, and conglomerates with subordinate argillite, chert, and limestone is exposed over wide areas in the Wallowa Mountains and farther east in Snake River canyon. Exposures in the northern part of the Baker quadrangle were named the Clover Creek Greenstone by Gilluly (1937) and the name was adopted for similar rocks in the Wallowa Mountains (Ross, 1938; Smith and Allen, 1941; and Prostka, 1962). Gilluly assigned the group to the Permian but the later works cited (see also Wetherell, 1960; Bostwick and Koch, 1962; and Vallier, 1967) have shown that the formation is largely Late Triassic in age.

Martin Bridge and Hurwal Formations: Structurally conformable above the Upper Triassic greenstones and volcaniclastic rocks in the Wallowa Mountains - Snake River canyon area are the Martin Bridge Formation of Late Triassic age and the Hurwal Formation of Late Triassic-Early Jurassic age. The Martin Bridge Formation, about 1500 feet thick, consists mainly of massive limestone with associated shales. The Hurwal Formation, about 4000 feet thick, is made up mainly of graywacke, siltstone, and shale with minor chert, conglomerate, and limestone.

Unnamed Upper Triassic-Jurassic rocks in the Huntington area: About 30 miles south of the Wallowa Mountains, undifferentiated Upper Triassic-Jurassic sedimentary and volcanic rocks are exposed in several places within a broad belt extending from the Snake River near Huntington westward to the vicinity of Ironside Mountain southwest of Unity. Farther west are the better known Mesozoic beds of the John Day country (Brown and Thayer, 1966). The lower part of the sequence along Snake River near Huntington is made up mainly of massive greenstones, tuffs, and volcaniclastic sedimentary rocks with subordinate interbedded argillite, chert, and limestone. These rocks are correlative with Upper Triassic greenstones in the southern Wallowa Mountains. The upper part of the sequence comprises several thousand feet of sheared graywacke, slate, and phyllite with intercalated volcanic and conglomeratic layers and scattered limestone lenses. Late Triassic and Jurassic fossils are present. Contiguous rocks in the Ironside Mountain area were named the Rastus Formation by Lowry (in press). Lithologic similarity and stratigraphic position suggest partial correlation of these rocks with the Hurwal Formation to the north.

Intrusive rocks

Pre-Late Triassic: The older of the two groups of plutons includes a wide variety of ultramafic rocks, gabbro, diorite, quartz diorite, and albite granite. The rocks have been strongly deformed, fractured, and chemically altered; commonly they show a gneissic banding. Exposures are widespread and are associated with Paleozoic formations in many parts of the gold belt. Locally, masses of the plutonic rocks have been remobilized and squeezed upward into younger, Mesozoic formations.

The older intrusives are typified by the Canyon Mountain magma series of the John Day area, which has been dated as Lower to Middle Triassic (Thayer, 1963; Thayer and Brown, 1964; and Brown and Thayer, 1966). There is evidence of a similar age for occurrences in the Sumpter quadrangle (Pardee, Hewett, and others, 1941), in the Baker quadrangle (Gilluly, 1937), and in the Sparta quadrangle (Gilluly, 1933; Prostka, 1962).

In the John Day area the plutonic rock types are dominantly peridotite and gabbro. In the Sumpter and Baker quadrangles, altered gabbros are the most plentiful, although albite granite and quartz diorite are common. A large part of the Sparta quadrangle is underlain by albite granite.

Jurassic-Cretaceous: The younger group of plutonic rocks is mainly granodiorite, although in the larger bodies more basic and acid differentiates are present. Distribution of the larger exposures is shown on figure 13. In addition there are many small masses and related dikes too small to show to scale. The dominant exposures are the Wallowa batholith and the Bald Mountain batholith, which underlie the highest, most rugged mountain ranges of the region. The Wallowa intrusive covers more than 225 square miles and the Bald Mountain batholith more than 170 square miles. Other important exposures are the Pedro Mountain and Lookout Mountain stocks and the Greenhorn Mountain intrusive. The plutons are probably related to the great Idaho batholith and are of Late Jurassic or Early Cretaceous age.

Since the Jurassic-Cretaceous intrusive rocks are the source of the gold mineralization of the region, the descriptions of the individual plutonic bodies are presented later in this bulletin under the discussion of the mining areas associated with them.

Tertiary and Quaternary Rocks

Overlying the pre-Tertiary rocks with profound angular discordance is a wide variety of lavas, pyroclastics, and loosely consolidated fresh-water sediments and gravels of Cenozoic age. The most widespread are the Miocene lavas of the Columbia River Group; also occupying large areas are Eocene-Oligocene rhyolitic and dacitic lavas and Miocene-Pliocene lake beds and tuffs. The Tertiary rocks have been deformed into broad, open folds and are cut by northwest-trending faults.

Quaternary gravels and alluvium fill present drainages. A few small alpine glacial moraines occur in the Wallowa Mountains and in the Bald Mountain-Elkhorn Ridge area.

ORE DEPOSITS

Most of the gold deposits in the Blue Mountains appear to be genetically related to the Jurassic-Cretaceous granitic rocks, for they occur in greatest profusion near the contacts of these intrusives with the older rocks. The majority occupy fissures in argillite or in the granodiorite itself. Important veins and replacement deposits were also formed in metavolcanic rocks and in the sheared, highly altered gabbros, diorites, and albite granites in the older group of intrusives. Deposits in serpentinites and other ultrabasic rocks are numerous in some areas, particularly in the Greenhorn district, but production has been relatively small, most of it having come from discontinuous lenses and pockets of high-grade ore.

The Cenozoic volcanic and sedimentary rocks in the region are not known to contain lode-gold deposits. It may well be that in some areas these rocks mask important gold deposits in the older formations.

The lode-gold deposits in the Blue Mountains are predominantly narrow, quartz-rich fissure veins, breccia fillings, and associated replacement bodies along faults and shear zones. Probably the majority were formed by the filling of open spaces with accompanying replacement of the fissure walls and intervening fragments of broken rock. Most of the veins strike northeast, but several of major importance strike northwest -- notably the Virtue and Connor Creek veins.

Some of the zones of fracturing and mineralization that contain ore bodies, such as the North Pole-Columbia lode in the Cracker Creek district, are quite wide and contain parallel, branching, and overlapping veins. Individual ore shoots, however, rarely exceed 10 feet in width; probably most of those mined had an average width of 1½ to 5 feet.

The Union-Companion vein at Cornucopia had a productive strike length of about 2500 feet. It has been developed to a vertical depth of 1400 feet below the outcrop with little decrease in gold content. Other nearby productive veins crop out 1000 to 1600 feet higher than the Union-Companion vein. The productive vertical range in this district has, therefore, been demonstrated to be about 3000 feet. The North Pole-Columbia lode has been mined almost continuously through a horizontal distance of about 12,000 feet and to a depth of about 2500 feet below the highest point on the outcrop. Several other deposits have been developed to depths of 750 to 1000 feet. However, the great majority of mine workings have attained vertical depths of less than 500 feet. Work on many veins ceased with the exhaustion of ore above a working level or of a particular shoot; for some, there is good reason to suspect that ore of similar character is present, either below existing levels or laterally along strike.

Most of the ore mined in the Blue Mountains has averaged from ½ to 1 ounce per ton in gold and varying amounts of silver, although many ore bodies locally yielded bunches of high-grade gold. One small ore body at the North Pole mine yielded 1115 tons of ore averaging $500 per ton.

E and E Mill, Bourne (top). **Stock certificate** (lower) **for investment in the Highland Gold Mines Company of Sumpter, issued in May 1905.**
—Archives, Baker County Library (top); Jim Evans collection (lower)

INCORPORATED UNDER THE LAWS OF OREGON

No 680 — 500 Shares

HIGHLAND GOLD MINES COMPANY

FULLY PAID — CAPITAL STOCK, $3,000,000 — NON-ASSESSABLE

This Certifies that Calista Andrews is the owner of Five Hundred Shares of the Capital Stock of HIGHLAND GOLD MINES COMPANY transferable only on the books of the Corporation by the holder hereof in person or by Attorney upon surrender of this Certificate properly endorsed.

In Witness Whereof the said Corporation has caused this Certificate to be signed by its duly authorized officers and to be sealed with the Seal of the Corporation AT SUMPTER, OREGON this day of May 1905

Secretary — President

$1 EACH

AMERICAN PRINT, SUMPTER, OREGON

GOLD MINING AREAS
IN THE GOLD BELT OF THE BLUE MOUNTAINS

Descriptions of the gold-producing districts and mines of the gold belt of the Blue Mountains are herein divided into eight areas, each of which contains one or more districts as shown in figure 13. The area divisions segregate the dominant Jurassic-Cretaceous intrusive bodies and their associated gold deposits. In the order of their productive rank they are as follows:

1. The Elkhorn Mountains area includes the Baker, Cable Cove, Cracker Creek, Granite, and Rock Creek districts, which lie in and along the edges of the Bald Mountain batholith.

2. The Wallowa Mountains area encompasses the Wallowa batholith and satellitic intrusives, the most prominent of which is the Cornucopia stock. This area contains the Cornucopia, Eagle Creek, Homestead, Medical Springs, and Sparta districts.

3. The Greenhorn Mountains area, which is in part underlain by the Greenhorn batholith, contains the Greenhorn and Susanville districts.

4. The Lookout Mountain-Pedro Mountain area with its several exposures of Jurassic-Cretaceous granodiorite contains the Connor Creek, Mormon Basin, and Weatherby districts.

5-8. The Virtue, Canyon, Quartzburg, and Unity areas each include one district wherein only small intrusive bodies are exposed.

ELKHORN MOUNTAINS AREA

Location

The area lies in the Elkhorn Mountains west of Baker and north of Sumpter (figure 18). It embraces several groups of gold-silver deposits which appear to be genetically associated with the Bald Mountain batholith, a large granodiorite body which has an outcrop area of more than 170 square miles.

The mining districts in which the lode mines occur are, in the order described: Baker, Cable Cove, Cracker Creek, Granite, and Rock Creek. The Baker, Cracker Creek, and Rock Creek districts are in northwestern Baker County; the Cable Cove district straddles the Baker-Grant County line; and the Granite district is in eastern Grant County. These five districts lie along the southern edge of the batholith within an area about 12 miles wide and extending from Baker Valley westward through Sumpter and across the John Day River divide to the vicinity of Granite, a distance of 28 miles. On the northern edge of the batholith, in southern Union County, is the Camp Carson district (not shown on index map), which is known primarily for its placer deposits.

Described separately are the placers of the Baker, Granite, Sumpter, and Camp Carson districts, which were important producers in the past but are now largely of historical interest.

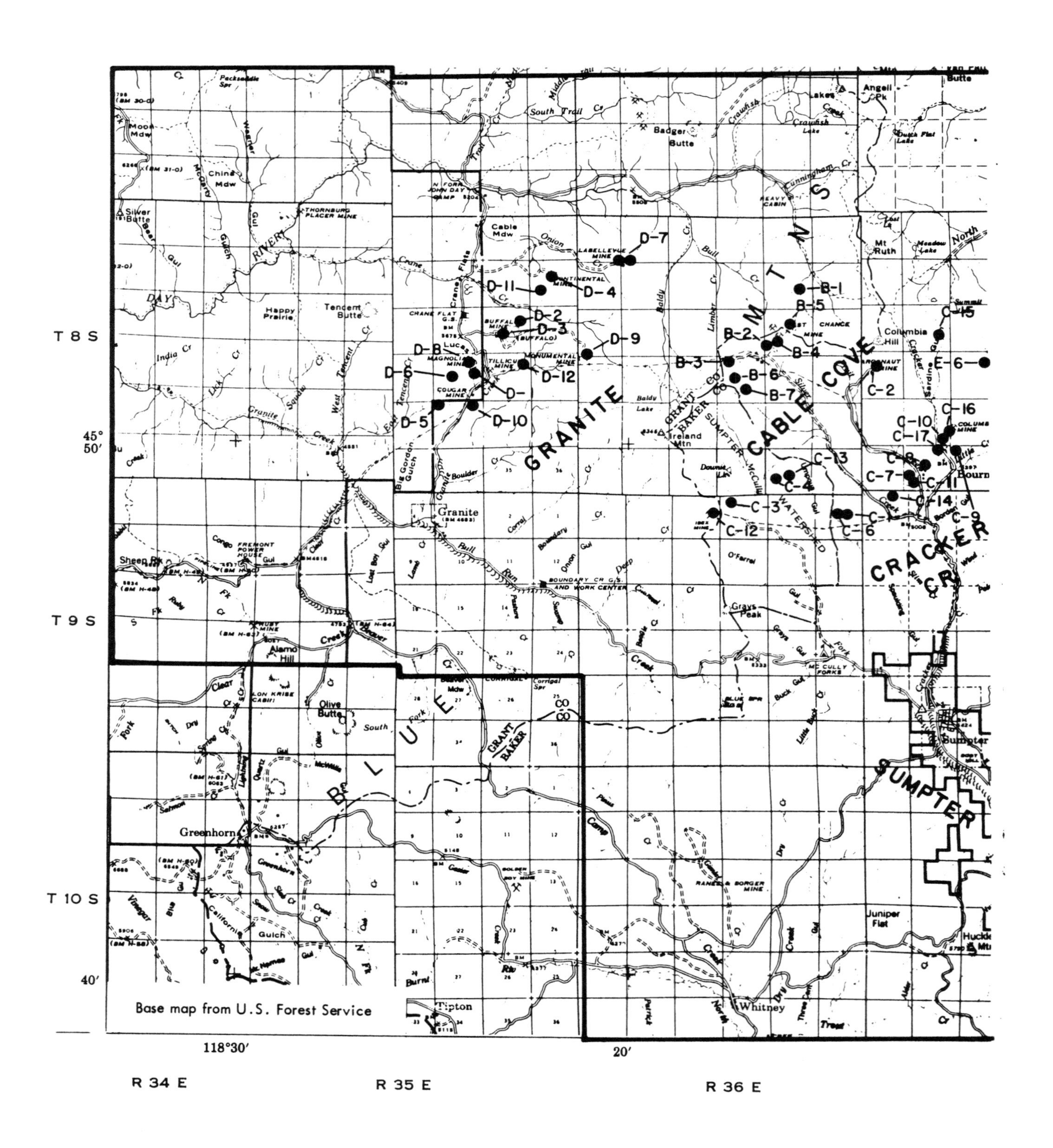

Figure 18. Index map of the Elkhorn Mountains area.

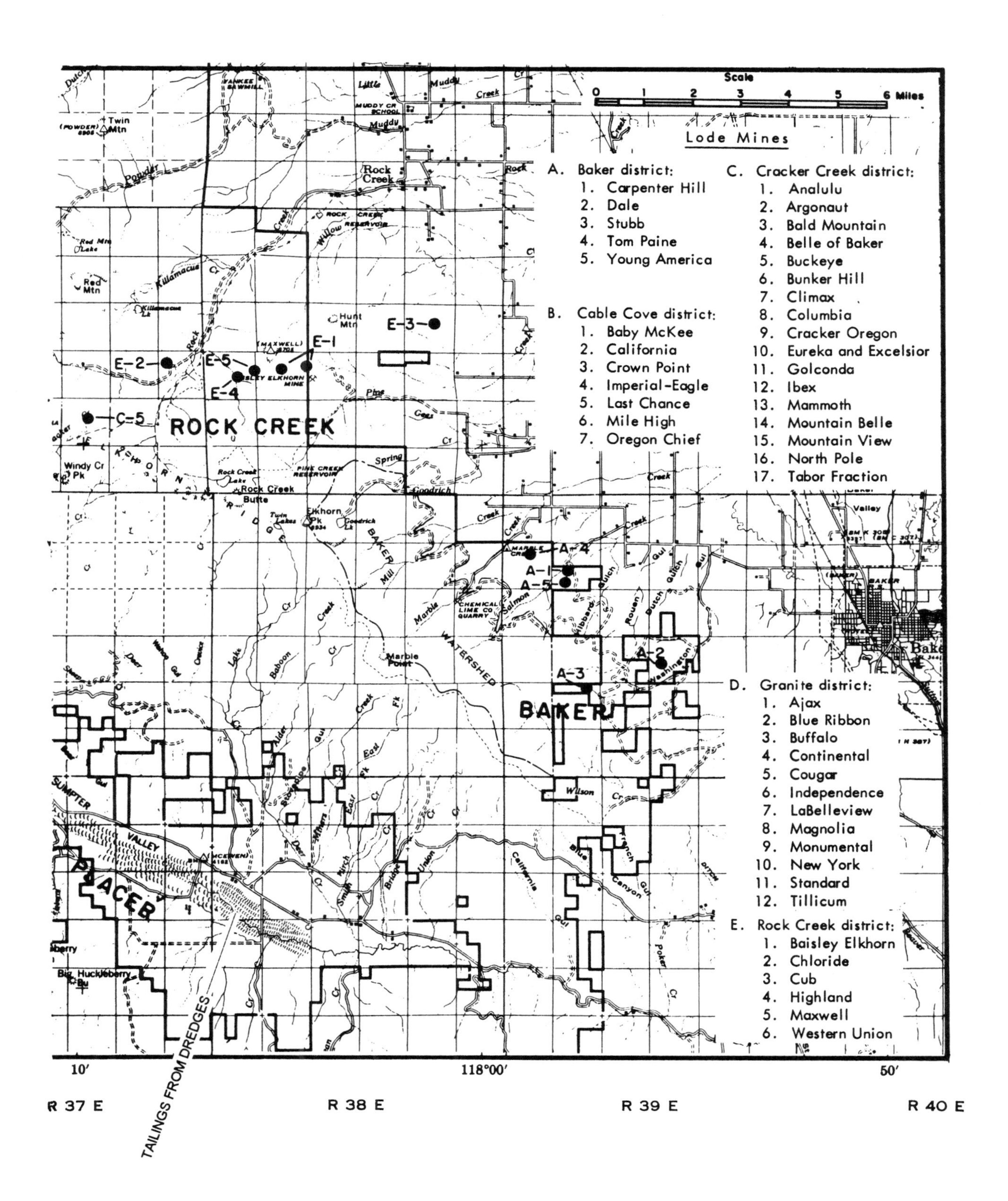
Scale
0 1 2 3 4 5 6 Miles
Lode Mines
A. Baker district:
1. Carpenter Hill
2. Dale
3. Stubb
4. Tom Paine
5. Young America
B. Cable Cove district:
1. Baby McKee
2. California
3. Crown Point
4. Imperial-Eagle
5. Last Chance
6. Mile High
7. Oregon Chief
C. Cracker Creek district:
1. Analulu
2. Argonaut
3. Bald Mountain
4. Belle of Baker
5. Buckeye
6. Bunker Hill
7. Climax
8. Columbia
9. Cracker Oregon
10. Eureka and Excelsior
11. Golconda
12. Ibex
13. Mammoth
14. Mountain Belle
15. Mountain View
16. North Pole
17. Tabor Fraction
D. Granite district:
1. Ajax
2. Blue Ribbon
3. Buffalo
4. Continental
5. Cougar
6. Independence
7. LaBelleview
8. Magnolia
9. Monumental
10. New York
11. Standard
12. Tillicum
E. Rock Creek district:
1. Baisley Elkhorn
2. Chloride
3. Cub
4. Highland
5. Maxwell
6. Western Union
ROCK CREEK
BAKER
PLACER
E-1
E-2
E-3
E-4
E-5
C-5
A-1
A-2
A-3
A-4
A-5
Twin Mtn
Rock Creek
Hunt Mtn
Elkhorn Pk
Marble Point
WATERSHED
SUMPTER VALLEY
TAILINGS FROM DREDGES
10′
118°00′
50′
R 37 E
R 38 E
R 39 E
R 40 E

Geography

The Elkhorn Mountains are topographically rugged, heavily timbered except at high elevations, and contain many permanent streams. Elkhorn Ridge, a prominent southeast-trending prong of the range, forms the bold and rugged skyline west of Baker Valley and north of Sumpter Valley. For much of its length, this ridge stands above 8000 feet in elevation, one of the highest points being Rock Creek Butte (9105 feet). The streams and gulches draining the ridge are all tributary to Powder River. North and a little west of Sumpter, Elkhorn Ridge intersects the straggling, generally north-south trending, divide between the Powder River drainage to the east and the John Day River drainage to the west. North along the divide is the Anthony Lakes country. To the south is Mount Ireland, formerly known as Bald Mountain, a prominent peak 8346 feet in elevation. For several miles this divide marks the boundary between Baker and Grant Counties.

Geology and Mineralization

The Bald Mountain batholith is a composite intrusive of early Cretaceous age composed of at least eight distinct rock types, ranging from norite to quartz monzonite with tonolite and granodiorite comprising about 97 percent of the rocks exposed (Taubeneck, 1957, p. 181-238). The batholith intrudes Permian Elkhorn Ridge Argillite and a metagabbro-serpentine complex of post-Permian and pre-Late Triassic age. Both of these older units were tightly folded prior to emplacement of the granitic intrusives. Near the edges of the batholith widespread faulting has resulted in intricate, almost unsolvable disruption of strata (Pardee, 1909, p. 88; Pardee and Hewett, 1914, p. 34; and Taubeneck, 1957, p. 187). These writers considered the faulting as later than the folding but earlier than, or contemporaneous with, the emplacement of the intrusives.

Most of the gold deposits along the southern contact of the batholith occur in the argillite, although several are wholly in granodiorite and some cut both. The veins characteristically strike northeast to east and dip steeply. They are arranged in extensive groups or systems. The veins in the Cracker Creek and Rock Creek districts produce an en echelon or overlapping system that is nearly continuous for about 12 miles. Included in this system is the important North Pole-Columbia lode that is traceable for about 5 miles. To the west, the veins of the Cable Cove district form a similar but much shorter group. The vein system of the Granite district farther west is about 6 miles long.

Hewett (1931) has recognized a zonal arrangement of the ores along the southern edge of the batholith. He points out that the ores of the outer zone contain relatively fewer sulfides and more free gold than the ores of the inner zone. He also states (p. 315): "Almost all the veins, and especially those which have been the source of most of the production, are composite; they are the result of several epochs of mineralization....the general order of deposition of the minerals is as follows: (1) Quartz in several epochs of deposition, the first generally replacing the country rock; (2) sulfides, sulfarsenides, and sulfantimonides; (3) gold. In most of the veins examined that contain quartz, sulfides and gold, the relations of these minerals reveal at least three epochs of mineralization, each distinguished by an uncommonly high proportion of that mineral or minerals."

Each new epoch of mineralization was preceeded by brecciation of earlier material. Hewett (p.345) further states: "From both the extent of underground explorations in length and depth and the relations of the fractures to the geologic features of the region, it is clear that the veins of the Sumpter region have more than average persistence horizontally, and it seems that they should persist vertically much deeper than they have been explored thus far, especially those veins that are grouped around the Bald Mountain batholith."

Production

From the standpoint of total output and number of mines with significant past production, the Elkhorn Mountains area ranks first among the gold- and silver-mining areas of Oregon. Incomplete records and estimates of production presented by various authorities indicate that total output exceeds 1,200,000 ounces gold and $700,000 in silver. About 55 percent of the gold came from lode mines and the rest from placers. Dredge operations account for more than half of the placer production. Roughly 65 percent of the production from lode deposits came from mines along the North Pole-Columbia lode in the Cracker Creek district. Other mines with important past production are the Baisley-Elkhorn and Highland-Maxwell in the Rock Creek district and the Buffalo, Cougar-Independence, and La Belleview in the Granite district.

Placer Mines

The most productive placers in the Elkhorn Mountains were in the vicinity of Sumpter and Granite and on the east flank of Elkhorn Ridge south and west of Baker.

Sumpter placers

The many streams and gulches draining the Cracker Creek district join waters to form the Powder River at Sumpter. For approximately 8 miles along the Powder River below Sumpter almost the entire width of the mile-wide valley has been mined by bucketline dredges. Operations were nearly continuous during 1913-1924 and 1935-1954. All of the dredges combined covered 2603 acres and handled 60,625,514 yards of gravel. The first bucketline dredge was constructed and operated by Powder River Dredging Co. during 1913-1917 and 1920-1924. A second boat began work in 1915 and operated through 1922. In 1935 operation of a third bucketline dredge was begun by Sumpter Valley Dredging Co. and was continued by successor companies - Baker Dredging Co., 1948-1950, and a second Powder River Dredging Co., 1950-1954. The dredge was a Yuba-Electric equipped with 72 nine-cubic-foot buckets and dug to an average depth of 18 feet. Between 20 and 30 men were employed. Except for a three-year shutdown during World War II, this dredge operated almost continuously through September 1954. During 1946-1954 its annual production often exceeded the output of all other gold mines, both lode and placer, in the state. Throughout its 20-year history the dredge produced more gold than any other placer in the state. Its output exceeded that of all lode mines except during the years 1939-1941, when the Cornucopia mine was the largest producer. Recorded output from all of the bucketline dredges totals 296,906 ounces gold and 70,983 ounces silver, which at today's prices is equal to about $10,483,000.

For a mile or two above Sumpter the gravels of both Cracker Creek and McCully Fork have been worked in part by dragline dredge and in part by dry-land washing plant. Upper portions of these creeks were earlier exploited by hand and hydraulic methods, as were parts of some of the streams and gulches tributary to them, such as Buck Gulch and Mammoth Gulch. On the adjacent slopes are bench gravels and scattered remnants of an early Tertiary drainage system which has been the source of considerable placer gold (Pardee and Hewett, 1941).

Granite placers

Granite Creek and the streams and gulches draining into it, especially Bull Run and Clear Creeks, have yielded a large amount of placer gold. Pardee and Hewett (1914, p. 10) computed the minimum output to 1914 at $1,033,000. During 1938-1942 and 1946-1951 a bucketline dredge was operated almost continuously on Granite, Clear, and Bull Run Creeks. Output from the dredge was large, but statistics are not available for publication.

To the north, along the North Fork of the John Day River, are other placers of past importance. Pardee and Hewett (1914, p. 10) estimated production of placer gold from this area as being not less than $893,000. Placering, mainly on a small scale, has continued off and on up to the present time. Some of the more important deposits of early days were the Klopp placer mine and adjacent diggings in the vicinity of the junction of Trail Creek with the North Fork, the French Diggings about 6 miles up the South Fork of Trail Creek from the Klopp mine, and the Thornburg placers on the North Fork about 5 miles below the Klopp mine. These deposits were worked mainly with hydraulic equipment. Some of the operations were of large scale. The Davis and Calhoun and Howell placers several miles farther down the North Fork were worked by washing plant and dragline during 1940-1942 and 1947-1950. Several thousand ounces of gold was produced.

Elkhorn Ridge placers

Nearly every stream and gulch on the east flank of Elkhorn Ridge south and west of Baker show evidence of early-day placer operations. Some of the diggings are thought to have been very rich locally, but much of the best ground was worked before records began to be kept and so no reliable estimates of production are available.

Some of the more noted placers are those of French Gulch and Blue Canyon near the old townsite of Auburn. The Auburn Ditch, which was built in 1863, took water from the head of Pine Creek and intermediate water courses and carried it down to Auburn a distance of more than 30 miles. The ditch is now part of the Baker city water system. Little is left to mark the old townsite of Auburn.

North of Auburn are the placers of Elk Creek, Griffin Gulch, Washington Gulch, and Salmon Creek. Production from the Nelson placers, an alluvial fan at the mouth of Salmon Creek, is believed to have exceeded $400,000 by 1900 (Lindgren, 1901, p. 652). West of Auburn, old placer diggings are found in Poker Gulch, California Gulch, Union Creek, Miners Creek, and several tributaries that drain southwestward into Deer Creek, a tributary of Powder River.

Camp Carson placers

The Camp Carson district (not shown on index map) lies near the head of the Grande Ronde River toward the north end of the Elkhorn Range and near the north boundary of the Bald Mountain batholith. The district is about 20 miles air line north of Sumpter, but is usually reached from La Grande. The district is best known for the Camp Carson hydraulic placers at the head of Tanners Gulch, which were first worked in the early 1860's. Although the gravels are extensive and much work was done, no production records are available. Several quartz veins have been prospected, but production records are scarce.

Baker District

The Baker district embraces the southeastern end of Elkhorn Ridge southwest of Baker (figure 18). The area is bordered on the north by Baker Valley and on the south and east by Powder River. Elevations range from about 4000 to 7000 feet, although the main ridge is considerably higher a few miles to the northwest. The north and south flanks of Elkhorn Ridge are 4 to 5 miles wide, and are steep and deeply incised. The eastern slope below 5000 feet elevation is more gentle. The streams and gulches draining the area are tributary to the Powder River. Parts of the district have been referred to in the past as the Pocahontas, Auburn, and Minersville districts.

The oldest rocks in the district are siliceous argillites with subordinate greenstones and small limestone lenses of the Elkhorn Ridge Argillite. These rocks are cut by numerous irregularly shaped masses of gabbro and related rocks. Younger porphyry dikes probably related to the Bald Mountain batholith intrude both the argillite series and the gabbroic rocks.

The nature of the quartz deposits is best summarized by Lindgren (1901, p. 650-651): "Though... nearly every creek and gulch heading in this part of the range has carried more or less placer gold and a few have been enormously rich, there is throughout a very marked absence of important vein systems to which the origin of these placers could be attributed. In part this may be due to insufficient prospecting, but in most cases I believe that the placer gold was here rather derived from small seams and veinlets than from prominent fissure veins."

A few lode prospects are located in the drainage area of Salmon Creek and Marble Creek. This area was formerly known as the Pocahontas district and is credited by Lindgren (1901, p. 651) with a total output exceeding $100,000; one mine, the Tom Paine, produced $70,000 from "one small chimney" in about 1882. Small quartz prospects are also known in the Washington Gulch and Auburn areas and on the south slope of the range in the drainage of Deer Creek.

Cable Cove District

Cable Cove lies at the head of Silver Creek about 12 miles northwest of Sumpter (figure 18). This amphitheater-shaped feature is a glacial cirque cut into the southeast-facing side of the Powder River-John Day River divide. At its widest part Cable Cove is roughly a mile and a half from rim to rim and about 800 feet deep. The inner slopes are steep and rocky. The outer slopes descend more gently north and west to the North Fork of the John Day River. Elevations range from about 6500 to more than 7000 feet above sea level. The road up Silver Creek is ordinarily closed by snow from November through May.

The gold deposits of the Cable Cove district are in granodiorite of the Bald Mountain batholith. Basalt dikes are present but not abundant. Shearing in a northeasterly direction has developed a broad system of faults and shear zones in which the gold veins were formed.

Most of the mines worked veins made up of brecciated and intensely altered granodiorite interspersed with lenses and streaks of quartz and a little calcite. Pyrite and arsenopyrite are the chief metallic minerals. Lead and zinc sulfides are present locally. The gold values were confined mainly to portions of veins rich in pyrite and arsenopyrite. Much of the ore exceeded 10 percent sulfides. The Imperial-Eagle vein is traceable for more than 2 miles, and is as much as 25 feet wide. Ore shoots were rarely more than 2 feet wide, short, and unpredictably located.

The district was first worked in the early 1870's. The main period of development was between 1899 and 1910. Production has been small, probably less than $200,000, although several veins are extensively developed. The bulk of production is credited to the Imperial-Eagle, California, and Last Chance mines in the order of their productive rank. The workings of the Imperial Eagle mine are said to aggregate about 10,000 feet.

Cracker Creek District

General information

The Cracker Creek district encompasses the area drained by Cracker Creek and McCully Fork, headwaters of the Powder River north of Sumpter (figure 18). The district includes the highly productive North Pole-Columbia lode and numerous lesser veins of a system which extends southwesterly from the head of Rock Creek on the Elkhorn Ridge divide to the Ibex mine at the head of Deep Creek on the John Day River divide. In the past, parts of the area have been referred to as the Bourne district and McCully Fork district, but there is now little reason for such division.

The North Pole-Columbia lode crosses Cracker Creek near Bourne, a ghost town 6 miles by good forest road north of Sumpter. The Bald Mountain and Ibex mines are reached by road up McCully Fork from Sumpter. Branch roads, some of which are steep and poorly maintained, extend to most of the other mines in the district. Elevations at the mines range between 5500 and 8000 feet. The area is heavily timbered except at very high elevations.

The district is underlain mainly by dark-colored siliceous to tuffaceous argillites of the Elkhorn Ridge Argillite. Interbedded greenstones and small limestone pods are seen in places. Also present locally are large bodies of metagabbro. The southern contact of the Bald Mountain batholith extends along the northern and western edges of the district. Many dikes related to the latter intrusive are found in the argillite series. Most of the gold deposits occur along steep, northeast-trending faults in argillite. Some, such as the Mountain View, Argonaut, and Mammoth veins, are very near the granodiorite contact and may locally cut the granodiorite.

Production from lode mines in the district is estimated at more than $9,000,000, most of which was produced by the mines along the North Pole-Columbia Lode. The Belle of Baker and Mountain View mines are credited by Hewett (1931, p. 318 and 321) with an estimated output of $400,000 and $90,000 respectively. Other mines from which production has been reported include the Bald Mountain, Ibex, Mammoth, Argonaut, and Climax. Placer production from gravels along creeks and gulches draining the district is probably equal to or perhaps greater than output from the lode mines.

Principal lode mines

North Pole-Columbia lode: The North Pole-Columbia lode is traceable for nearly 5 miles from Elkhorn Ridge southwesterly across Cracker and Fruit Creeks to McCully Fork. The more productive part of the lode is more than 2 miles in length, extending southwest and northeast from the old town of Bourne on Cracker Creek. In this stretch the mines are, from southwest to northeast, the Golconda, Columbia, Taber Fraction, Eureka and Excelsior (better known as the E & E), and the North Pole.

Claims were located on the lode in the 1870's but the main periods of activity of each of the mines fell between 1894 and 1916. Since 1916 there have been numerous ill-fated attempts to reactivate some of the mines and comparatively small output has been made periodically by lessees.

An unpublished report by William W. Elmer dated June 30, 1930 contains the statement: "The gross production of the mines has been not less than $8,000,000 of which amount $7,782,005.01 is shown in available records." Production figures presented by Elmer and the principal periods of activity of each of the mines follows:

North Pole mine	$2,485,006.96	1895-1908
Eureka and Excelsior	1,064,833.57	1894-1898; 1903-1905; 1920-1922
Taber Fraction	445,255.34	1903-1905
Columbia mine	3,638,959.60	1897-1916
Golconda mine	147,949.50	1903-1904
Total gross output	$7,782,005.21	

The Golconga mine was operated during 1897-1904. Pardee and Hewett (1914, p. 92) state: "According to J. A. Howard of Baker the total production to 1904 was $550,000."

The following statement of production for the North Pole mine was taken from records of Emil Melzer, manager of the mine during 1895-1908.

	Dry tons treated	Gold contents per ton	Gold Value	Ounces recovered and sold	
				Gold	Silver
Milling ore	157,801.84	$ 12.21	$1,927,836.06	72,186.496	78,268.758
Shipping ore	1,115.55	499.45	557,170.96	27,858.545	25,347.437
Total production	158,917.40	15.63	$2,485,006.96	100,045.041	103,616.195

Records concerning the grade and quantity of ore mined and treated at the other mines are presented in the individual mine reports at the end of this chapter.

Old maps (figure 19) indicate that ore shoots were nearly continuous for about 9000 feet along strike through the North Pole, E & E, Taber Fraction, and Columbia mines. About 1500 feet south of the Columbia is the Golconda mine with nearly 1200 feet of continuous stoping. Therefore, the over-all productive length of the lode has been close to 12,000 feet.

South from the Golconda mine the lode splits into two branches, along which a considerable amount of development work has been done with little apparent success. To the northeast beyond the North Pole mine, extensive developments along the lode on the adjoining South Pole ground also failed to yield an appreciable amount of ore.

The country rock along the North Pole-Columbia lode is in most places dark-gray to black argillite with rare exposures of greenstone. Granodiorite dikes are abundant. A highly altered dike that was probably originally a granodiorite porphyry has been encountered at several places along the more productive parts of the lode.

The North Pole-Columbia lode is a composite vein, in that it consists of several bands of quartz and silicified argillite breccia separated by gouge or sheared country rock. Between the Golconda and

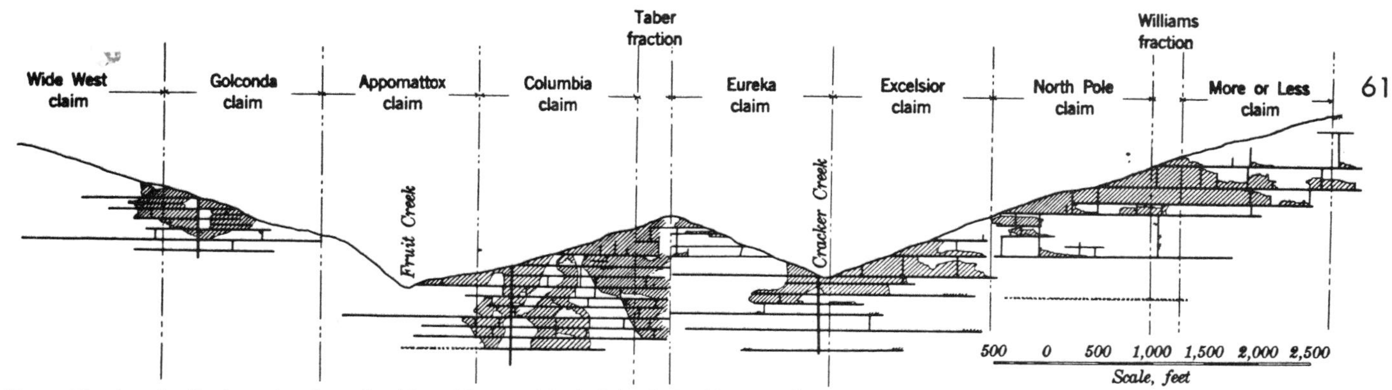

Figure 19. Longitudinal section through old workings on North Pole-Columbia vein (reprinted from Lorain, 1938).

Armstrong nugget at 80.4 ounces, is exhibited in U.S. National Bank Baker City.

the North Pole mines, the over-all trend of the lode is N. 30° E. It varies in dip from $86\frac{1}{2}$° SE. at the Columbia to 78° SE. at the North Pole. In width the lode ranges from 7 feet to a maximum of about 300 feet, averaging perhaps about 25 feet. Its walls are fairly well defined in most places. According to Pardee (1909), this wide zone is a normal fault which has a vertical displacement of 400 feet and a horizontal displacement of approximately 1800 feet. As a result of this movement, which was probably accomplished by a series of small displacements over a long period of time, the rocks were shattered and in places thoroughly crushed. Thus a great amount of open space for the circulation of the ore-forming solutions was provided.

Much silica was introduced as replacement of the argillite and as quartz filling in open spaces. From place to place along and across the lode all gradations are found from fractured argillite impregnated with sulfides and cut by seams of quartz to almost completely silicified zones more than 100 feet wide. On the whole, vein quartz occupies a large part of the lode. The chief sulfides are pyrite, arsenopyrite, and chalcopyrite. Small amounts of tetrahedrite, stibnite, marcasite, and some tellurides have been found.

Swartley (1914, p. 151) states: "Most of the massive quartz does not contain to exceed $1 per ton in gold, while much of the less altered argillite is of low grade. The best values are more frequently contained in highly replaced argillite, and often bear a close relation to a gouge streak. The gold occurs chiefly in fine arsenopyrite....The ore is usually in a series of overlapping lenses, which make up the several shoots found in the developed part of the lode. These lenses vary from a mere seam to 25 feet in width." The over-all width of ore mined probably averages about 4 feet.

Ore shoots were usually found along or very near the footwall of the lode, although hanging wall shoots were not uncommon, and in at least one place ore was mined from a shoot near the middle of the lode.

Continuous stopes range up to as much as 2500 feet in length in the North Pole mine and ore was mined to a depth of 1000 feet in the Columbia mine. The horizontal length of most stoped areas, except at the Columbia mine, greatly exceeded the depth to which mining was carried on. Vertical distance from the highest point at which ore was mined in North Pole ground to the low point in Columbia ground is about 2100 feet. The horizontal interval between these points is about 6000 feet. A considerable block of ground within this zone has not been fully explored and there is incomplete development between the walls of the lode where drifting has been done.

During early productive days (figures 20 and 21) the Golconda, Columbia, and E and E mines were equipped with 20-stamp mills and the North Pole with a 30-stamp mill and cyanide plant. Sulfide concentration was accomplished mainly with tables and vanners. Bulk roasting and cyanidation was tried at the North Pole mine for a time, but later only the tailings were cyanided. Swartley (1914, p. 155) states that the combined average mill recovery from the several plants did not exceed 67 percent. The tailings were released into flowing streams and irretrievably scattered. A 50-ton-per-day flotation plant was erected in 1942 but has been little used.

Granite District

General information

The Granite district lies in the upper reaches of the North Fork of the John Day River, which drains the west flank of the Elkhorn Mountains (figure 18). The principal lode mines occury a northeast-trending belt about 2 miles wide and 5 miles long that extends from the Cougar mine on the southwest to the La Belleview mine on the northeast. The old ghost town of Granite, one of the earliest settlements in eastern Oregon, is about 3 miles south of the Cougar mine and 14 miles by gravelled road west of Sumpter. The southern part of this area is drained by Granite Creek and the northern part by Crane and Onion Creeks. Timber is abundant. Elevations range from about 5000 feet at the Cougar-Independence mill on Granite Creek to 7000 feet at the La Belleview mine.

The lode mines of the Granite district lie along the southwestern edge of the Bald Mountain batholith. The veins are mainly in older argillite of the Elkhorn Ridge Argillite, but a few also cut granodiorite of the batholith. Dikes related to the batholith have been observed in several of the mines. Most of

Figure 20. Columbia mine near Sumpter about 1914.

Figure 21. E & E mine mill about 1900. (Photograph courtesy of Baker County Historical Society.)

them are intensely altered near the veins. With very few exceptions, the veins of the district strike northeast and dip steeply either east or west.

Production from lode mines in the district exceeds $2.2 million, most of which has come from the Buffalo, Cougar-Independence, and La Belleview mines. Several other lode mines have had small production.

Pardee and Hewett (1914) estimate that the placers along Granite Creek and the North Fork of the John Day River and their tributaries produced nearly $2,000,000 to 1914. Production figures are not available for a dredge that operated for most of the period between 1938 and 1951 on Granite, Clear, and Bull Run Creeks. Although total placer output for the district is not recorded, it is estimated to be well in excess of $5,000,000.

Principal lode mines

Cougar and Independence mines: The two mines are located on the ridge west of Granite Creek about 3 miles north of Granite; they are jointly owned. During 1939-1942 ore from both mines was treated in a large mill which stood adjacent to the Granite Creek road below the mines. Since 1907, according to figures presented by Koch (1959, p. 30-31), the mines have collectively produced 22,509 ounces gold; 27,629 ounces silver; 8,032 pounds of copper; 866 pounds of lead; and 19 pounds of zinc from 61,125 tons of ore. More than 75 percent of this came from the Cougar mine during 1938-1942. Koch states: "There are two principal veins, the Cougar and Independence....These veins are more or less parallel, striking about N. 50° E. and dipping about 70° E. It is entirely possible that the Independence and Magnolia veins are the same and that the Cougar and Ajax veins are the same, but to determine this would require examining the intervening ground in detail and trenching the surface. In addition to the two principal veins there are numerous small veins and stringers....All of the veins are in argillite that strikes consistently northwest and dips southwest." One of the largest Cougar ore shoots was 475 feet long, 3 to 4.5 feet wide, and was worked to a depth of about 450 feet.

Buffalo mine: The Buffalo is the largest mine in the district; it is 5 miles by road north of Granite. Development began here in the mid-1880's but production records are not available for the years prior to 1903. Records since then indicate that the mine has been almost continuously active, although in some years there was little or no production. Since 1958 activity at the mine has been confined to exploration and development of the veins to greater depths. This work has been only periodically pursued and mill operations have been limited to the treatment of development ore from the lower levels. Recorded production from 1903 through 1964 totals 33,418 ounces of gold and 239,305 ounces of silver from 42,246 tons of ore. In addition, small amounts of copper, lead, and zinc are recovered from the ores but add little or nothing to their value. James Jackson, operator of the mine from 1951 through 1965, suggests that, considering the total stope area, production may actually be about double the recorded amount.

The mine is developed by about 10,000 feet of drifts and crosscuts divided among four adit levels, the 200, 400, 500, and 600 levels. The lowest, or 600, level is 450 feet below the highest, or 200-foot level. Present development is confined to the two lower levels.

The Buffalo mine works five roughly parallel veins. These are, from west to east, the Monitor, No. 1, No. 2, No. 3, and Constitution. The Monitor vein is a silicified breccia zone in argillite which has yielded little or no ore. The other four are roughly parallel composite quartz veins having similar structure and mineralogy. All have been productive, although the No. 2 and Constitution veins have been the most important. Since the early 1940's, mining has been confined to the Constitution vein. Probably no less than 4000 feet of drifting had been done previously on the No. 1, No. 2, and No. 3 veins on the 200 and 400 levels and, according to old maps, much of the ground above the drifts was stoped. These workings have long been abandoned.

The four quartz veins range from 80 to 220 feet apart, strike N. 15° to 30° E., and dip 60° to 80° W., except the north end of the Constitution vein, which has rolled in dip to steeply east.

The Constitution vein cuts back and forth across the contact between argillite and granodiorite. The other veins are said to lie largely in argillite. The argillite strikes N. 30° to 40° E. and dips 45° SW. The veins, formed partly by open-space filling and partly by replacement, contain pyrite with minor amounts of arsenopyrite, chalcopyrite, galena, sphalerite, and tetrahedrite in a gangue of quartz, calcite, and incompletely replaced fragments and streaks of wall rock. The veins range in thickness from 1 to 6 feet, averaging probably about 20 inches. The sulfide minerals, which make up on the average about 10 percent of the ore and contain nearly all of the values, are concentrated by bulk flotation and the concentrates shipped to smelters. Free gold is rarely visible and no silver minerals other than tetrahedrite and a very little pyrargyrite have been identified.

During 1953-1958 treatment of 7938 tons of ore yielded 8760 ounces of gold and 62,004 silver. Mill recovery averaged 90 to 94 percent. Ratio of concentration was 8:1 to 10:1. Concentrates averaged 9 ounces gold and 68 ounces silver.

La Belleview mine: The La Belleview mine is at the head of Onion Creek about 3 miles air line northeast of the Buffalo. This mine produced close to $500,000 during the periods 1878-1892, 1927-1929, and 1939-1941. About 6000 feet of drifting and crosscutting has been done from four adits over a vertical range of about 600 feet.

The total production up to 1911, including ore shipped crude and ore treated in a mill on Onion Creek, amounted to 8000 tons having a gross value of $200,000. Concentrates averaged 1.20 ounces gold and 55 ounces silver to the ton and shipping ore was worth $60 to $300 to the ton. The 4800 tons of ore treated in 1940 yielded 1001 tons of concentrates containing 811 ounces gold; 40,444 ounces silver; 8243 pounds of copper; and 33,732 pounds of lead.

The country rock is quartz-biotite gneiss and granodiorite. The workings expose several veins, but most of the development follows one vein which trends N. 35° to 45° E., dips 65° to 70° NW., and varies from a narrow seam of chloritic gouge to 4 feet of crushed rock containing streaks of quartz with pyrite, arsenopyrite, and minor galena, chalcopyrite, and tetrahedrite. Silver minerals include argentiferous tetrahedrite, pyrargyrite, native silver, and possibly proustite. The massive pyrite rarely carried much gold. The ores were richest where galena, chalcopyrite, and antimony minerals were present. In the poorest zones the pyrite is dense and the other sulfides are only sparingly present.

Two shoots were mined, one pitching nearly vertically; the other had a rake of 30° SW. The intersection of the two shoots is said to have been very rich, with ore running $500 to $600 to the ton. One stope attained a length of 280 feet. The lower portion of this shoot yielded ore containing 0.40 ounce gold and 15 ounces silver.

Rock Creek District

General information

The Rock Creek district includes the upper reaches of Rock and Pine Creeks, which head close together in the high mountains northwest of Baker (figure 18). Here Elkhorn Ridge culminates in a straggling network of barren sawtooth ridges and peaks, some of which tower more than 5000 feet above Baker Valley. Elevations range from 5500 to 8500 feet. The principal mine workings lie between 6000 and 7000 feet, although some of the veins apex at much higher elevations.

The intrusive contact between rocks of the Elkhorn Ridge Argillite to the south and granodiorite of the Bald Mountain batholith to the north extends easterly through the district. The intruded rocks are mainly dark-colored argillites which, near the contact, have been altered to fine-grained crystalline hornfels. A number of veins occur along or close to the argillite-granodiorite contact. Some are in argillite, some are in granodiorite, and others cross the contact. This system of veins strikes northeast and connects with the Cracker Creek system on the southwest.

Past production of the Rock Creek district is credited largely to the Baisley-Elkhorn and Highland-Maxwell mines, whose combined output totals more than $1,560,000. Placer output from the district has been small.

Principal lode mines

Baisley-Elkhorn mine: The Baisley-Elkhorn mine is on the north fork of Pine Creek about 18 miles from Baker by road. The last few miles are in poor condition.

The Baisley-Elkhorn vein was discovered in about 1882 and for several years was worked by two different mines, the Baisley-Elkhorn and the Robbins-Elkhorn. Around 1900 the two properties were consolidated. Since 1907 there has been very little production, although during the 1920's considerable development work was done. A small output was made in the late 1930's. Pardee and Hewett (1914, p. 74) presented the following production statistics but indicated that the statement may be incomplete, since there is no record of production from 1901 to 1905.

Prior to Jan. 1, 1898:	$ 342,861.07
1898 to Dec. 1, 1900:	
26,095 tons crude ore (bullion)	84,591.64
3759 tons concentrates	239,529.84
472 tons shipped at $45.03 per ton	21,254.04
1905: 20,000 tons crude ore, yielding 3000 tons concentrates	210,000.00
1907: 7680 tons crude ore, yielding 1280 tons concentrates	38,481.00
1912: Small production	?
Total	$ 936,717.59

The vein is developed by about 10,000 feet of workings. Most of the mining was done through a 626-foot adit crosscut to the vein and a 400-foot shaft sunk on the vein from the adit level. The adit level is about 6700 feet in elevation. The lowest of four levels turned from the shaft is about 665 feet below the outcrop.

The vein has an average strike of N. 40° E., a nearly vertical dip, and is partly in granodiorite and partly in argillite. The ore shoots varied from 2 to 10 feet in width and were chiefly in granodiorite. Lindgren (1901, p. 647) describes the ore as "...a soft mixture of coarse sulfides with much crushed diorite and occasional streaks of quartz which may show comb structure; in one place a 2-foot ore streak was adjoined by 10 inches of white, barren quartz.

"...The sulfides in order of their abundance are pyrite, black zinc blende, galena, and chalcopyrite....Ruby silver is occasionally found."

Pardee and Hewett (1914, p. 77) state: "The vein is reported to be traceable for 1800 feet on the surface and has been explored for 1400 feet on the second level from the shaft. Within this distance two shoots have been found, the Baisley-Elkhorn 850 feet long, and the Robbins-Elkhorn 150 feet long. Both appear to have pitched directly down the dip of the vein. Though the former was stoped continuously to the third level 515 feet below the outcrop, the fourth level, 150 feet lower, appears to have found only sporadic masses of ore."

A crosscut adit, driven during the 1920's, intersects the vein about 750 feet east of the old shaft and about 285 feet below the lower level. A drift extends westward for about 1000 feet on the vein which, according to Hewett (1931, p. 326), "...contains one or two strands of quartz 6 to 12 inches wide, containing coarse sulfides in the midst of quartz-diorite gouge." No assay data are available.

Highland and Maxwell mines: The Highland and Maxwell mines, which work portions of the same vein, lie in the Rock Creek drainage about 2 miles west of the Baisley-Elkhorn mine. The old camp and lower-level adits of the Highland are on the east fork of Rock Creek at an elevation of about 6100 feet. The Maxwell claims cover the higher eastern extension of the vein; the uppermost tunnel on Maxwell ground is about 200 feet below the crest of the high ridge west of the Baisley-Elkhorn mine.

Principal periods of production were 1900-1905, 1909-1914, and 1936-1938. Records of production prior to 1935 are incomplete, but estimates contained in old reports are $375,000 for the Highland and $100,000 for the Maxwell. The Highland produced more than $100,000 in 1913 but was idle in 1914. The properties were consolidated during the 1936-1938 operations. Ore produced during that period had a gross value of approximately $150,000 and was treated in a 100-ton flotation plant.

The Highland-Maxwell vein crosses the east fork of Rock Creek just above the old Highland camp and is developed on both sides of the gulch line. It has been prospected by numerous tunnels and cuts for more than a mile along strike and over a vertical range of about 2000 feet. The vein is mostly in argillite, but in the upper eastern workings it is partly in granodiorite. The west extension of the vein strikes N. 75° E.; the east end strikes N.55° E. The dip is nearly vertical. The vein in most places is an incoherent mass of crushed and generally silicified argillite, clay, quartz, calcite, and ore minerals. The ore minerals in order of abundance are pyrite, sphalerite, galena, arsenopyrite, chalcopyrite, and tetrahedrite. The vein material is bounded by well-defined walls from which it breaks freely. Crosscuts locally show a width of as much as 28 feet between the walls, but the ore is seldom more than 3 feet thick. An old map of the property contained in Department files indicates that ore shoots were scattered. Stoped areas vary from less than 100 to more than 1000 feet in length. The larger shoots mined were in the lower levels.

During the period April 1936 to March 1938, about 7036 tons of ore having a gross value of $18.60 per ton were mined and treated. Ratio of sulfide concentration ranged from 4.5 to 1, to 7.5 to 1, with a recovery ratio of 75 to 98.7 percent. Net smelter returns averaged about $12.30 per ton of ore. From April 1936 to September 1, 1937 production amounted to 5400 tons of ore containing an average of 0.42 ounces gold and 3.65 ounces silver. Most of this ore came from the east end of the lowest level. The mine closed in the fall of 1938.

STAMP MILL: An implement or device used to bring down the (foot) forcibly as in smashing big rocks to make little rocks, or in the process of reducing chunks of ore to smaller pieces, from which gold and other valuable metals can be extracted. Throughout the gold areas, hundreds of stamp mills were used. Some single machines had a many as 25 heads (stamps). Pictured is the 5 stamp mill used at Custer, Idaho.

—Bert Webber photo June 1994

Lode Mines of the Elkhorn Mountains Area

A. BAKER DISTRICT

Carpenter Hill mine — Baker District, A-1

Location: Baker County, SW$\frac{1}{4}$ sec. 8, T. 9 S., R. 39 E., on Salmon Creek above Nelson placers.

Development: 1200-foot adit.

Geology: Several quartz veins in greenstone, the largest seldom wider than 6 inches.

Production: Records not available. Small amount of ore treated in 5-stamp mill in early days.

References: Pardee and Hewett, 1914:146; Swartley, 1914:162; Parks and Swartley, 1916:51.

Dale mine — Baker District, A-2

Location: Baker County, W$\frac{1}{2}$ sec. 22, T. 9 S., R. 39 E.

Development: Several adits and pits; one adit is more than 400 feet long.

Geology: Country rock is mostly gabbro. Fractured zone strikes N. 23° E. and dips 70° W. Quartz seams locally show free gold.

References: Grant and Cady, 1914:150; Parks and Swartley, 1916:85; Gilluly, Reed, and Park, 1933:83; Gilluly, 1937:103.

Stub (Kent) mine — Baker District, A-3

Location: Baker County, secs. 20 and 29, T. 9 S., R. 39 E., in upper Washington Gulch.

Development: Tunnel several hundred feet long, with a short winze and some raises.

Geology: Country rock is argillite, greenstone, and chert. Quartz-impregnated shear zone up to 15 feet wide. Strike NE; dip nearly vertical.

Production: No record.

References: Grant and Cady, 1914:150; Swartley, 1914:162; Parks and Swartley, 1916:135; Gilluly, Reed, and Park, 1933:83; Gilluly, 1937:102; Department Bulletin 14-A, 1939:16.

Tom Paine - Old Soldier Group (Yellowstone) — Baker District, A-4

Location: Baker County, NE$\frac{1}{4}$ sec. 7, T. 9 S., R. 39 E. in McCord Gulch.

Development: Three adits, longest 700 feet.

Geology: Two veins 600 feet apart in argillite and limestone. Both pre-mineral and post-mineral dikes are present. Tom Paine vein strikes N. 25° W. and dips 30° to 50° SW. Old Soldier vein strikes N. 68° W. and dips 42° S. Both vary from a few inches to about 4 feet in width and consist largely of quartz and a little calcite with a scattering of fine pyrite.

Production: $36,000 in early days.

References: Lindgren, 1901:651; Grant and Cady, 1914:147-149; Swartley, 1914:162; Parks and Swartley, 1916:240; Gilluly, Reed, and Park, 1933:81; Gilluly, 1937:101; Department Bulletin 14-A, 1939:16.

Young America mine — Baker District, A-5

Location: Baker County, SW$\frac{1}{4}$ sec. 8, T. 9 S., R. 39 E. on Salmon Creek.

Development: 940 feet of tunnels.

Geology: Massive quartz vein, up to 5 feet wide, in greenstone. Strike NW.

Production: No record.

References: Grant and Cady, 1914:147; Swartley, 1914:162; Parks and Swartley, 1916:241; Department Bulletin 14-A, 1939:17.

B. CABLE COVE DISTRICT

Baby McKee mine — Cable Cove District, B-1

Location: Baker County, E$\frac{1}{2}$ sec. 11, T. 8 S., R. 36 E.

Development: Poor record. Said to include 1800-foot crosscut which cuts vein at depth of 900 feet.

Geology: Mineralized shear zone similar to that developed at California mine.

Production: No record.

References: Newspaper accounts circa 1900-1904.

California mine — Cable Cove District, B-2

Location: Baker County, secs. 14 and 15, T. 8 S., R. 36 E.

Development: Six adits over vertical range of 800 feet; longest 1750 feet.

Geology: Shear zone about 3 feet wide in granodiorite contains streaks and lenses of heavy sulfides, quartz, and calcite. Strikes NE. Sulfides mostly pyrite and arsenopyrite with galena, chalcopyrite, and sphalerite.

Production: About $40,000 mostly from sporadic output between 1873 and 1910. Mill erected in 1879, but little used. Estimated ratio of sulfides to gangue, 1:10.

References: Lindgren, 1901:675; Swartley, 1914:140; Parks and Swartley, 1916:49; Hewett, 1931:318; Lorain, 1938:19; Department Bulletin 14-A, 1939:18.

Crown Point mine

Cable Cove District, B-3

Location: Baker County, NW¼ sec. 22, T. 8 S., R. 36 E.

Development: 1800 feet of tunnels.

Geology: Shear zone as much as 4 feet in width in granodiorite contains quartz seams and scattering of pyrite and arsenopyrite.

Production: No record, property located in 1930.

References: Department Bulletin 14-A, 1939:18.

Imperial - Eagle mine

Cable Cove District, B-4

Location: Baker County, SW¼ sec. 14, T. 8 S., R. 36 E.

Development: Totals about 10,000 feet from six adits.

Geology: Three veins in granodiorite - Imperial, Eagle, and Winchester - made up of more or less completely crushed and altered granodiorite interrupted locally by streaks and lenses of heavy sulfides and quartz. Pyrite, arsenopyrite, chalcopyrite, galena, sphalerite. Vein zones said to vary up to 25 feet in width. Eagle vein traceable for two miles. Ore mined rarely exceeded 2 feet in width, with stope and pitch lengths of usually less than 50 feet.

Production: Estimated at $100,000 mostly during 1900-1915. Mill erected about 1900 and intermittently operated up to about 1910. Poor recovery. Estimated ratio of sulfides to gangue 1:10. Gold less than 10 percent free. Silver to gold ratio 10:1.

References: Lindgren, 1901:673; Pardee and Hewett, 1914:98; Swartley, 1914:140; Parks and Swartley, 1916:128; Hewett, 1931:314,317-8; Lorain, 1938:19; Department Bulletin 14-A, 1939:19-20.

Last Chance mine

Cable Cove District, B-5

Location: Baker County, sec. 14, T. 8 S., R. 36 E.

Development: Two adits and 150-foot shaft; 400-foot drift on vein.

Geology: Shear zone in granodiorite, strikes N. 50° E., dips 80° SE.; contains heavy sulfides locally. Ore shoots about 18 inches wide and less than 50 feet long.

Production: Estimated at $5,000. Inactive since about 1910.

References: Lindgren, 1901:675; Pardee and Hewett, 1914:102; Swartley, 1914:142; Parks and Swartley, 1916:139; Hewett, 1931:318; Lorain, 1938:19; Department Bulletin 14-A, 1939:20.

Mile High mine — Cable Cove District, B-6

Location: Baker County, sec. 22, T. 8 S., R. 36 E. on south wall of Cable Cove.

Development: Drift adit 640 feet long with two short raises and a small stope near face.

Geology: Shear zone in granodiorite. Strikes N. 45° E. and is nearly vertical. Vein splits 400 feet from portal. North branch thickens to 2 feet and contains streaks and bunches of pyrite with small amounts of chalcopyrite, galena, and sphalerite.

Production: It is said that several cars of crude ore were shipped during 1934-1939. Two shipments totaling 38 tons averaged about 0.84 ounces gold, 1.8 ounces silver, and 0.9 percent copper.

References: Department mine file report, 1963.

Oregon Chief mine — Cable Cove District, B-7

Location: Baker County, SE$\frac{1}{4}$ sec. 22, T. 8 S., R. 36 E.

Development: Several adits total 1400 feet.

Geology: Quartz vein as much as 4 feet wide in granodiorite.

Production: No record.

References: Parks and Swartley, 1916:171; Department Bulletin 14-A, 1939:21.

C. CRACKER CREEK DISTRICT

Analulu mine — Cracker Creek District, C-1

Location: Baker County, E$\frac{1}{2}$ sec. 1, T. 9 S., R. 36 E., about 9 miles from Sumpter on a branch of Silver Creek.

Development: About 600 feet of tunnels and shafts.

Geology: Vein as much as 10 feet in width in argillite is said to be the southwest extension of the North Pole-Columbia lode.

Production: No record.

References: Lindgren, 1901:667; Parks and Swartley, 1916:15; Hewett, 1931:321; Department Bulletin 14-A, 1939:34.

Argonaut mine

Cracker Creek District, C-2

Location: Baker County, NW$\frac{1}{4}$ sec. 19, T. 8 S., R. 37 E., about 4.5 miles by road northwest of the Columbia mine.

Development: Includes about 2000 feet of drifts from two adit levels 160 feet apart vertically.

Geology: Country rocks are argillite and granodiorite, cut by premineral aplite and granodiorite porphyry dikes. Two veins mostly in argillite; one strikes N. 30° E. and dips 65° to 70° E., the other strikes N. 80° E. and dips about 60° S. Ore zones are abruptly discontinuous and range from a few inches to about 6 feet in width.

Production: Small shipments made during 1937-1941 period.

References: Lorain, 1938:23; Department Bulletin 14-A, 1939:34; Department mine files.

Bald Mountain mine

Cracker Creek District, C-3

[See Ibex mine.]

Belle of Baker

Cracker Creek District, C-4

Location: Baker County, S$\frac{1}{2}$ sec. 35, T. 8 S., R. 36 E., on the divide between McCully Fork and Silver Creek; a few hundred feet west of the Mammoth mine and on the same vein.

Development: Shaft 385 feet deep and about 2000 feet of drifts distributed between four levels.

Geology: Vein cuts granodiorite and locally argillite; strikes N. 45° E.; dips 70° SE.; attains maximum width of 35 feet. Splits on the southwest and pinches rapidly on the northeast and below the 100-foot level. Gangue is mainly granodiorite and argillite breccia and gouge that has been irregularly silicified. Wire gold is present in thin quartz seams and along fractures. Roscoelite is also present. Pyrite and arsenopyrite form a small percentage of the ore.

Production: Estimated at $400,000 prior to 1931. Gold largely free; average grade of ore is low.

References: Lindgren, 1901:669; Pardee and Hewett, 1914:97; Parks and Swartley, 1916:148; Hewett, 1931:321; Department Bulletin 14-A, 1939:34.

Buckeye mine

Cracker Creek District, C-5

Location: Baker County, sec. 27, T. 8 S., R. 37 E., about 2 miles northeast of Bourne on the Rock Creek-Cracker Creek divide.

Development: Seven adits with an aggregate length of about 4000 feet attain depth of 900 feet below outcrop.

Geology: Country rock is argillite. The main vein, consisting of argillite breccia cemented with quartz, strikes N. 60° to 70° E., dips 70° SE. and averages about 4 feet in width for considerable distances. Branch fissures locally contain quartz. Percentage of sulfides is low.

Production: Estimated at $6,000 to 1931.

References: Pardee and Hewett, 1914:94; Swartley, 1914:160; Parks and Swartley, 1916:204; Hewett, 1931:18 and 32; Department Bulletin 14-A, 1939:35.

Bunker Hill mine

Cracker Creek District, C-6

Location: Baker County, sec. 1, T. 9 S., R. 36 E. On southwest extension of the North Pole-Columbia lode.

Development: Poor record. Exceeds 1000 feet, including 300-foot crosscut to vein.

Geology: Vein in argillite is a quartz-replacement breccia said to be 25 feet wide locally. See description of North Pole-Columbia lode.

Production: No record.

References: Lindgren, 1901:667; Department Bulletin 14-A, 1939:36.

Climax mine

Cracker Creek District, C-7

Location: Baker County, SW$\frac{1}{4}$ sec. 32, T. 8 S., R. 37 E. On a vein that is 600 to 800 feet northwest of the Columbia and Golconda mines. The vein approximately parallels the North Pole-Columbia lode.

Development: Includes two adits, one a crosscut 550 feet long, with several hundred feet of drifts.

Geology: Similar to North Pole-Columbia lode, though the vein here is much narrower, ranging from a few feet to 10 or 12 feet in width.

Production: Small. An ore shoot 50 feet long and 100 feet deep was worked out in early days.

References: Swartley, 1914:159; Parks and Swartley, 1916:56; Hewett, 1931:17; Lorain, 1938:20; Department Bulletin 14-A, 1939: 36.

Columbia mine

Cracker Creek District, C-8

Location: Baker County, sec. 32, T. 8 S., R. 37 E. on the North Pole-Columbia lode.

Development: About 50,000 feet of drifts, crosscuts, and raises from a shaft 918 feet deep and three adits.

Geology: See general description of North Pole-Columbia lode under Cracker Creek district. On Columbia ground the lode has an average strike of N. 34° E. and dip of 86½°SE. The walls are generally well defined and are rarely less than 25 feet apart. Maximum width is about 100 feet. The ore shoots ranged from 3 to 8 feet in width.

Production: $3,638,959.60 total during 1897-1916 period, according to report by W.W. Elmer dated June 30, 1930. Bullion shipments by way of First National Bank of Baker from 1897 to 1916 totaled 112,066.51 ounces, which contained 58,062.66 fine ounces gold and $26,365.92 in silver. Total bullion value was $1,226,521.15.

When closed in 1916, the mine was equipped with a 20-stamp amalgamation and concentration mill and a 60-ton cyanidation plant. Bullion averaged 518 fine in gold. Gold 40 percent free. Ratio of concentration 10:1 to 15:1. Recovery averaged about 75 percent. Mill ore averaged $11.04 per ton. Shipping ore averaged $212.72 per ton.

References: Lindgren, 1901:659-663; Pardee and Hewett, 1914:81-94; Swartley, 1914:146-159; Parks and Swartley, 1916:59-65; Hewett, 1931: 312 to 336; Lorain, 1938:19.

Cracker-Oregon Mine

Cracker Creek District, C-9

Location: Baker County, NW¼ sec. 33, T. 8 S., R. 37 E. On a vein about 1800 feet southeast of and paralleling the North Pole-Columbia lode.

Development: About 3000 feet mostly from adit levels.

Geology: Vein in argillite similar to North Pole-Columbia lode, though narrower and less well mineralized.

Production: Small output from 10-stamp mill in early days.

References: Swartley, 1914:159; Department Bulletin 14-A, 1939:36.

E and E mine

Cracker Creek District, C-10

Location: Baker County, NE¼ sec. 32, T. 8 S., R. 37 E. on a segment of the North Pole-Columbia lode and a short distance up Cracker Creek from Bourne.

Development: A shaft 760 feet deep on the north bank of Cracker Creek and eight adits, three extending north and five south from Cracker Creek; total about 20,000 feet.

Geology: See general description of North Pole-Columbia lode under Cracker Creek district.

Production: Main periods of activity were 1891, 1894-1898, 1903-1905, and 1920-1922, according to Elmer report of 1930. Total output was $1,064,833.57. During 1894-1898, mill ore averaged $9.28 per ton. Ratio of gangue to sulfides, 25:1 to 20:1. Free gold less than 5 percent. Gold to silver ratio about 1 to 2. 100-ton flotation mill on property.

References: See Columbia mine.

Golconda mine — Cracker Creek District, C-11

Location: Baker County, SW$\frac{1}{4}$ sec. 32, T. 8 S., R. 37 E. on the North Pole-Columbia lode immediately southwest of the Columbia mine. The property embraces about 3000 feet of the lode.

Development: Totals about 7000 feet, including a 510-foot underground shaft with several levels which is located near the portal of a 1300-foot adit.

Geology: The North Pole-Columbia lode is here as much as 200 feet wide. Ore has been mined from three veins which are nearly parallel in strike (N. 45° E.) but differ greatly in dip. The western and intermediate veins have not been found below the adit level. The eastern-most vein is more persistant, dipping steeply NW. in the upper levels and SE. in the lower levels.

Production: Estimated at $550,000. Main period of activity was 1897-1904. Ratio of concentration 7:1 to 15:1. Free gold 40 to 50 percent. Gold-silver ratio in bullion 720 to 220.

References: Lindgren, 1901:665; Pardee and Hewett, 1914:92; Swartley, 1914:146; Parks and Swartley, 1916:100; Hewett, 1931: 321, 339; Department Bulletin 14-A, 1939:36.

Ibex mine (including Bald Mountain mine) — Cracker Creek District, C-12

Location: Baker County, sec. 4, T. 9 S., R. 36 E., near the head of Deep Creek and McCully Fork on the John Day River divide. The Ibex and Bald Mountain are contiguous mines on the same vein.

Development: Four adits and a shaft total about 10,000 feet; lowest adit cuts vein about 1000 feet below outcrop.

Geology: Vein in argillite near contact with younger granodiorite; strikes N. 25° to 60° E., dips 60° to 80° SE.; varies from 5 to 25 feet in width. Gangue is crushed argillite cemented by quartz and a little calcite. Chief sulfides are pyrite and arsenopyrite with tetrahedrite, marcasite, and pyrargyrite present locally.

Production: Meager record. Ratio of sulfides to gangue is less than 5 percent. Ratio of gold to silver about 1:10 or less. Gold 30 percent free. Small production.

References: Lindgren, 1901:667; Pardee and Hewett, 1914:95; Swartley, 1914:144; Parks and Swartley, 1916:127; Hewett, 1931: 312 to 330; Lorain, 1938:22.

Mammoth mine

Cracker Creek District, C-13

Location: Baker County, SE$\frac{1}{4}$ sec. 35, T. 8 S., R. 36 E., a few hundred feet east of the Belle of Baker mine and on the same vein.

Development: Shaft 300 feet deep with drifts.

Geology: See Belle of Baker mine.

Production: Estimated at $40,000 prior to 1931. Bullion 500 to 600 fine.

References: See Belle of Baker mine.

Mountain Belle mine

Cracker Creek District, C-14

Location: Baker County, NE$\frac{1}{4}$ sec. 6, T. 9 S., R. 37 E.

Development: Shaft 300 feet deep and tunnel of unknown length.

Geology: Breccia zone in argillite, contains quartz and a little pyrite.

Production: No record.

References: Hewett, 1931: 321.

Mountain View mine

Cracker Creek District, C-15

Location: Baker County, SE$\frac{1}{4}$ sec. 17, T. 8 S., R. 37 E., 4 miles north of Bourne.

Development: Two adits and a 100-foot shaft.

Geology: Vein in argillite near contact with younger granodiorite. Vein strikes N. 45° E.; dips steeply SE. and is a maximum of 4 feet wide. Filling is silicified argillite breccia and gouge cut by quartz seams.

Production: $100,000 between 1904 and 1906. Small shipments made in mid-1930's.

References: Pardee and Hewett, 1914:93; Swartley, 1914:159; Parks and Swartley, 1916:158; Lorain, 1938:23; Department Bulletin 14-A, 1939:37.

North Pole mine

Cracker Creek District, C-16

Location: Baker County, secs. 28 and 29, T. 8 S., R. 37 E. on North Pole-Columbia lode.

Development: About 13,000 feet distributed among five adits and an intermediate level.

Geology: See general description of North Pole-Columbia lode under Cracker Creek district.

Production: Total recovery 1895-1908 -- 100,045.04 fine ounces gold and 103,616.19 fine ounces silver from 158,917.40 tons of ore.

References: See Columbia mine.

Taber Fraction mine — Cracker Creek District, C-17

Location: Baker County, sec. 32, T. 8 S., R. 37 E. Property embraces 300-foot segment of North Pole-Columbia lode between the Columbia and E. & E. mines.

Development: Levels above 500 interconnect with Columbia and E. & E. workings. Lower levels worked through Columbia shaft.

Geology: See general description of North Pole-Columbia lode under Cracker Creek district.

Production: During 1903-1905, 24,910.9 tons of ore were treated at Columbia mill. Total assay value was $295,881.54 or $11.88 per ton. Total recorded output, according to the Elmer report, was $445,255.34 including shipments prior to 1903.

References: See Columbia mine.

D. GRANITE DISTRICT

Ajax mine — Granite District, D-1

Location: Grant County, E½ sec. 22, T. 8 S., R. 35½ E., in Lucas Gulch, 3 miles north of Granite.

Development: 5 short adits; longest contains 500 feet of drift on Ajax vein.

Geology: Intersecting shear zones up to 6 feet wide in argillite contain discontinuous quartz seams and lenses. A small amount of pyrite is present. Ajax vein strikes about N. 70° E. and dips 47° to 63°SE. Snowbird vein strikes about N. 10° E. and dips westerly.

Production: Five-stamp mill on property at one time. Mill record unavailable. Gold-silver ratio about 1:1. A shoot 90 feet long in Ajax vein produced $40,000 in 1905-1906.

References: Pardee and Hewett, 1914:106; Parks and Swartley, 1916:7; Hewett, 1931:320; Department Bulletin 14-B, 1941:41; Koch, 1959:13.

Blue Ribbon mine — Granite District, D-2

Location: Grant County, NE¼ sec. 14, T. 8 S., R. 35½ E., about half a mile northeast of the Buffalo mine.

Development: Two adits about 250 feet apart vertically, and 40-foot underground shaft from upper adit. There is a total of about 2000 feet of workings, attaining a depth of 190 feet below outcrop.

Geology: Two or more subparallel veins in argillite intruded by basic dikes. Veins strike about N. 60° and dip 85° SE. Width 2 feet or less. Gangue is quartz and calcite. Sulfides are chiefly pyrite with some tetrahedrite.

Production: Small.

References: Lindgren, 1901:685; Swartley, 1914:137; Parks and Swartley, 1916:36; Hewett, 1931:319,339; Department Bulletin 14-B, 1941:42-43; Koch, 1959:14.

Buffalo mine

Granite District, D-3

Location: Grant County, sec. 14, T. 8 S., R. 35½ E., about 5 miles north of Granite.

Development: About 10,000 feet of crosscuts and drifts divided among four adits; attain depth of about 650 feet below highest point on outcrop.

Geology: Five roughly parallel veins mostly in argillite but partly in granodiorite. Four have been productive. The veins are 80 to 220 feet apart. They are from 1 to 6 feet wide, averaging about 20 inches. Strike N. 15° to 30° E., dip 60° to 80° W. Gangue is mainly quartz, calcite, and altered country rock containing pyrite, arsenopyrite, chalcopyrite, galena, sphalerite, tetrahedrite, and rare ruby silver. Gold rarely visible.

Production: Bulk sulfide flotation. Concentration ratio 8:1 to 10:1. Silver to gold ratio averages about 7.5:1. Recorded output from 1903 to 1965 was 33,142 ounces gold and 252,893 ounces silver from 42,246 tons of ore. The 1880-1903 output unknown.

References: Lindgren, 1901:685; Swartley, 1914:137; Pardee and Hewett, 1914:106; Parks and Swartley, 1916:46; Hewett, 1931:312 to 346; Department Bulletin 14-B, 1941:43-45; Koch, 1959:15-25.

Continental mine

Granite District, D-4

Location: Grant County, sec. 12, T. 8 S., R. 35½ E., about 8 miles by road northeast of Granite.

Development: Four short adits and several cuts. Longest adit contains about 900 feet of crosscuts and drifts.

Geology: Several short veins and mineralized fracture zones in quartz-mica schist derived from thermal alteration of argillite. Pegmatites numerous. Strike of veins is within a few degrees of east. Dips average about 45° S. Gangue is quartz. Sulfides are chiefly pyrite with minor amounts of arsenopyrite, sphalerite, and galena.

Production: About $50,000 said to have been produced from high-grade ore about 1915.

References: Department Bulletin 14-B, 1941:46-47; Koch, 1959:27-29.

Cougar mine

Granite District, D-5

Location: Grant County, NW¼ sec.27, T. 8 S., R. 35½ E., about 3 miles north of Granite.

Development: Four drift adits and 270-foot underground shaft with two connecting levels; total about 5000 feet of drifts and 2000 feet of raises. Vertical extent of workings about 440 feet.

Geology: Vein in argillite varies from a few inches to 9 feet in width; traced for about

2000 feet; strikes N. 43° to 50° E., dips 70° to 83° SE. Filling is largely argillite breccia, partly recemented by quartz and locally dolomite. Pre-mineral dikes are numerous, some being cut by the vein. Chief sulfides are pyrite and arsenopyrite. Ore shoots averaged 3 to 4 feet thick.

Production: 1938 to 1942 - 19,126.24 ounces gold and 10,976.30 ounces silver from roughly 51,500 tons of ore. Bulk flotation; sulfide concentration ratio 10:1. Little free gold. Ratio of gold to silver about 1.7:1.

References: Lindgren, 1901:683; Swartley, 1914:135; Pardee and Hewett, 1914:103-104; Parks and Swartley, 1916:81-82; Department Bulletin 14-B, 1941:47-51; Koch, 1959:30-35.

Independence mine

Granite District, D-6

Location: Grant County, sec. 22, T. 8 S., R. 35½ E., about 3½ miles north of Granite.

Development: 1100-foot crosscut adit and three drift levels totaling about 2500 feet explore vein to depth of 350 feet below outcrop.

Geology: Vein in argillite. Numerous light-colored dikes 2 to 4 feet wide. Vein strikes N. 50° E., dips 65° SE. Filling 3 to 6 feet wide is argillite breccia and gouge, cemented by dolomite and quartz. Sulfides mainly pyrite and arsenopyrite, with minor chalcopyrite and sphalerite; oxidized in the upper levels where manganese stains are abundant.

Production: Sporadic output from 1907 to 1940 totals 3,202.19 ounces gold, 14,582 ounces silver, and 4,724 pounds copper from about 9,500 tons of ore.

References: Swartley, 1914:135-136; Pardee and Hewett, 1914:104-105; Parks and Swartley, 1916:129; Department Bulletin 14-B, 1941:47-51; Koch, 1959:30-35.

La Belleview mine

Granite District, D-7

Location: Grant County, secs. 6 and 7, T. 8 S., R. 36 E. on Onion Creek about 10 miles by road northeast of Granite.

Development: 4 adits; total about 6000 feet of drifts and crosscuts reaching 500 feet below outcrop.

Geology: Vein in quartz-biotite gneiss, and granodiorite; varies from a few inches to 4 feet thick; strikes N. 35° to 45° E., dips 65° to 70° NW. Vein comprised mainly of crushed rock with seams and lenses of quartz and containing pyrite, arsenopyrite, galena, chalcopyrite, and silver-bearing tetrahedrite. Silver sulfides and native silver are rare.

Production: Estimated at about $500,000. Work began in 1878. 1939-1941 operations utilized 50-tons-per-day flotation plant. Ore runs 15 to 20 percent sulfides. Silver-to-gold ratio about 50:1.

References: Lindgren, 1901:685; Swartley, 1914:138; Pardee and Hewett, 1914:109; Parks and Swartley, 1916:137; Hewett, 1931:311 to 336; Lorain, 1938:18; Department Bulletin 14-B, 1941:53-55.

Magnolia mine — Granite District, D-8

Location: Grant County, NE¼ sec. 22, T. 8 S., R. 35½ E., in Lucas Gulch about 3½ miles north of Granite.

Development: Three adits, lowest and longest about 1000 feet; attains 280 feet below outcrop.

Geology: Quartz lenses and stringers in zone of crushed and locally silicified argillite. Vein zone said to average 4½ feet, strike N. 60° E., dip 70° SE. Sulfides are pyrite, marcasite, and arsenopyrite. Three stopes; longest 205 feet with average width of 4 feet.

Production: Ten-stamp mill erected in 1899; ore averaged less than $10 per ton. Gold 15 to 20 percent free. Small production.

References: Lindgren, 1901:684; Pardee and Hewett, 1914:105; Swartley, 1914:136; Parks and Swartley, 1916:148; Hewett, 1931: 312-320; Koch, 1959:35-36; Department Bulletin 14-B, 1941:55.

Monumental mine — Granite District, D-9

Location: Grant County, secs. 18 and 19, T. 8 S., R. 36 E., about 2 miles by road east of the Buffalo mine.

Development: Two tunnels, a shaft, and several raises total about 4000 feet, reach 700 feet below outcrop.

Geology: Twelve narrow veins in granodiorite. Strike N. to N. 20 E.; dip 65° NW. Most of the work done on four. Gangue is crushed granodiorite, with quartz stringers and lenses and later calcite. Sulfides are arsenopyrite, pyrite, galena, tetrahedrite, pyrargyrite. Ore lenses mined are a maximum of 18 inches wide and stope lengths are less than 100 feet.

Production: Mine located in 1870. Ore shipped to San Francisco in 1874. Total output estimated at $100,000 to 1928. Gold-to-silver ratio 1:20.

References: Lindgren, 1901:685; Swartley, 1914:139; Pardee and Hewett, 1914:108; Parks and Swartley, 1916:154; Hewett, 1931:314 to 340; Lorain, 1938:18; Department Bulletin 14-B, 1941:56-57.

New York mine — Granite District, D-10

Location: Grant County, NE¼ sec. 27, T. 8 S., R. 35½ E., opposite old Cougar-Independence Camp, 2½ miles north of Granite.

Development: Five adits; about 900 feet total.

Geology: Two veins, 3 to 7 feet wide, consisting of silicified argillite breccia and gouge.

One strikes about N. 30° E., and dips 70° E. to vertical; the other is nearly vertical and stikes N. 40° W. Ore oxidized. Pyrite encountered at face of lower adit 150 feet below outcrop.

Production: Most of the development work done between 1937 and 1941. Ore cyanided. More than 1000 tons treated. Millheads averaged $8.00 per ton.

References: Department Bulletin 14-B, 1941: 57-58; Koch, 1959:36-37.

Standard mine — Granite District, D-11

Location: Grant County, SW$\frac{1}{4}$ sec. 12, T. 8 S., R. 35$\frac{1}{2}$ E., on the divide between Crane and Onion Creeks; about 8 miles by road northeast of Granite and a quarter of a mile south of Continental mine.

Development: 400-foot adit and shallow shaft.

Geology: Two quartz- and gouge-filled fissure zones in granodiorite and quartz-mica schist. One about 8 inches wide strikes NE.; the other, up to 6 feet wide, strikes E. Values best at intersection. Sulfides mostly pyrite and arsenopyrite with some galena.

Production: No record.

References: Department Bulletin 14-B, 1941: 64; Koch, 1959:37.

Tillicum mine — Granite District, D-12

Location: Grant County, NE$\frac{1}{4}$ sec. 23, T. 8 S., R. 35$\frac{1}{2}$ E., on the north side of Granite Creek 3$\frac{1}{2}$ miles north of Granite.

Development: Two short adits 50 feet apart vertically.

Geology: Two parallel veins 40 to 50 feet apart in granodiorite. Strike N. 30° E., dip 50° SW. Ore consists of quartz-impregnated limonitic gouge. Gold is largely free.

Production: No records.

References: Department Bulletin 14-B, 1941:64-65; Koch, 1959:38.

E. ROCK CREEK DISTRICT

Baisley-Elkhorn mine — Rock Creek District, E-1

Location: Baker County, secs. 20 to 21, T. 8 S., R. 38 E., near the head of Pine Creek, about 15 miles from Baker.

Development: About 10,000 feet. Mining in early days was done chiefly through a 626-foot crosscut adit and a 400-foot underground shaft. Later work includes a 2300-foot crosscut adit and long drift about 950 feet below the outcrop.

Geology: Nearly vertical vein in granodiorite and argillite. Strikes N. 40° E. and is traceable for 1800 feet. The ore shoots ranged from 2 to 10 feet wide, mostly in granodiorite. Ore consisted of crushed granodiorite containing coarse sulfides and occasional streaks of quartz. Sulfides in order of abundance: pyrite, sphalerite, galena, and chalcopyrite with a little ruby silver.

Production: Estimated at $950,000, mostly between 1890 and 1907. Ratio of concentration, 5:1 to 7:1. Gold 20 to 25 percent free. 1898-1900 recovery averaged $12.30 from 26,095 tons of mill ore and $45.03 per ton from 472 tons of shipping ore. The value for 1905 and 1907 was $7 and $5 respectively.

References: Lindgren, 1901:646; Pardee and Hewett, 1914:74; Swartley, 1914:161; Parks and Swartley, 1916:20; Hewett, 1931: 312-317; Lorain, 1938:25; Department Bulletin 14-A, 1939:85.

Chloride mine

Rock Creek District, E-2

Location: Baker County, NW¼ sec. 24, T. 8 S., R. 37 E.

Development: Poor record. Includes 3 adits with several hundred feet of drifts.

Geology: Vein in argillite near granodiorite contact. Strike N. 45° to 50° E.; dip 55° to 75° S. Ore contains galena, sphalerite, arsenopyrite, pyrite, chalcopyrite, and argentite.

Production: No record; some milling done in early days.

References: Lindgren, 1901:648; Department Bulletin 14-A, 1939:86.

Cub mine

Rock Creek District, E-3

Location: Baker County, sec. 14, T. 8 S., R. 38 E.

Development: No record.

Geology: Vein in granodiorite 6 inches to 2½ feet thick consists of crushed and sericitized granodiorite, with kidneys and streaks of quartz and sulfides. Strike N. 22° E.; dip 55° W. High-grade gold assays reported.

Production: No record.

References: Department Bulletin 14-A, 1939:86.

Highland and Maxwell mines

Rock Creek District, E-4, 5

Location: Baker County, secs. 19 and 20, T. 8 S., R. 38 E., on the east fork of Rock Creek. The two mines are contiguous properties on the same vein.

Development: About 15,000 feet distributed among numerous adits and a shaft below the lowest level. Vertical range is about 2000 feet.

Geology: Vein mostly in argillite near contact with granodiorite. Strike N. 75° to 55° E.; dip nearly vertical. Vein described as incoherent mass of crushed and generally silicified argillite, clay, quartz, calcite, and ore minerals: pyrite, sphalerite, galena arsenopyrite, chalcopyrite, and tetrahedrite. Walls well defined and locally as much as 28 feet apart. Ore shoots were seldom more than 3 feet thick, and varied from less than 100 to more than 1000 feet long. Largest shoots found in lower levels.

Production: Estimated at $625,000; operated 1900-1905, 1909-1914, and 1936-1938. Ratio of sulfide concentration -- 4.5:1 to 7.5:1. Smelter returns averaged $12.30 per ton of ore. Gold-silver ratio about 1:10.

References: Lindgren, 1901:648; Swartley, 1914:161; Pardee and Hewett, 1914:77; Parks and Swartley, 1916:121, 150; Hewett, 1931: 314 to 319; Lorain, 1938:24; Department Bulletin 14-A, 1939:87.

Western Union mine

Rock Creek District, E-6

Location: Baker County, sec. 21, T. 8 S., R. 37 E., near the head of Rock Creek, 2 miles above Chloride mine.

Development: About 2000 feet of tunnels.

Geology: Country rocks are argillite and granodiorite. Vein strikes northeast.

Production: No record. Property located in 1887.

References: Department Bulletin 14-A, 1939:90.

WALLOWA MOUNTAINS AREA

Location and Geography

This area includes the high Wallowa Mountains and the adjoining foothills south and southeast of the range (figure 22). The region lies in the northeastern part of Baker County, the southwestern part of Wallowa County, and the southeastern part of Union County. Gold production from the region has come almost entirely from deposits on the south slope and adjoining foothills of the range, but a few prospects occur high in the mountains. Mining districts in the order described are: Cornucopia, Eagle Creek, Homestead, Medical Springs, and Sparta.

The Wallowa Mountains, which present some of the most scenically rugged terrain in eastern Oregon, are a northwest-trending, arched uplift about 50 miles long and about 20 miles wide. Many peaks and ridges in the central and northern parts of the range are more than 8000 feet in elevation. The Matterhorn, 9845 feet above sea level, is the highest peak in the eastern part of the state. Local relief exceeds 4000 feet. The foothills south and southeast of the range comprise a dissected plateau whose surface is characterized by rolling hills and bench lands cut by steep-sided gulches. The region is bordered on the south by Powder River, which flows easterly to Snake River following a winding course through an area of low relief. Eagle Creek, with several branches heading in the high reaches of the Wallowa range, flows generally southeastward entering Powder River at Richland. Pine Creek, whose upper tributaries drain the southeastern part of the range, enters Snake River at Oxbow.

Geology

General features

The Wallowa range is composed mainly of granitic rocks of the Wallowa batholith, a composite intrusive of Late Jurassic-Early Cretaceous age. Older rocks exposed in the range and along its flanks comprise volcanic, sedimentary, and plutonic rocks of Permian, Triassic, and Early Jurassic age. These older rocks have been regionally metamorphosed to the greenschist facies. Within one mile of the batholith they are generally altered to schists and hornfels.

The granitic rocks of the Wallowa batholith were deroofed and in places deeply dissected during Cretaceous and early Tertiary time, after which most, possibly all, of the area was buried under flow upon flow of Columbia River Basalt. The range was subsequently uplifted and most of the lava has since been stripped, but evidence of its former presence is seen in the remnants of nearly flat-lying flows perched on some of the highest peaks and ridges and the multitude of basalt dikes which intrude the granitic and older rocks. Removal of the lava was accelerated by vigorous Pleistocene glaciation, of which there is much evidence. Uplift of the Wallowa Mountains during the Pliocene was greatly accentuated by major faults on the north, south, and east (Taubeneck, 1963). The northern scarp in the vicinity of Enterprise is particularly impressive, being locally more than 4000 feet high.

The foothill belt south and southeast of the range is underlain chiefly by Tertiary basalt and by lake and stream deposits; rocks older than the Wallowa batholith are exposed in irregularly shaped areas where erosion has stripped the Tertiary cover. The Tertiary rocks dip generally southward and thus reflect the uplift of the Wallowa range. South of Powder River the Tertiary rocks dip to the north. Lower Powder River, therefore, roughly follows the axis of a syncline.

Pre-Tertiary sedimentary and volcanic rocks

Greenstones and associated sedimentary rocks: The oldest strata comprise a heterogeneous assemblage of altered lavas, pyroclastics, and coarse-to-fine-grained clastic sedimentary rocks consisting largely of volcanic detritus. Greenstones and tuffs predominate, but conglomerate, breccia, sandstone, shale, mudstone, and argillite layers locally attain considerable thicknesses and collectively make up a large part of the assemblage. Small limestone lenses are common. These rocks are partly Permian and partly Late Triassic in age but, because both sequences contain similar rocks and fossils are scarce, stratigraphic boundaries have been recognized only locally. Exposures of these rocks in the northeastern part of the Baker quadrangle were named the Clover Creek Greenstone by Gilluly (1937). In Idaho, equivalent rocks along Snake River and in the Seven Devils Mountains have been called the Seven Devils Volcanics.

Martin Bridge Formation: Comformable above the Upper Triassic part of the greenstone sequence is the Martin Bridge Formation, a lithologically distinctive series of thin-bedded to locally massive limestones and calcareous shales of widespread exposure in the Wallowa Mountains. Some of the best and most readily accessible exposures are along Eagle, East Eagle, and Paddy Creeks in the northern part of the Sparta quadrangle (Prostka, 1962). Because of strong folding over much of its extent, the thickness of the formation is not accurately known. Estimates range from 1000 to 2000 feet for different areas. Fossils of Karnian (Upper Triassic) age are locally abundant.

Hurwal Formation: The Hurwal Formation, which rests conformably on the Martin Bridge, is a sequence of gray, purplish, and black argillite, laminated siltstone, sandstone, and shale with minor limestone, conglomerate, and chert. The top of the formation has not been recognized, but a partial section in the Sparta quadrangle was estimated to be 4000 feet thick. Nolf (1967) found Lower Jurassic fossils in the upper part of the formation in the northern Wallowas.

Plutonic rocks

Exposures of Lower Triassic plutonic rocks are widespread in the southern foothills of the Wallowa Mountains, particularly in the Sparta quadrangle (Prostka, 1962). Albite granite and gabbro are the most prevalent representatives of the complex. Quartz diorite, peridotite, and serpentine are found locally. Contacts between the various rock types are transitional. In places Upper Triassic conglomerate and greenstones have been deposited directly upon eroded surfaces of these Lower Triassic plutonic rocks.

Granitic rocks of the Wallowa batholith of Late Jurassic-Early Cretaceous age are exposed over an area of about 225 square miles in the Wallowa Mountains. According to Taubeneck (1964, p. 1064), "The batholith...contains at least five small gabbroic units, four large zoned intrusions of tonalite-granodiorite and many small felsic bodies that include a sequence of unique cordierite trondhjemites and cordierite trondhjemite porphyries."

Satellitic bodies are numerous. Largest among them is the Cornucopia stock, which Taubeneck describes in the report cited above. The Cornucopia stock, 3 to 9 miles southeast of the parent batholith, is cut by the gold-quartz veins of the Cornucopia district.

Mineralization

According to their mode of occurrence, the gold deposits of the Wallowa Mountains area may be divided roughly into three classes--contact metamorphic deposits, replacement deposits, and fissure veins.

The contact metamorphic deposits contain copper, tungsten, and molybdenum minerals with small amounts of gold and silver. Known deposits of this type are found in the high central and northern parts of the range. Most are within the Eagle Cap Wilderness Area, where access is difficult and use of mechanized equipment is prohibited. The deposits occupy zones of thermal alteration along the contacts between granitic rocks and calcareous sediments into which the granitic rocks were intruded. Such zones are characteristically rich in epidote, garnet, calcite, and quartz. Metallic minerals present locally include magnetite, pyrite, chalcopyrite, galena, sphalerite, molybdenite, scheelite, hematite, and

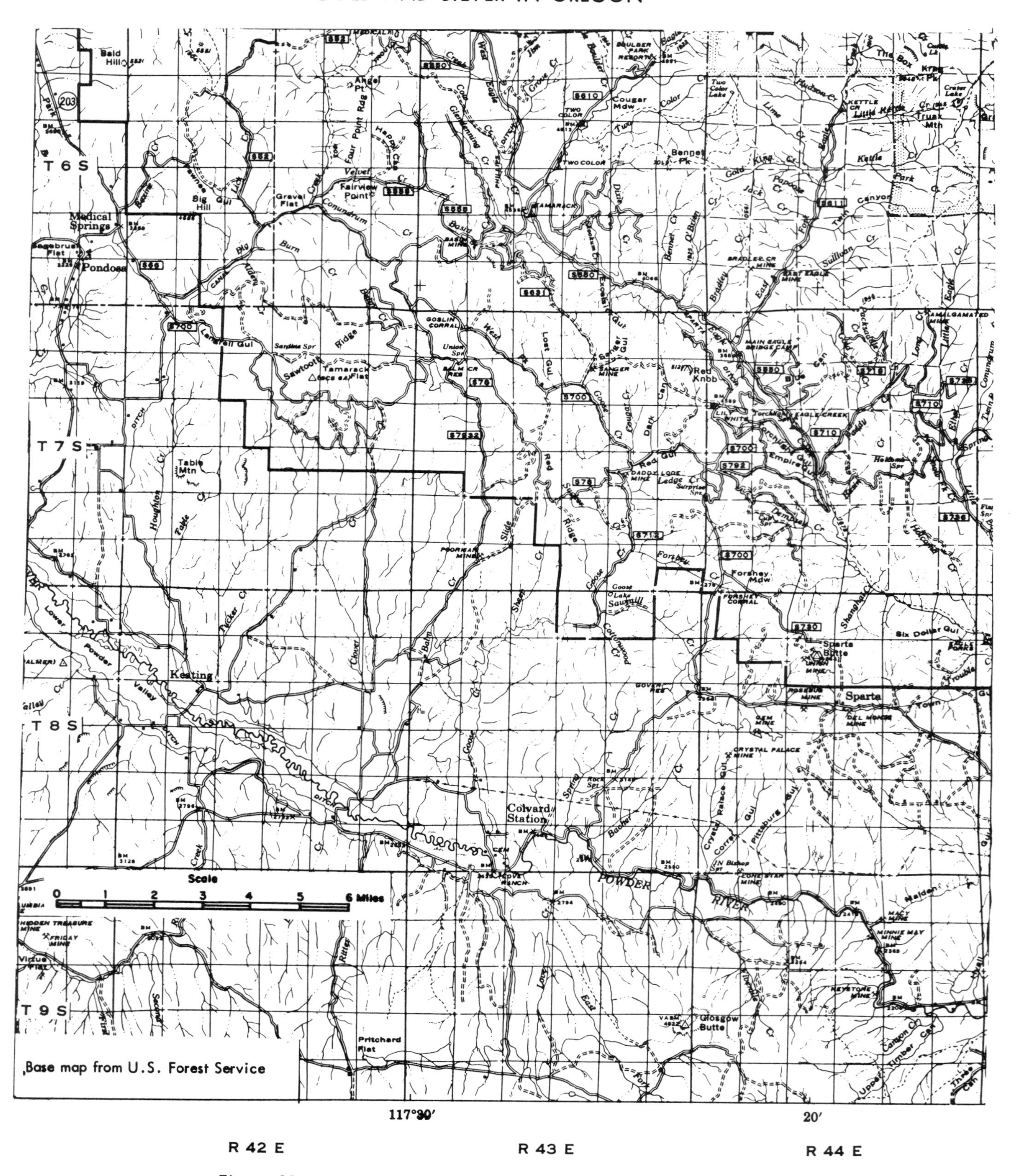

Figure 22. Index map of the Wallowa Mountains area.

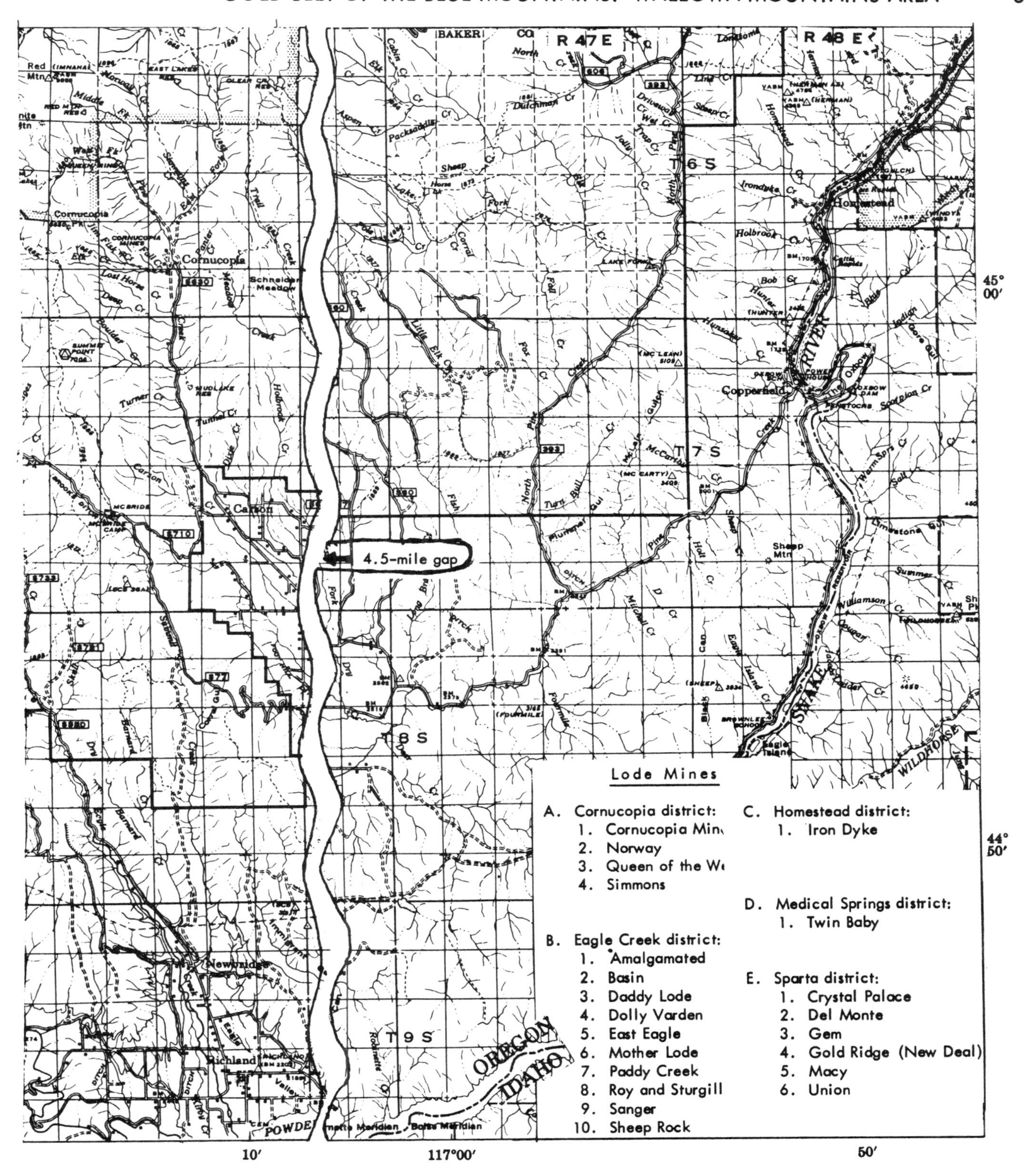

Lode Mines
A. Cornucopia district:
1. Cornucopia Min
2. Norway
3. Queen of the W
4. Simmons
B. Eagle Creek district:
1. Amalgamated
2. Basin
3. Daddy Lode
4. Dolly Varden
5. East Eagle
6. Mother Lode
7. Paddy Creek
8. Roy and Sturgill
9. Sanger
10. Sheep Rock
C. Homestead district:
1. Iron Dyke
D. Medical Springs district:
1. Twin Baby
E. Sparta district:
1. Crystal Palace
2. Del Monte
3. Gem
4. Gold Ridge (New Deal)
5. Macy
6. Union
4.5-mile gap
R 47 E
R 48 E
T 6 S
T 7 S
T 8 S
T 9 S
BAKER
Cornucopia
Carson
Newbridge
Richland
Copperfield
Homestead
Schneider Meadow
SNAKE
RIVER
OREGON
IDAHO
45° 00′
44° 50′
10′
117°00′
50′

R 45 E

pyrrhotite. Gold and silver are rarely present in more than trace amounts. The several metalized contact zones range from a few inches to about 20 feet in width (Oregon Dept. Geology and Mineral Industries, 1939, p. 113-120). Proven metallic mineral concentrations are of limited extent. These deposits are little developed and are not further described here.

The replacement deposits which contain copper minerals and associated gold are widespread in the southern foothills of the Wallowa Mountains and along the Snake River to the east. The deposits occur as irregular replacement bodies in and along the edges of faults and shear zones in greenstones and related volcaniclastic sedimentary rocks. The gold and silver values appear to be independent of the copper content. The mineralized bodies are abruptly discontinuous and difficult to outline. None, so far as known, are amenable to open-pit mining. Many of the occurrences were known before 1900. Of the replacement-type deposits, only the Iron Dyke mine in the Homestead district and the Mother Lode mine in the Eagle Creek district have yielded significant production.

The fissure-vein deposits account for the bulk of gold output from the Wallowa Mountains region. The most productive deposits have been the narrow, steeply dipping quartz veins of the Cornucopia district. These cut granitic rocks of the Cornucopia stock, an offshoot of the Wallowa batholith. In the Eagle Creek district there are a number of gold-quartz veins in sandstones and shales of the Hurwal Formation. Most notable are those of the old Sanger mine. The East Eagle mine and a few other prospects develop copper- and gold-bearing fissure veins in old greenstones and volcaniclastic rocks. The deposits in the Sparta district occupy narrow, discontinuous quartz-filled fractures in Lower Triassic albite granite and diorite.

Production

Combined estimates of production from lode mines in the Wallowa Mountains area total more than $12,500,000, of which $10,000,000 was produced by the Cornucopia mines north of Halfway. The remainder was produced chiefly by the Sanger mine in the Eagle Creek district and the Iron Dyke mine in the Homestead district. Lesser producers include the Mother Lode and Basin mines in the Eagle Creek district and the Macy and Gem mines in the Sparta district.

Placer Mines

Few records exist upon which to base estimates of placer output from the Wallowa Mountains area. Considerable placer mining has been done in the vicinity of Sparta and the Sanger mine, along Eagle Creek and some of its tributaries, and along Pine Creek below Cornucopia.

Sparta

Near Sparta several gulches have been placered and in parts of the area the soil and decomposed granite on the dividing ridges contained placer gold. Among the better placers were those of Shanghai and Town Gulches, which drain north and east into Eagle Creek, and Pittsburg and Maiden Gulches, which drain southward into Powder River. During the few years immediately following completion of the 22-mile-long Sparta Ditch from West Eagle Creek in 1873, a good deal of placer mining was done in the Sparta area.

Sanger

The area in the vicinity of the Sanger mine constituted the old placer camp of Hog'em, and from the gulches leading up to the mine the sum of $500,000 is reported to have been extracted (Lindgren, 1901, p. 738-739). Fir Gulch below the Sanger mine is said to have been one of the richest placers in the district.

Eagle Creek

All along Eagle Creek there are benches of heavy gravels as high as 100 feet above present stream level. These benches have been placered to some degree from a point below the mouth of Paddy Creek to a few miles upstream above the mouth of East Eagle Creek. Parts of both Paddy and East Eagle Creeks have also been worked to some extent.

Pine Creek

Placer yield from Pine Creek below Cornucopia is unknown, but it probably was not large because of the excessive amount of large boulders in the gravels and their thickness, which is said to be 60 feet. Attempts to sink shafts through the gravels to bedrock have been hampered by water. The canyon is too narrow, the boulders too large, and the values too deep for dredging.

Cornucopia District

General information

The Cornucopia district is situated near the head of Pine Creek in northeastern Baker County (figure 22). The nearly deserted mining camp of Cornucopia on Pine Creek 12 miles north of Halfway is in the south-central part of the district. The workings of numerous mines and prospects are scattered along both slopes of the creek within 4 miles of the old camp site. The area is very rugged topographically. Elevations range from about 4700 feet at Cornucopia townsite to 8650 feet at the top of Cornucopia Peak about 2 miles to the west.

The gold deposits of this district lie within and along the irregular southeastern edge of the Cornucopia stock, a small offshoot of the Wallowa batholith. Country rocks are mainly hornblende-biotite schist and hornfels derived from thermal metamorphism of graywackes, shales, conglomerates, and greenstones. The greenstones are well exposed on Simmons Mountain northeast of Cornucopia. A description of the Cornucopia stock and country rocks in the thermal aureole is presented by Taubeneck (1964). Goodspeed (1939 and 1956) discusses the geology of the Cornucopia district and suggests that the stock was emplaced as a result of static granitization.

Estimates of the value of production range from $10,000,000 to about $18,000,000 in gold and byproduct silver, copper, and lead. The latter figure is probably excessive. Mining operations, which began about 1880, have been periodic. In 1930 Cornucopia Gold Mines Co. acquired the important mines and worked them collectively from 1933 through October 1941. Since that time little mining has been done in the district. Regarding production, Swartley (1914, p. 22) states: "According to Bernard MacDonald's report upon the property, the Union-Companion, Red Jacket, and Last Chance claims produced $1,008,000 previous to 1903." Recorded production since 1903 amounts to 272,776.64 ounces gold and 1,088,051 ounces silver recovered from 983,927 tons of ore. Gross content of copper in ores produced during 1933-1941 was 671,778 pounds and of lead 121,983 pounds.

Principal lode mines

Cornucopia mines group: The most important veins crop out 1 to 2 miles northwest and 1000 to 3000 feet above Cornucopia on the west side of Pine Creek. From east to west these are the Whitman, Union-Companion, Last Chance, Wallingford, and Valley View veins. The veins are from 1500 to 2500 feet apart, strike N. 20° to 40° E., and dip about 45° W. The bulk of production from the district is credited to the Union-Companion and Last Chance veins. The Union-Companion, the more extensively developed of the two, is a maximum of 2500 feet horizontally east of the Last Chance and it crops out roughly 1000 feet lower on the mountainside. The Union Companion vein is said to be traceable on the surface for 6800 feet.

About 36 miles of underground work has been done. The principal access during the last few years of operation was the Coulter tunnel (portal elevation 4805 feet) which was driven in 1936. This adit enters the mountain at the north edge of Cornucopia, trends N. 60° W., and crosscuts the Union-Companion vein about 6200 feet from its portal. Ore was treated in a 150-ton flotation plant erected in 1936 at the portal of the Coulter tunnel. About 4700 feet nearly due west and 985 feet in elevation above Coulter tunnel portal is the Clark tunnel, a long crosscut which taps both the Union-Companion and Last Chance veins. This tunnel continues in a northwesterly direction and taps the Wallingford and other small veins. During the early days, the upper part of the Last Chance vein was worked mainly through the Lawrence tunnel at 6910 feet elevation and 4100 feet north and a little west of the Clark tunnel portal. Several other adits cut either the Union-Companion or Last Chance veins, but development has progressed mainly from the three mentioned. Mine maps (plate 1, in pocket) are cross-sectional views of the workings and stopes on the Union-Companion and Last Chance veins. Figure 23 is a photograph of the Union-Companion mill built in 1913 at the portal of the Clark tunnel.

The veins are partly in granodiorite of the Cornucopia stock and partly in the older hornfels and schist. Dikes of granodiorite porphyry, aplite, and Tertiary basalt are numerous. Basalt dikes cut the veins, often causing offsets of several feet. In places the veins are split by aplite dikes.

The veins consist mainly of quartz with intermixed streaks of wall rock that, in general, has been thoroughly crushed and altered. Two stages of quarts are recognized. The first was probably introduced shortly after the fracturing and was contemporaneous with the alteration of the wall rocks; it contains no values. After the original quartz was deposited the vein was sheared and fractured, thus providing openings for the second stage of quartz deposition; this contains the ore values. The original quartz had deposited with it quite coarse pyrite, some of which was later fractured; second-stage quartz and ore minerals were then deposited in the fractures. Microbrecciation of the later quartz provides a key for mine development.

Scattered sulfides make up, on the average, less than 5 percent of the ore and contain the bulk of the gold values. They consist chiefly of pyrite, with smaller quantities of arsenopyrite, chalcopyrite, galena, and sphalerite. Ordinarily the richness of the ore was in direct proportion to the amount of copper minerals contained.

At the surface the ore contained free gold and small amounts of petzite and hessite. The ore was generally oxidized to a depth of about 300 feet, with local oxidation extending down as much as 2000 feet below the outcrop.

According to H. F. Anderson (written communication, 1965) who was staff geologist at Cornucopia during 1939-1940, "The ore shoots of both the Union-Companion and Last Chance veins were abruptly localized by structural controls, and comprised in the aggregate less than 25 percent of the total planar area of the vein systems. The tenor of the shoots commonly changed markedly in grade within a few feet from high-grade ore containing an ounce or more gold to the ton to barren vein material consisting of early-stage quartz and non-auriferous pyrite. Mining costs were above average, due to the excessive amount of development work required to explore and mine the localized shoots.

"The Union-Companion vein looked healthy and showed no diminution of structural strength where exposed by drifting on the Coulter tunnel level. Average stoping width ranged from 4 to 9 feet in the different ore shoots. Where last developed by drifts on the Clark tunnel level the Last Chance vein appeared less impressive than in upper levels. Stoping widths averaged 3 to 7 feet."

During 1938-1941, the last 4 years of operation, 156,388 tons of ore were concentrated by bulk flotation and 36 tons of crude ore were shipped directly to smelters. The 7170 tons of concentrates plus the 36 tons of crude ore contained 75,268 ounces of gold, 343,626 ounces of silver, 356,781 pounds of copper, and 82,563 pounds of lead. Thus the ore before treatment contained, in recoverable metals, approximately 0.48 ounces gold, 2.2 ounces silver, 0.1 percent copper, and 0.025 percent lead. The ratio of sulfide concentration was about 20 to 1.

Queen of the West mine: This mine is located high on the west side of Pine Creek about 3 miles northwest of Cornucopia and a little less than a mile north of, and on the opposite side of Bonanza Basin from, the Lawrence tunnel portal. Parks and Swartley (1916, p. 186) state: "The vein has the usual strike of N. 20° E. and a dip near the surface of about 45° but at depth this decreases to about 30°. The average width of the vein near the surface is between 3 and 4 feet, but generally speaking it decreases in width with the decrease in dip.

"The gangue minerals are quartz and calcite containing pyrite, chalcopyrite, galena, and sphalerite in bunches. It is said that the zinc, lead, and copper minerals carry most of the gold values. In many places the vein shows included fragments of altered granodiorite, and the granodiorite on each side of the vein for about 2 feet is badly altered and impregnated with pyrite which is said to contain some values in gold and silver. This vein can be traced for a long distance, reported to be as much as 3000 feet."

Simmons mine: The Simmons mine vein, which has yielded a relatively small production, is exposed near the top of Simmons Mountain on the east side of Pine Creek and about 2½ miles north of Cornucopia. The country rocks are mostly greenstones. The vein strikes N. 30° W., dips gently eastward, and is traceable for about 2000 feet. In 1941 a lessee shipped 1046 tons of ore from the Simmons for treatment at the Cornucopia mill; concentrates recovered contained 264 ounces gold and 1840 ounces silver.

Eagle Creek District

General information

This district includes the upper drainage of Eagle Creek and the adjoining area on the Powder River slope northeast of Keating that is drained by Clover, Balm, and Goose Creeks (figure 22). Roads enter the district from Sparta, Keating, and Medical Springs. Elevations at the mines range from 3500 to 7000 feet. Most of the area is timbered and there are several permanent streams.

Pre-Tertiary rocks exposed in the district include greenstones, tuffs, and related volcaniclastic sedimentary rocks of the Clover Creek Greenstone; limestone of the Martin Bridge Formation; and sandstones, mudstones, and shales of the Hurwal Formation. Granitic rocks of the Wallowa batholith underlie the northern part of the area. A large part of the district is blanketed by Tertiary basalt. The Sparta quadrangle geologic map (Prostka, 1963) covers most of the district.

Lode-gold production from the Eagle Creek district is dominated by the Sanger mine, whose output has been estimated at $1,500,000. The Mother Lode copper mine produced a limited amount of gold during 1935-1938. Production from other lode deposits in the Eagle Creek district has been small. Some of the better known prospects are the Basin, East Eagle, Amalgamated, Lilly White, and Dolly Varden.

Principal lode mines

Sanger mine: This old mine is located on a branch of Goose Creek near the top of the Powder River-Eagle Creek divide. The following description of the deposit is taken from Lindgren (1901, p. 738-739) and a Department mine-file report compiled in the early 1900's by Charles P. Berkey.

The principal vein, called the Summit lode, was discovered in 1870, and was actively worked during the following years. In 1874 the production was $60,000 from ore containing $16 to the ton. The total production to 1887 is unknown, though probably small, but a mill was then built, and in 1889 production began to increase rapidly. The Mint reports for the 4 years 1889-1892 give $813,000 as the production of the mine. Production ceased in 1897. Total output is estimated at about $1,500,000.

The rocks at the Sanger mine are dark-colored, medium-to-fine-grained mudstones and shales of the Upper Triassic Hurwal Formation. The rocks are pyritic near the veins. The Summit vein strikes nearly due east, dips 30° N. and has been worked to a depth of 400 feet on the dip from several adits and an inclined shaft. An old map dated January 1, 1901 seems to indicate that drifting was done on at least two other veins or fault zones, one paralleling the summit vein and the other crossing it at nearly right angles. The ore shoot in the upper stopes of the Summit vein was 600 feet long, about 15 inches wide and averaged $20 to $25 a ton in gold; below the zone of oxidation the vein widened to 2 to 4 feet and the value dropped to $12 a ton. The gangue is coarse quartz with a little calcite and about 3 percent sulfides, consisting mostly of pyrite with a little sphalerite and galena. Much of the gold was free. In its easterly extension on all levels the vein bends in a broad curve to the south and appears to blend with the strike of

Figure 23. Union-Companion mill, Cornucopia district. Built at the portal of Clark tunnel in 1913.

Figure 24. Iron Dyke copper mine at Homestead, Baker County, prior to 1914.

the host rocks, losing its characteristic size and value. Toward the west the vein has be
ing, at which point mining ceased. Other veins on the property include the Packwood,
and Knight. The Big Vein at the head of Fir Gulch may have been the source of the rich ... Fir
Gulch.

Mother Lode mine: The Mother Lode mine, which consolidates the old Balm Creek and Poorman workings, is located on Balm Creek about 8 miles by road northeast of Keating. Development, which began prior to 1900, led to the construction of a 100-ton flotation plant and the production of 8,108.8 ounces of gold and 1,047,015 pounds of copper between June 1935 and January 1, 1938. Yield in 1937 amounted to 3517 ounces of gold, 1665 ounces of silver, and 520,000 pounds of copper from 20,380 tons of ore. Mining ceased in December 1938 and there has been no further production. Sporadic promotion and exploration have continued to the present time.

Development comprises more than 15,000 feet of drifts, crosscuts, shafts, and raises. Principal access includes two shafts, one 700 and the other 420 feet deep; and three adits. Rocks cut by the workings are almost exclusively greenstones which have been sheared and hydrothermally altered. Gilluly (1931, plates 2 and 3) indicates that the rocks are cut by a great number of small faults and fractures, the majority of which strike west-northwest and dip southwest. He states (p. 26-28): "The mineralized masses appear, in general, to be elongated a little north of west, roughly parallel to the strike of the formation and the most prominent fractures, but in detail they are exceedingly irregular....the distribution of the metals is very irregular. Little copper is present on the upper levels, what was formerly there having been leached out. However, oxidation has not seriously affected the minerals at lower levels, and here too the distribution of gold and copper is erratic. Most of the material is highly silicified and carries sporadically distributed specks and veinlets of chalcopyrite and pyrite. The gold and chalcopyrite do not appear to be closely associated, as grab samples show wide variations in gold content although running about the same in copper...Pyrite and chalcopyrite are essentially the only primary metallic minerals. The gangue is dominantly quartz, with some sericite, ankerite, calcite, chalcedony, and barite."

The 1935-1937 mining operations proved uneconomic because of extreme irregularity of mineralization and structural disturbance of the ore bodies.

Homestead District

The Iron Dyke copper mine at Homestead on Snake River (figure 22) has produced a considerable amount of gold and silver in addition to copper. The deposit was discovered in 1897 and a 50-ton mill was erected before 1908 but little production was made prior to 1915. A 150-ton flotation plant was placed in operation in 1916 and for several years thereafter the Iron Dyke was one of the most productive mining enterprises in Oregon (figure 24). Concentrates and crude ore were shipped to the International Smelter at Salt Lake City, Utah. Production figures since 1910, as recorded by the U.S. Bureau of Mines, are shown in table 8. No mining has been done since 1928, and important workings are inaccessible. Reports by Swartley (1914, p. 107-109) and Parks and Swartley (1916, p. 130-132) were written when the mine was in early stages of development. Little data has since been published regarding the ore deposits and what was learned about them during mining.

Development includes a 650-foot vertical shaft, a 500-foot inclined shaft, and four adits. The lower adit is about 300 feet above Homestead and about 450 feet in elevation below the outcrop of the ore body on the south wall of Irondyke Creek.

Rocks exposed in the near vicinity of the deposit are mostly greenstones derived mainly from andesitic to rhyolitic lavas, breccias, and tuffs. Elsewhere in the adjacent area conglomerates and limestones form thin interbeds in the series. Permian fossils have been identified from nearby localities. A considerable amount of shearing and faulting has taken place and the rocks have been further disturbed by the intrusion of numerous dikes.

The main ore body of the mine is associated with a broad zone of shearing that strikes about N. 20° E. and dips steeply eastward. Swartley (1914, p. 109) states: "The best ore in the lower tunnel is massive chalcopyrite and pyrite with but little quartz as a gangue in a lens-shaped body dipping 60° E. with a maximum width of about 6 feet which is said to extend from the lower to the upper tunnel...On either side of this high grade ore, which is said to average 15 to 20 percent copper, is a much larger body of disseminated pyrite and chalcopyrite in the chloritic greenstone, in which are abundant quartz seams, veinlets, and nodules that contain pyrite. There is often a silicification of the rock itself. Statements are made that it contains about $2.00 in gold, and 6 to 30 ounces in silver, regardless of the percent of copper present. This deposit, both high and low grade, is in a zone of crushing in which copper-bearing solutions have deposited their contents largely by replacement." State of Oregon Department of Geology and Mineral Industries Bulletin 14A (1939, p. 62) states: "Since the above was written...the mine was developed by shaft to levels below the lower crosscut. On the lowest level the ore body was cut off by a nearly horizontal fault. The ore body here was egg shaped, about 140 feet wide, and 210 feet long, carrying good grades of copper and about $\frac{1}{4}$ ounce in gold."

Table 8. Production of gold, silver, and copper from the Iron Dyke mine, Homestead district, Baker County, Oregon 1910 to 1934.

Year	Ore smelted, tons	Ore milled, tons	Concentrates produced, tons	Gold, ounces	Silver, ounces	Copper, pounds
1910	68			1	535	13,861
1915	3,565			55	9,803	396,972
1916	1/ 23,225			377	80,856	2,230,729
1916			1,673	58	8,337	290,971
1917		36,676	7,522	1,279	31,256	1,372,110
1918		33,583	6,734	3,794	24,212	1,602,145
1919		27,618	7,044	10,753	17,624	2,087,276
1920		34,804	7,910	8,322	18,890	2,353,276
1921		2,398	573	434	1,339	174,300
1922	2,047			513	4,167	198,320
1922		15,070	3,570	2,259	10,238	813,869
1923	369			26	862	57,345
1923		17,980	5,117	3,141	21,244	1,176,144
1924		14,746	3,418	1,879	12,039	757,440
1925		2,740	548	375	1,938	105,600
1926	27			81	97	6,519
1926		5,155	1,031	510	3,512	227,691
1927	185			7	729	43,356
1927		16,018	1,236	805	7,513	439,696
1928		2,800	223	148	1,283	70,300
1934		2/ 1		150	15	-
	29,486	209,589	46,599	34,967	256,489	14,417,920

1/ Total ore milled and smelted. 2/ Bullion produced.

Medical Springs District

The Medical Springs district lies about 18 miles northeast of Baker (figure 22). Old prospect workings are found here and there among the hills surrounding the village of Medical Springs, but records of production from the district are meager. Much of the area is covered by Tertiary basalt. The prospects are in small exposures of Permian or Upper-Triassic greenstones and related sedimentary rocks.

The Twin Baby mine (including the Grull prospect) is 4 miles by road northeast of Medical Springs on the timbered divide between two tributaries of Big Creek. Development which began prior to 1900 includes numerous shallow adits, shafts, and open cuts scattered over several claims. One shaft is said to be 335 feet deep with drifts on the 100, 150, 230, and 335-foot levels. Near this shaft a 40-ton amalgamation and flotation mill was erected and operated for a short time during the mid-1930's.

The country rocks are mostly greenstones. Several narrow veins and mineralized fault zones of diverse trend have been exposed. The most persistent vein averages about 2 feet in width, strikes N. 5° W., dips steeply east, and has been traced along strike for more than 2000 feet. Where visible this vein consists mainly of brecciated and altered greenstone with thin lenses and stringers of brecciated quartz. In surface exposures, sulfide minerals are sparse, although their former presence is indicated by iron oxides; gold, where present, is largely free and occurs in very fine particles. Specimens found on the dump adjacent to the 335-foot shaft contain pyrite and chalcopyrite. Gold mineralization is spotty. Small pockets of high-grade ore have been won from near-surface workings.

Sparta District

General information

The Sparta district encompasses a small area between Eagle Creek and Powder River in the vicinity of Sparta, a ghost town about 40 miles by road east of Baker (figure 22). Elevations range from about 2300 feet on Powder River along the southern edge of the district to 4944 feet at the top of Sparta Butte in the northern part. Timber is absent on the Powder River slope of the divide but plentiful to the north.

The host rocks of the district are albite granite and quartz diorite of Early Triassic age (Prostka, 1963). The Triassic intrusive rocks are sheared and deeply decomposed. Patches of Cenozoic basalt, tuffs, and lake and stream sediments are present locally.

The Sparta district is noted mainly for its placers, although no production records exist. The several quartz veins that have been developed are small and nonpersistent, but locally contained pockets of rich ore. Total lode production is believed to be small. Lindgren's (1901, p. 736) figure of $677,000 for the four years 1889-1892 is said to be greatly exaggerated by local historians.

Principal lode mines

Macy mine: The Macy mine in Maiden Gulch, a quarter of a mile above Powder River, was discovered in 1920. Intermittent operations have produced in the neighborhood of $90,000. About 2000 feet of development work has been done, mainly from adit levels. Early work was mapped by Gilluly, Reed, and Park (1933, p. 61). The workings explore several small quartz veins of diverse trend, some of which intersect. The veins range in width from about an inch to 4 feet, and consist chiefly of quartz with a little calcite, sericite, and chlorite. Ore minerals are pyrite, sphalerite, and free gold. Ore shoots were small and irregular. Stope widths averaged 1 to $1\frac{1}{2}$ feet.

Gem mine: Probably the most extensive underground development is at the Gem mine where, according to Lorain (1938, p. 33), there is an inclined shaft 550 feet deep from which short drifts have been turned on eight different levels. The vein strikes north, dips 30° to 40° E., and ranges from 1 to 4 feet in width between sharply defined walls. It consists of crushed and altered albite granite with streaks and lenses of quartz as much as 2 feet wide. The ore ordinarily consists of coarse quartz containing free gold with pyrite and sphalerite.

Crystal Palace mine: At the Crystal Palace mine, about 2000 feet of workings develop a curving vein which strikes from N. 30° W. to N. 60° E. and dips about 25° E. The vein consists of pyrite, arsenopyrite, and free gold in a gangue of quartz and minor sericite, chlorite, and ankerite.

Lode Mines of the Wallowa Mountains Area

A. CORNUCOPIA DISTRICT

Cornucopia Mines Group — Cornucopia District, A-1

Location: Baker County, secs. 27 and 28, T. 6 S., R. 45 E., on steep west wall of Pine Creek 12 miles north of Halfway.

Development: About 36 miles of workings over a vertical interval of more than 3000 feet, mostly from three adit levels. The lowest is near creek level.

Geology: Two principal veins, the Union-Companion and Last Chance, in granodiorite and along the contact between granodiorite and intruded greenstones and metasediments. Strike N. 20° to 40° E.; dip about 45° W. Veins range up to 20 feet in width, average 4 or 5 feet in the mined shoots which were abruptly discontinuous. Gangue is mostly quartz and altered wall rock. Ore minerals include pyrite, arsenopyrite, chalcopyrite, galena, and sphalerite with a little free gold and tellurides.

Production: About $10,000,000. Latest mill was a 150-ton bulk flotation plant. Ratio of concentration about 20:1. Mined ore produced during 1938-1941 averaged 0.48 oz. gold, 2.2 oz. silver, 0.1 percent copper, and 0.025 percent lead.

References: Lindgren, 1901:743; Swartley, 1914:36; Parks and Swartley, 1916: 21,74; Goodspeed, 1939:1-18; Lorain, 1938:38; Department Bulletin 14-A, 1939:26.

Norway mine — Cornucopia District, A-2

Location: Baker County, E½ sec. 9, T. 6 S., R. 45 E.

Development: 1000-foot drift adit with short cross cuts.

Geology: Quartz lenses along shear zone in greenstone. Trend nearly north. Dips steeply east. Lenses range from a few inches to a few feet in thickness.

Production: Small.

References: Swartley, 1914:61; Parks and Swartley, 1916:213; Ross, 1938:215; Department Bulletin 14-A, 1939:30.

Queen of the West mine — Cornucopia District, A-3

Location: Baker County, W½ sec. 21, T. 6 S., R. 45 E.

Development: Unknown.

Geology: Quartz vein in granodiorite traceable for 3000 feet. Strike N. 20° E., dip 30°

to 45° W. Vein minerals are quartz, calcite, pyrite, chalcopyrite, galena, and sphalerite.

Production: No record of production. Some milling done in early days.

References: Lindgren, 1901:745; Swartley, 1914:54; Parks and Swartley, 1916:186; Department Bulletin 14-A, 1939:30.

Simmons mine — Cornucopia District, A-4

Location: Baker County, secs. 15 and 16, T. 6 S., R. 45 E., on Simmons Mountain on the east side of Pine Creek opposite Cornucopia.

Development: Several short adits, shafts, and cuts. Total length unknown.

Geology: Vein in greenstone 18 inches to 4 feet wide; strikes N. 30° W., dips gently east, and is traceable for about 2000 feet. The vein consists chiefly of quartz with scattered sulfides.

Production: Small. Concentrates produced in 1941 from 1046 tons of ore contained 246 ounces gold and 1840 ounces silver. Previous output unknown.

References: Lindgren, 1901:745; Swartley, 1914:59; Parks and Swartley, 1916:203; Department Bulletin 14-A, 1939:31.

B. EAGLE CREEK DISTRICT

Amalgamated mine — Eagle Creek District, B-1

Location: Baker County, N$\frac{1}{2}$ sec. 2, T. 7 S., R. 44 E., at the head of Paddy Creek.

Development: Several tunnels, two of which aggregate 1500 feet.

Geology: Crushed zones along faults in greenstone; locally, 20 to 30 feet wide; strike NW.; dip steeply SW. and NE. Locally contain quartz seams and lenses.

Production: Small; 25-ton Chile mill operating in 1937.

References: Parks and Swartley, 1916:14; Gilluly, Reed, and Park, 1933:63.

Basin mine — Eagle Creek District, B-2

Location: Baker County, SW$\frac{1}{4}$ sec. 29, T. 6 S., R. 43 E.

Development: Two or more adits, total about 2000 feet.

Geology: Country rock is quartz diorite. Numerous shear and joint openings of widely different attitude contain quartz lenses as much as 2 feet thick. One stoped vein strikes N. 60° W. and dips 35° S. Ore oxidized. Relict sulfides include pyrite, chalcopyrite, and sphalerite. Gold free.

Production: Small.

References: Lindgren, 1901:738; Gilluly, Reed, and Park, 1933:68; Department Bulletin 14-A, 1939:44-45.

Daddy Lode (Blue Bell mine)

Eagle Creek District, B-3

Location: Baker County, sec. 23, T. 7 S., R. 43 E.

Development: Shaft 210 feet deep and several adits; total about 2000 feet.

Geology: Copper-gold mineralization along fractures in argillite, chert, and greenstone. Light-colored dikes are present.

Production: No record.

References: Department Bulletin 14-A, 1939:45.

Dolly Varden mine

Eagle Creek District, B-4

Location: Baker County, secs. 19 and 30, T. 7 S., R. 44 E.

Development: Small.

Geology: Large outcrop of quartz-impregnated volcanic tuffs.

Production: $115,000 prior to 1900 from surface ore.

References: Lindgren, 1901:739; Parks and Swartley, 1916:87; Department Bulletin 14-A:47-48

East Eagle (McGee and Woodard Groups)

Eagle Creek District, B-5

Location: Baker County, NW$\frac{1}{4}$ sec. 32, T. 6 S., R. 44 E., just above level of East Eagle Creek.

Development: Two drift adits 600 and 350 feet in length respectively.

Geology: Quartz vein in argillite and minor greenstone. Pinches and swells rapidly from less than a foot to more than 4 feet in width; much gouge included locally. Strike averages N. 47° W.; dip 22° S. Chief ore minerals are pyrite and chalcopyrite with copper oxides and a little free gold.

Production: Small.

References: Department file report dated December 13, 1947.

Mother Lode (Balm Creek, Poorman) mine

Eagle Creek District, B-6

Location: Baker County, NW$\frac{1}{4}$ sec. 32, T. 7 S., R. 43 E.

Development: About 15,000 feet of drifts, crosscuts, and raises, including two shafts 700 and 420 feet deep respectively, and three adits.

Geology: Scattered copper-gold mineralization along small faults and fractures in sheared greenstone. Zone of shearing and mineralization trends a little north of west. Gangue is dominantly quartz with some sericite, ankerite, calcite, chalcedony, and barite. Sulfides are pyrite and chalcopyrite.

Production: Output during 1935-1938 amounted to 8108.8 ounces gold and 1,407,015 pounds of copper. No production since. Yield in 1937 was 3517 ounces gold, 1665 ounces silver, and 520,000 pounds of copper from 20,380 tons of ore treated in a 100-ton flotation plant.

References: Lindgren, 1901:732; Swartley, 1914:121; Parks and Swartley, 1916:181; Gilluly, 1931:24-28; Lorain, 1938:29; Department Bulletin 14-A, 1939:43.

Paddy Creek mine

Eagle Creek District, B-7

Location: Baker County, sec. 15, T. 7 S., R. 44 E.

Development: Several hundred feet of underground work.

Geology: Lens-like veins in sedimentary rocks.

Production: Small. Ten-stamp mill in early days; no mill records.

Reference: Parks and Swartley, 1916:77.

Roy and Sturgill mines

Eagle Creek District, B-8

Location: Baker County, SE$\frac{1}{4}$ sec. 3, T. 7 S., R. 43 E.

Development: Roy, 1000 feet of drift; Sturgill, 500 feet of drift on two levels.

Geology: Country rock is blocky argillite cut by diorite dikes. Vein strikes N. 10° to 35° E.; dips 35° to 45° E.; iron-stained quartz. No sulfides; gold free.

Production: No records.

References: Department Bulletin 14-A, 1939:49 and 50.

Sanger mine

Eagle Creek District, B-9

Location: Baker County, SW$\frac{1}{4}$ sec. 2, T. 7 S., R. 43 E.

Development: Inclined shaft and several adits work Summit vein through horizontal distance of 650 feet and inclined depth of 400 feet.

Geology: Country rock is shale and argillite. Several veins exposed; only one, the Summit, has been extensively worked. It strikes E., dips 30° N. Gangue is coarse quartz and about 3 percent sulfides. Average thickness, 2 feet.

Production: $1,500,000. Little output since 1900. Ore averaged $20 to $25 per ton, mostly free gold, in upper levels and $12 in lower levels as sulfides increased.

References: Lindgren, 1901:738; Swartley, 1914:118; Parks and Swartley, 1916:198; Gilluly, Reed, and Park, 1933:67; Lorain, 1930:32; Department Bulletin 14-A, 1939:50.

Sheep Rock mine (including Summit property) — Eagle Creek District, B-10

Location: Baker County, NW¼ sec. 32, T. 6 S., R. 44 E.

Development: Several short adits as much as 200 feet long, and many cuts and pits scattered about several claims.

Geology: Several narrow quartz veins in greenstone cut by basic dikes.

Production: $30,000 in early days.

References: Lindgren, 1901:738-739; Swartley, 1914:117; Parks and Swartley, 1916:200; Gilluly, Reed, and Park, 1933:64.

C. HOMESTEAD DISTRICT

Iron Dyke mine — Homestead District, C-1

Location: Baker County, sec. 21, T. 6 S., R. 48 E., on bank of Snake River.

Development: Includes 650-foot vertical shaft, 500-foot inclined shaft, and four adits.

Geology: Copper-gold mineralization associated with broad zone of shearing in greenstones. Zone trends N. 20° E. and dips steeply east. The rocks are partly silicified and contain abundant quartz seams, veinlets, and nodules that contain pyrite. Main copper mineral is chalcopyrite.

Production: Main period of operation was 1916-1928. Total recorded output since 1910 has been 34,967 ounces of gold, 256,489 ounces of silver, and 14,417,920 pounds of copper. (See table 8.)

References: Lindgren, 1901:749; Swartley, 1914:107; Parks and Swartley, 1916:124; Department Bulletin 14-A, 1939:60.

D. MEDICAL SPRINGS DISTRICT

Twin Baby mine — Medical Springs District, D-1

Location: Union County, sec. 20, T. 6 S., R. 42 E., about 4 miles by road northeast of Medical Springs.

Development: Shaft 335 feet deep; several short adits and cuts.

Geology: Country rock is mostly greenstone with some argillite. Principal vein strikes N. 5° W., dips steeply east and averages about 2 feet wide. On the surface the vein consists of brecciated country rock with stringers and nodules of quartz. Sparse pyrite. Gold largely free. Several other veins and mineralized fault zone have been meagerly explored.

Production: A few thousand dollars. Mill erected in 1930's but little used.

References: Gilluly, Reed, and Park, 1933:70; Lorain, 1938:31; Department Bulletin 14-A, 1939:112.

E. SPARTA DISTRICT

Crystal Palace mine — Sparta District, E-1

Location: Baker County, E½ sec. 19, T. 8 S., R. 44 E., 2½ miles west-southwest of Sparta.

Development: Two adits 80 feet apart vertically aggregate 2000 feet.

Geology: Quartz vein and silicification along shear planes in albite granite. Main vein ranges up to 7 feet thick, but averages about 1 foot. Strike varies from N. 35° W. to N. 60° E. Dip is shallow. Sulfides present are pyrite and arsenopyrite.

Production: Small.

References: Gilluly, Reed, and Park, 1933:59; Department Bulletin 14-A, 1939:47.

Del Monte mine — Sparta District, E-2

Location: Baker County, E½ sec. 15, T. 8 S., R. 44 E.

[No other data available.]

Gem mine — Sparta District, E-3

Location: Baker County, secs. 17 and 20, T. 8 S., R. 44 E.

Development: 550-foot inclined shaft with eight levels.

Geology: Vein 1 to 4 feet wide in albite granite; strike N.; dip 30° to 40° E. Vein matter is quartz and gouge with little pyrite and sphalerite and free gold.

Production: Probably small. Some milling done.

References: Lindgren, 1901:737; Swartley, 1914:127; Parks and Swartley, 1916:98; Gilluly, Reed, and Park, 1933:58; Lorain, 1938:33; Department Bulletin 14-A, 1939:92.

Gold Ridge (New Deal) mine — Sparta District, E-4

Location: Baker County, SE$\frac{1}{4}$ sec. 16, T. 8, R. 44 E.

Development: Many cuts, and pits and 140-foot incline with short drifts on 100-foot level.

Geology: Quartz vein in albite granite; surface croppings on Gold Ridge yielded rich pockets in early days. Strike N. 30° W.; dip 40° E. Mostly free gold.

Production: $124,000 between 1889 and 1892.

References: Lorain, 1938:34; Department Bulletin 14-A, 1939:94.

Macy mine — Sparta District, E-5

Location: Baker County, SW$\frac{1}{4}$ sec. 2, T. 9 S., R. 44 E. in Maiden Gulch, about half a mile from Powder River.

Development: About 1500 feet, mostly from adit levels. There are two short, inclined shafts.

Geology: Branching quartz veins, paper thin to about 4 feet thick, in albite granite. The main vein strikes N. 50° W. and dips 40° to 75° S. Gangue is chiefly quartz, locally brecciated and vuggy, containing pyrite, sphalerite, and free gold. Stope widths average 1 to 1$\frac{1}{2}$ feet.

Production: Approximately $90,000. Discovered about 1920.

References: Gilluly, Reed, and Park, 1933:59; Department Bulletin 14-A, 1939:93.

Union mine — Sparta District, E-6

Location: Baker County, sec. 9, T. 8 S., R. 44 E.

Development: Short adit.

Geology: Vein in albite granite.

Production: No record.

References: None.

GREENHORN MOUNTAINS AREA

Location and Geography

The area lies in the Greenhorn Mountains in Grant and Baker Counties (figure 25). A granitic intrusion named the Greenhorn Mountain batholith by Hewett (1931, p. 340) underlies the area. Numerous gold-silver lode deposits occur within the intrusive and also in older rocks along its edges. The deposits are grouped in two separate districts: the Greenhorn, which straddles the Baker-Grant County line, and the Susanville to the west in Grant County.

The Greenhorn Mountains trend generally northwestward from near Whitney to a few miles beyond Susanville, a total distance of about 30 miles. The range is bounded on the northeast and southwest by the North and Middle Forks of the John Day River, respectively. The east slope is drained by the North Fork of Burnt River. Elevations range from 8131 feet on Vinegar Hill, the highest point in the central part of the range, to 4200 feet at Whitney and 3400 feet on the Middle Fork of the John Day River at Galena. The elevation at Susanville on Elk Creek about 2 miles above Galena is 3796 feet. At the old townsite of Greenhorn, about 3 miles east along the crest of the range from Vinegar Hill, the elevation is about 6400 feet. Most of the area is heavily timbered with only portions of the highest ridges bare.

Geology

The Greenhorn batholith is compositionally similar to the larger Wallowa and Bald Mountain batholiths. It intrudes meta-argillites and greenstones which appear to represent the westward continuation of the Elkhorn Ridge Argillite. Near Susanville, however, these rocks are somewhat more schistose than is typical of that formation. Also found are numerous small masses of gabbro, peridotite, serpentinite, and related rocks which are younger than the argillite series but older than the Greenhorn batholith.

Production

The principal lode mines in the Greenhorn area are the Bonanza, Red Boy, and Ben Harrison mines in the Greenhorn district and the Badger mine in the Susanville district. Output from these mines has been about $3,500,000. There are also several small mines and prospects, particularly in the Greenhorn district, that have produced small amounts of gold and silver. It seems unlikely that total lode production exceeds $4,000,000. Total placer output cannot be as closely estimated, since only sketchy records are available.

Placer Mines

Greenhorn placers

Pardee and Hewett (1914, p. 10) estimated the yield from placers in the Greenhorn district, including the Bonanza and Whitney areas, as not less than $1,568,000. Subsequent production is unknown. A bucketline dredge operating on Burnt River above Whitney during 1941-1942 and 1945-1946 produced 4457 ounces gold and 902 ounces silver. During recent years there has been a small annual output from the Winterville hydraulic placers, which are situated about a mile south of the Bonanza.

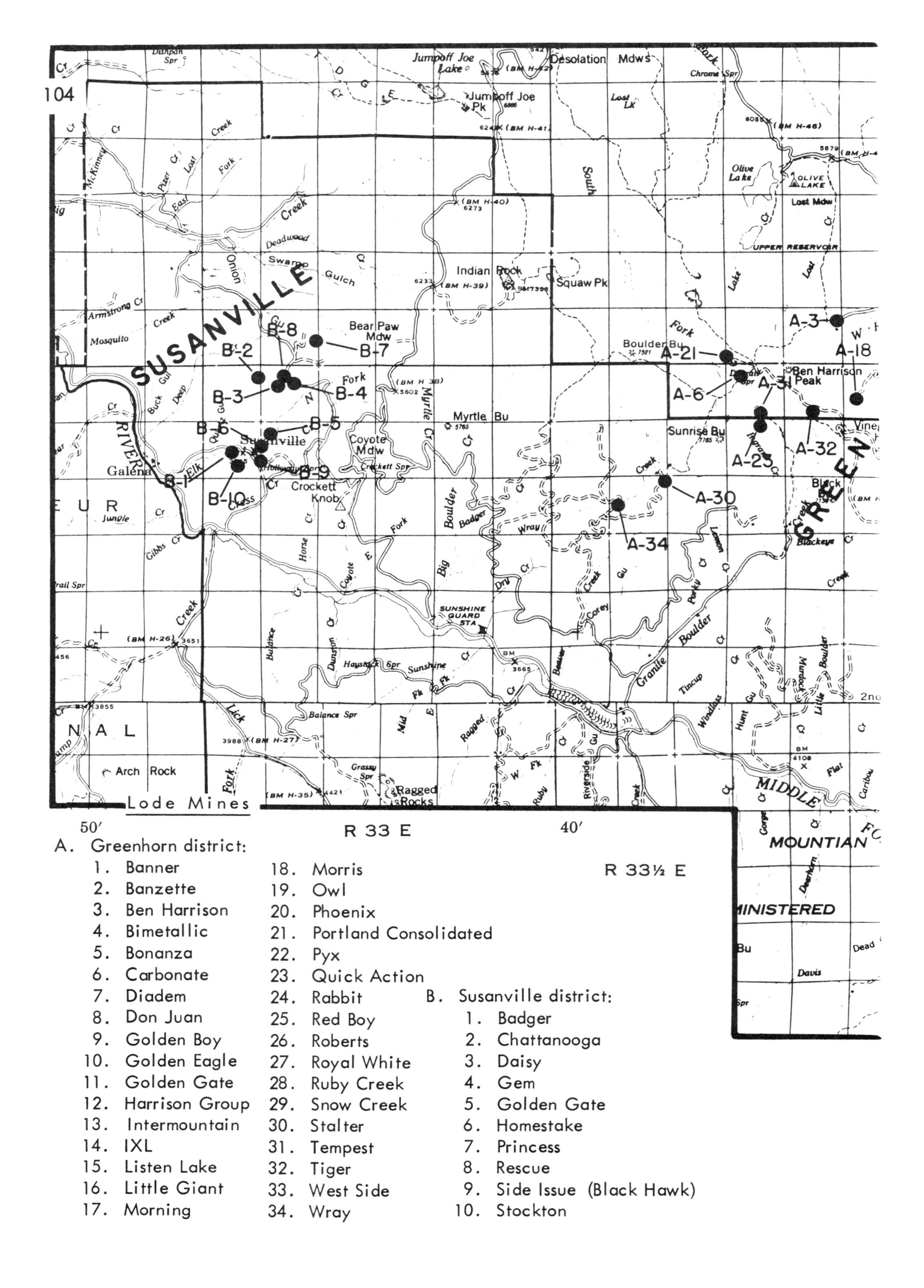
SUSANVILLE
GREEN
Jumpoff Joe Lake
Jumpoff Joe Pk
Desolation Mdws
Olive Lake
Lost Mdw
UPPER RESERVOIR
Indian Rock
Squaw Pk
Bear Paw Mdw
Boulder Bu
Ben Harrison Peak
Myrtle Bu
Sunrise Bu
Coyote Mdw
Crockett Knob
Susanville
Galena
SUNSHINE GUARD STA
Arch Rock
Ragged Rocks
MIDDLE
MOUNTAIN
MINISTERED
B-1
B-2
B-3
B-4
B-5
B-6
B-7
B-8
B-9
B-10
A-3
A-6
A-18
A-21
A-23
A-30
A-31
A-32
A-34
50′
R 33 E
40′
R 33½ E
Lode Mines
A. Greenhorn district:
1. Banner
2. Banzette
3. Ben Harrison
4. Bimetallic
5. Bonanza
6. Carbonate
7. Diadem
8. Don Juan
9. Golden Boy
10. Golden Eagle
11. Golden Gate
12. Harrison Group
13. Intermountain
14. IXL
15. Listen Lake
16. Little Giant
17. Morning
18. Morris
19. Owl
20. Phoenix
21. Portland Consolidated
22. Pyx
23. Quick Action
24. Rabbit
25. Red Boy
26. Roberts
27. Royal White
28. Ruby Creek
29. Snow Creek
30. Stalter
31. Tempest
32. Tiger
33. West Side
34. Wray
B. Susanville district:
1. Badger
2. Chattanooga
3. Daisy
4. Gem
5. Golden Gate
6. Homestake
7. Princess
8. Rescue
9. Side Issue (Black Hawk)
10. Stockton

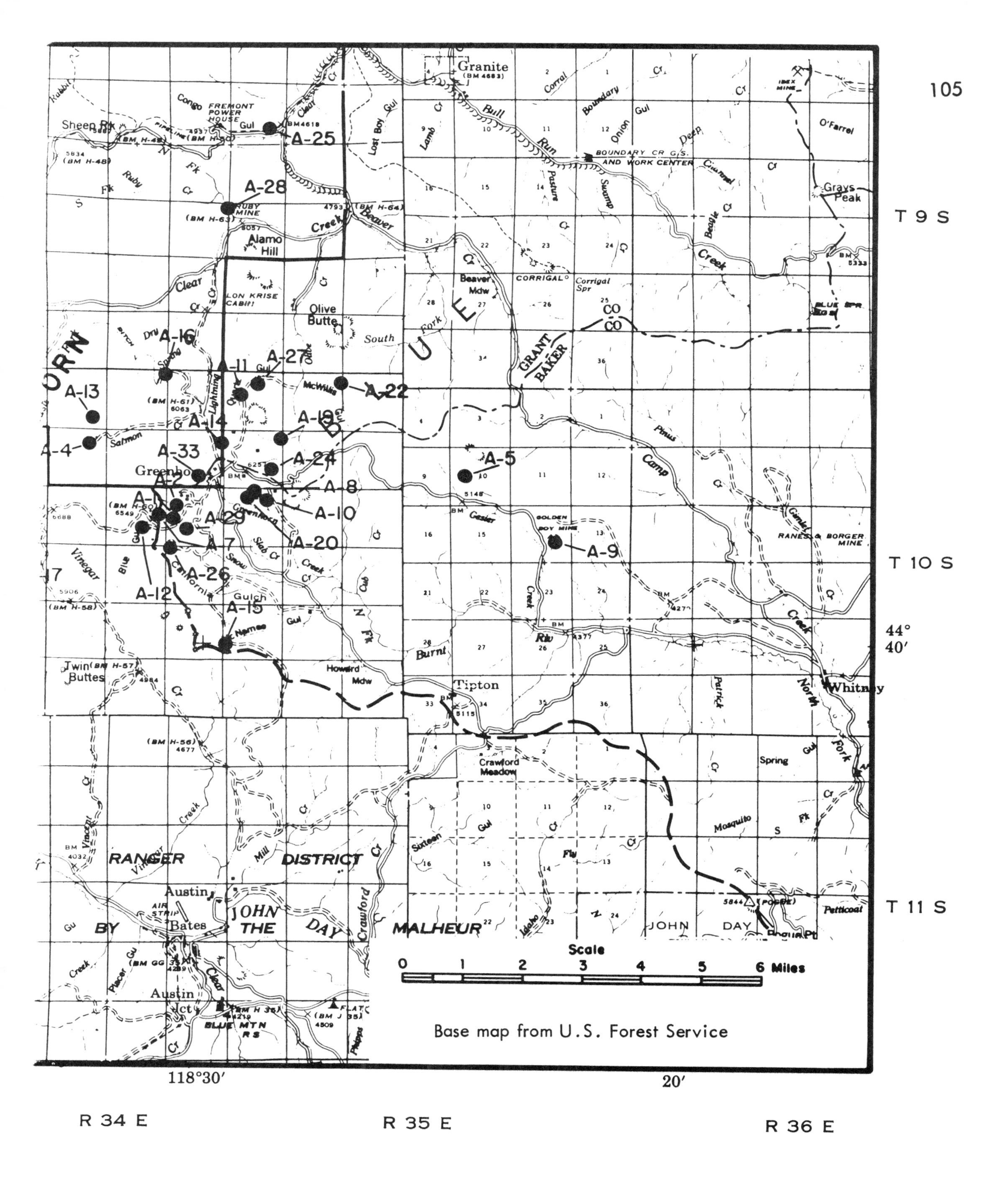

Figure 25. Index map of the Greenhorn Mountains area.

Susanville placers

The placer mines of the Susanville district were, according to Swartley (1914, p. 169-170), discovered in 1864 and by 1914 had produced about $600,000. Elk Creek was the most productive, but other creeks at a lower elevation along the north side of the Middle Fork were also important producers.

The Timms Gold Dredging Co. operated a bucketline dredge on the Middle Fork just below the mouth of Elk Creek from November 1933 until the spring of 1939. Placer yield from the district during this period was nearly $500,000, most of it from the Timms dredge. In 1939 the dredge was moved to the De Witt Ranch on the Middle Fork of the John Day River 10 miles below Bates, where it operated until 1942.

Greenhorn District

General information

The Greenhorn district embraces the eastern part of the Greenhorn Mountain range (figure 25) and incorporates all of the small subdistricts into which Swartley (1914, p. 173-194) divided this part of the range. The deposits are grouped around the old town of Greenhorn, which is about 50 miles by road from Baker by way of Whitney.

The mines and prospects in the Greenhorn district occur in a wide variety of host rocks. The Bonanza and Red Boy mines, several miles east and north of Greenhorn respectively, are in argillite. Closer to Greenhorn there are a great number of small veins in greenstone, argillite, and serpentinized intrusive rocks. Several of these small veins are associated with porphyritic dikes of intermediate to acid composition and they probably formed within the same zones of structural weakness (Allen, 1948, p. 21). Farther west the Ben Harrison, Tiger, Tempest, and several other small mines and prospects develop veins in granodiorite of the Greenhorn batholith. Still farther west in the drainage of Big Boulder Creek are several prospects in granodiorite and greenstone. Some of these, including the Wray and Reed prospects, contain copper minerals. The ores of several veins west of Greenhorn, including the Ben Harrison, Bi Metallic, Morris, and Intermountain, show a high silver-to-gold ratio.

Hewett found less definite evidence of mineral zoning in the veins around the Greenhorn batholith than around the Bald Mountain batholith, but suggested that the dissimilarities may be due to the fact that the deposits in the Greenhorn district occur in a wide variety of host rocks. Also only a few of the Greenhorn veins have been opened below the zone of oxidation.

The largest mines in the district and their estimated output are the Bonanza, $1,750,000; the Red Boy, $1,000,000; and the Ben Harrison, $425,000. In addition to these deposits the district embraces a large number of small mines and prospects. There is probably a greater concentration of small veins in the near vicinity of Greenhorn than is present in any other area of similar size in the Blue Mountains. Production records are available for very few of them, but while the combined production is probably small the presence of so great a concentration of small veins is indicative of widespread mineralization. Several of these veins have produced rich bunches of ore from near-surface workings and from residual placers.

Principal lode mines

Bonanza mine: The Bonanza mine is on the ridge north of Geiser Creek about 5 miles east of Greenhorn. It was discovered in 1877 and actively operated between 1892 and 1907 (figure 26). According to Pardee and Hewett (1914, p. 119), the lower levels were abandoned in 1904. From several adits and a shaft the mine is developed to a depth of 1250 feet. There are about 18,000 feet of workings. The vein, in argillite, strikes N. 55° W. and is nearly vertical. According to Lindgren (1901, p. 701): "The ore body as a whole forms a mass of clay slate traversed by quartz veins and seams of all sizes. The gold is low grade, being about 600 fine....Something like 70 percent is free, though it is said that as depth is increased more concentrates and less gold are obtained. The concentrates are said to vary from $20 to $60 per ton, chiefly in gold. The average ore is believed to run from $7 to $12 per ton, but lenses of ore 8 to 16 inches wide have been mined which ran as high as $1400 per ton, and several hundred tons are said to have yielded at the rate of $100 in free gold per ton.

"Though the pay streak averages only 5 to 6 feet, it swelled in places to 40 feet by the appearance of a vast number of quartz stringers."

Pardee and Hewett (1914, p. 120) presented the following statement of production for the period from 1899 to 1904.

1899		$ 146,419.47	a
1900		175,953.45	a
1901	14,885 tons ore	279,556.42	b
1902	5,371 tons ore	84,003.08	b
1903	11,495 tons ore	202,375.85	b
1904	3,887 tons ore	52,315.81	b
		$ 940,624.08	

a. probably net; b. gross.

The 1901-1904 output amounted to $636,251.16 from 35,638 tons of ore, indicating an average gross value of $17.85 per ton of ore treated.

Red Boy mine: The Red Boy mine is at the head of Congo Gulch 5 miles southwest of Granite. Development of this mine began about 1890. The period of greatest activity was from 1893 through 1903, although work was continued through 1914 (figure 27). The Red Boy mine is developed by about 5000 feet of drifts and crosscuts from three adits and a 300-foot shaft. Swartley (1914, p. 192) shows a plan of the workings.

The country rock is argillite that dips about 15° W. and is cut by numerous highly altered felsite dikes. The dikes are probably premineral but formed in zones of weakness where postmineral movement occurred.

At least five veins have been explored, the Red Boy, Monarch, Blaine, Concord, and Congo. The Red Boy and Monarch yielded nearly all of the ore produced by the mine. The other three, which were worked mainly during the later years of the mine's operation, are said to have been well defined, but few shoots that would pay to mill were found (Pardee and Hewett, 1914, p. 113).

The Red Boy and Monarch veins were explored for distances of 1000 feet and 900 feet respectively. Each has been stoped for a horizontal distance of about 800 feet. Good ore values are said to have held to a depth of only about 300 feet.

The Red Boy vein strikes nearly due north and dips about 80° W., whereas the Monarch strikes N. 30° E., and dips 50° to 55° W. The two join near the south end of the workings, and a short distance farther south are offset an unknown distance along a broad fault zone that contains one of the felsite dikes.

The veins consist of crushed argillite traversed by a great number of veins and stringers of quartz. Width of the crushed zones ranges from 3 to 15 feet. The values, which were generally best toward the footwall, were mainly in the quartz and consisted chiefly of free gold alloyed with much silver, the bullion being 515 to 525 fine. Sulfides, mainly fine pyrite, which made up about 5 percent of the mill ore, were largely contained in the argillite rather than in the quartz. According to data collected by the operators and presented by Pardee and Hewett (1914, p. 113), the combined areas of stopes on the Red Boy and Monarch veins up to January 1, 1902 was 437,000 square feet and the yield 83,000 tons, indicating an average stoping width of 28 inches. The return from this tonnage was $666,322.10 or $8.00 to the ton.

Ben Harrison mine: The Ben Harrison mine is near the head of the west fork of Clear Creek, about 28 miles by road from Sumpter by way of Granite, at an elevation of about 6500 feet. The mine was actively worked during several short periods including 1913-1914, 1916-1920, 1926-1928, and 1936-1937, producing roughly $425,000. The country rock is granodiorite cut by numerous aplite dikes. The vein strikes N. 3° E. and dips 80° E. in the upper levels, but flattens to 55° in the lower levels.

The vein and associated gouge is said to vary in width from 18 inches to a maximum of 21½ feet; the average stope width is probably 4 to 5 feet (Swartley, 1914, p. 176-180, and Lorain, 1938, p. 16-18). The ore shoot is a strong quartz vein which in most places contains numerous inclusions of incompletely replaced fragments of the granodiorite.

The ore minerals are pyrite, arsenopyrite, stibnite, a little chalcopyrite, tetrahedrite, and sphalerite. The silver sulfides are pyrargyrite and stephanite.

An ore shoot 300 to 400 feet long has been mined to a depth of 450 feet below the outcrop, or about 300 feet below the main adit level. Development was carried down at least another 100 feet to the 600-foot level, but little ore was removed. Swartley (1914, p. 176-180) states that the average value of ore in the upper workings was a little more than $10 a ton but that ore developed on the 600-foot level was 19 to 20 percent higher in grade. There was a wide variation in the gold-silver ratio of the ore mined from different parts of the shoot. In 1917 the mine produced 977 ounces gold and 22,534 ounces silver from 4600 tons of ore treated in a 20-stamp mill. Flotation concentrates produced in 1937 contained 1294 ounces gold and 10,823 ounces silver.

Other mines: The Snow Creek mine is said to have produced $52,000 between 1902 and 1905. Some very small shipments were made in 1925-1927 and in 1939. Total production of the mine is probably not more than $60,000. The vein in argillite and serpentine averages about 2 feet in width and strikes east. It is developed from a 240-foot shaft and connecting crosscut adit 1400 feet long.

At the Morning mine, the most important vein lies along the footwall of a diorite-porphyry dike which strikes N. 45° E. and dips 35° to 75° NW. This vein pinches and swells from 1 to 4 feet in width. Other narrower veins lie within or transect the dike. A stope on the footwall vein is 200 feet long, 65 feet high, and 3 to 6 feet wide. Production records are not available.

Susanville District

General information

The Susanville district lies in Grant County about 18 miles by road down the Middle Fork of the John Day River from Bates (figure 25). The several lode mines and prospects of this district are confined mainly to an area 4 miles long and 2 miles wide that extends northeasterly through the old town of Susanville, which is about 1½ miles up Elk Creek from its junction with the Middle Fork. Relief is about 1200 feet; slopes are fairly steep; and timber and water are abundant.

The rocks in the district are mainly schists with subordinate amounts of quartzite, slate, greenstone, serpentinized peridotite, and gabbro, all of which are cut by numerous aplite dikes. Quartz diorite of the Greenhorn batholith occupies the northern and eastern edges of the district. The trend of the principal partings in the schists is east to northeast with steep dips to the south. Faulting parallel to the schistosity is common throughout the district. Most of the veins of the district occur in schists and fill fissures paralleling the schistosity. Several of the ore bodies show evidence of being related to the aplite dikes which, in turn, are related to the quartz diorite. Some of the veins are in serpentinized rocks or along their contacts with the schists.

The dominant gangue of the veins is quartz; the metallic minerals include pyrite, marcasite, arsenopyrite, pyrrhotite, sphalerite, galena, stibnite, tetrahedrite-tennantite, and chalcocite. The percentage of the gold values extractable by amalgamation was small except from oxidized surface ores which were soon exhausted.

Gold output of the Susanville district has come mainly from placers. Lode production has been small, probably not much more than $500,000 or $600,000.

Principal lode mine

Badger mine: The most prominent mine of the district is the Badger, which was first operated during the late 1870's. Records are scarce but the gross value of production from this mine probably lies between $250,000 and $500,000. Development includes a 900-foot shaft, a 1600-foot crosscut adit on the 500-foot level, and several hundred feet of drifts. Greatest activity was between 1899 and 1905, after which the mine was closed. Little mining has been done since. According to limited records for the 1899-1905 period of operation (Oregon Dept. of Geology and Mineral Industries, 1941, p. 133-137), the ores contained between 8 and 11 percent recoverable sulfide concentrate; total recovery from mill ore was between 58 and 60 percent of assay value. At least two veins have been stoped. They are 6 feet apart and vary in width from 1 to 20 feet (Gilluly, Reed, and Park, 1933, p. 111).

Figure 26. Surface plant at the Bonanza mine, Greenhorn district.

Figure 27. Red Boy mine, Greenhorn district, prior to 1915.
(Photograph courtesy of Brooks Hawley.)

Lode Mines of the Greenhorn Mountains Area

A. GREENHORN DISTRICT

Banner mine — Greenhorn District, A-1

Location: Grant County, NW$\frac{1}{4}$ sec. 16, T. 10 S., R. 35 E., about 1$\frac{1}{2}$ miles southwest of Greenhorn.

Development: 200-foot shaft; 600 feet of drift.

Geology: Vein 4 inches wide in serpentine. Strike W.; dip S.

Production: No record.

References: Hewett, 1931:18; Department Bulletin 14-B, 1941:68.

Banzette mine — Greenhorn District, A-2

Location: Grant County, NW$\frac{1}{4}$ sec. 16, T. 10 S., R. 35 E., about 1$\frac{1}{2}$ miles southwest of Greenhorn.

Development: 1600-foot adit to shaft 100 feet deep.

Geology: Quartz veinlets and sulfide minerals associated with quartz monzonite porphyry dikes intruded into serpentine, argillite, and greenstone. Sulfides include pyrite, chalcopyrite, sphalerite, and galena.

Production: No record.

References: Hewett, 1931:19; Department Bulletin 14-B, 1941:68; Allen, 1948:41.

Ben Harrison mine — Greenhorn District, A-3

Location: Grant County, NE$\frac{1}{4}$ sec. 35, T. 9 S., R. 34 E., near the head of Clear Creek.

Development: More than 4000 feet from two adits and a shaft; workings attain a depth of about 550 feet below outcrop.

Geology: Country rock granodiorite with small inclusions of greenstone. Aplite dikes abundant. Vein 18 inches to 21$\frac{1}{2}$ feet wide; strikes N. 3° E, dips 67° E. Average stope width 77 inches. Filling is sericitized granodiorite breccia and gouge cemented and partly replaced by quartz. Calcite also present. Sulfides are pyrite, stibnite, a little chalcopyrite, sphalerite, pyrargyrite, and stephanite.

Production: Records scarce, output estimated at $425,000. Last operated 1937. Concentrate ratio 20:1; silver to gold ratio 5:1 to 50:1. Ore mined prior to 1914 averaged $10 per ton.

References: Lindgren, 1901: 694; Swartley, 1914:176; Parks and Swartley, 1916:29; Hewett, 1931:10; Lorain, 1938:16; Department Bulletin 14-B, 1941:68-72.

Bi Metallic mine

Greenhorn District, A-4

Location: Grant County, NE$\frac{1}{4}$ sec. 7, T. 10 S., R. 35 E., 2$\frac{1}{2}$ miles west of Greenhorn.

Development: Several adits; one includes 2150-foot crosscut and 410-foot drift.

Geology: Country rock is granodiorite intruded by quartz diorite and quartz monzonite porphyry dikes. Vein quartz and minor iron and copper sulfides fill narrow fissures in the dikes and in the granodiorite. Molybdenite present locally.

Production: No record.

References: Swartley, 1914:181-183; Parks and Swartley, 1916:37; Department Bulletin 14-B, 1941:73-74.

Bonanza mine

Greenhorn District, A-5

Location: Baker County, sec. 10, T. 10 S., R. 35$\frac{1}{2}$ E.

Development: Three adits and 1200-foot shaft; total about 18,000 feet.

Geology: Country rock mainly argillite with some greenstone; vein strikes N. 55° W. and is nearly vertical. Consists of sheared and brecciated country rock cemented by quartz veins and stringers. Ore shoots averaged 5 to 6 feet in width, but swelled in places to 40 feet.

Production: $1,750,000 estimated. Discovered 1877; little output since 1904. Gold average about 70 percent free and 600 fine. Sulfide concentrates varied from $20 to $60 per ton. Gross output during 1901-1904 totaled $636,251.16 from ore averaging $17.85 per ton.

References: Lindgren, 1901:700; Pardee and Hewett, 1914:119 (map); Swartley, 1914:188.

Carbonate mine

Greenhorn District, A-6

Location: Grant County, NW$\frac{1}{4}$ sec. 3, T. 10 S., R. 34 E., adjoining the Tempest.

Development: Crosscuts and drifts, amount unknown.

Geology: Several narrow quartz veins in a broad northeast-trending fracture zone in granodiorite. Veins contain arsenopyrite, pyrite, sphalerite, and a little galena.

Production: Some ore shipped; date, quantity, or value not recorded.

References: Lindgren, 1901:964; Swartley, 1914:175; Parks and Swartley, 1916:43-44; Department Bulletin 14-B, 1941:75.

Diadem mine

Greenhorn District, A-7

Location: Grant County, NE$\frac{1}{4}$ sec. 17, T. 10 S., R. 35 E., near Banzette, 1$\frac{1}{2}$ miles southwest of Greenhorn.

Development: Unknown.

Geology: Vein in greenstone, strike east to west; vertical. Ore minerals are pyrite and cinnabar.

Production: No record.

References: Lindgren, 1901:698; Swartley, 1914:185; Parks and Swartley, 1916:86; Hewett, 1903:19, 36; Department Bulletin 14-B, 1941:75.

Don Juan mine

Greenhorn District, A-8

Location: Baker County, secs. 10 and 15, T. 10 S., R. 35 E., one mile southeast of Greenhorn.

Development: Adit of unknown length.

Geology: Vein in serpentine and greenstone. Ore minerals are dolomite, chalcopyrite, and pyrite.

Production: Small.

References: Lindgren, 1901:696; Parks and Swartley, 1916:87; Hewett, 1931:20; Department Bulletin 14-A, 1939:54.

Golden Boy mine

Greenhorn District, A-9

Location: Baker County, sec. 14, T. 10 S., R. 35$\frac{1}{2}$ E.

[No further data available.]

Golden Eagle mine

Greenhorn District, A-10

Location: Baker County, E$\frac{1}{2}$ sec. 15, T. 10 S., R. 35 E.

Development: Three adits and an intermediate drift aggregating about 2600 feet; vertical range about 175 feet, plus a shaft 75 feet below lowest level.

Geology: Workings explore system of branching mineralized fractures in serpentine. Trend NW.; dip 40° -70° NE. Lenses of quartz and dolomite contain chalcopyrite, galena, and free gold. Average width 6 inches. Ore oxidized. High-grade pockets encountered. Gold 850 fine.

Production: Estimated at $75,000.

References: Pardee and Hewett, 1914:116; Parks and Swartley, 1916:105; Department

Bulletin 14-A, 1939:54.

Golden Gate mine — Greenhorn District, A-11

Location: Grant County, sec. 3, T. 10 S., R. 35 E., 2 miles north of Greenhorn.

Development: Three adits; the longest contains about 2400 feet of work, including raises.

Geology: Two veins, the Golden Gate and Belcher. The latter, being the most extensively developed, is in greenstone and argillite, strikes NNE. and dips steeply eastward. Vein filling is mostly quartz. Three small shoots have been stoped; maximum width 20 inches; low grade. The Golden Gate vein is said to be 40 feet wide, most of which is quartz.

Production: No record; output probably very small.

References: Lindgren, 1901:697; Pardee and Hewett, 1914:114; Swartley, 1914:187; Parks and Swartley, 1916:106; Hewett, 1931:20, 36; Department Bulletin 14-B, 1941:77-78.

Harrison Group (Windsor, Psyche, and Big Johnny) — Greenhorn District, A-12

Location: Grant County, sec. 17, T. 10 S., R. 35 E., 3 miles west of Greenhorn.

Development: Poor record; includes a 60-foot shaft and four adits 1200, 1100, 700, and 300 feet long.

Geology: North-northeast-trending shear zone in serpentine and altered gabbro contains several narrow quartz veins and porphyry dikes. High-grade ore occurs in scattered seams and lenses rarely more than 8 inches thick and containing calcite, dolomite, pyrite, and chalcopyrite.

Production: Small, no mill record. Psyche said to have produced $90,000 in 1905.

References: Department Bulletin 14-B, 1941:78.

Intermountain mine — Greenhorn District, A-13

Location: Grant County, SE$\frac{1}{4}$ sec. 6, T. 10 S., R. 35 E., half a mile north of Bi metallic.

Development: Unknown.

Geology: Vein in diorite and greenstone. Strike east-west and may be extension of Bi metallic vein. Ore consists of quartz and tetrahedrite rich in silver.

Production: Small. Property located in 1937. Latest shipments (crude ore) made in 1938 and 1940.

References: Lindgren, 1901:694; Swartley, 1914:183; Department Bulletin 14-B, 1941:79.

IXL mine

Greenhorn District, A-14

Location: Grant County, secs. 9 and 10, T. 10 S., R. 35 E., just east of Greenhorn.

Development: Two shafts, with drifts on the veins.

Geology: Three veins in argillite and greenstone.

Production: No record.

References: Hewett, 1931:20; Department Bulletin 14-B, 1941:79.

Listen Lake mine

Greenhorn District, A-15

Location: Baker County, SW$\frac{1}{4}$ sec. 27, T. 10 S., R. 35 E.

Development: Includes shaft 120 feet deep.

Geology: Silicified shear zone in gabbro reportedly attains width of 50 feet; contains pyrite, chalcopyrite, and associated gold along fractures.

Production: Small; no mill record.

References: Pardee and Hewett, 1914:118; Parks and Swartley, 1916:141; Hewett, 1931: 19, 36; Department Bulletin 14-A, 1939:55.

Little Giant mine

Greenhorn District, A-16

Location: Grant County, secs. 4 and 5, T. 10 S., R. 35 E., near head of Spring Creek.

Development: Includes shaft 40 feet deep and two short adits 1000 feet apart.

Geology: Little data available. Country rocks are argillite, diorite, and serpentine. Ore minerals include quartz and massive pyrite. Some malachite in upper workings.

Production: No record. Property discovered in 1898; 20-stamp mill built in 1899; little activity since 1906.

References: Department Bulletin 14-B, 1941:81-82.

Morning mine

Greenhorn District, A-17

Location: Grant County, SE$\frac{1}{4}$ sec. 13, T. 10 S., R. 34 E., about 5 miles southwest of Greenhorn.

Development: Two adits and an intermediate level. Lower level contains more than 1300 linear feet of workings and some small stopes.

Geology: The country rocks are argillite, greenstones, and serpentine intruded by quartz-diorite porphyry dikes. One dike ranges up to 100 feet in width, strikes N. 45° E. and dips 35° to 75° NW. Within this dike and along its walls are narrow, branching quartz veins which have produced small amounts of ore. A footwall shoot was 200 feet long, 65 feet in maximum height, and averaged 3 feet in width; values were spotty, ranging from $15 to $50 per ton. Sulfides are mainly pyrite with small amounts of arsenopyrite, sphalerite, and chalcopyrite.

Production: Little data available. Output probably small. Mine active during 1937-1942 and for a short time after World War II. Gold-silver ratio about 1:3. Sulfides comprise about 5 percent of the ore. Mine located in 1893.

References: Swartley, 1914:183; Parks and Swartley, 1916:155; Department Bulletin 14-B, 1941:82; Allen, 1948:33.

Morris mine

Greenhorn District, A-18

Location: Grant County, SW$\frac{1}{4}$ sec. 1, T. 10 S., R. 34 E., in Morris Basin.

Development: Two groups of workings about 800 feet apart comprise about 1000 feet of crosscuts and drift distributed among four adits.

Geology: Country rocks are argillite and quartz diorite cut by quartz-diorite porphyry dikes. Workings expose several quartz veins 6 inches to 15 inches wide, at least one of which has been stoped. Sulfides include pyrite, arsenopyrite, sphalerite, tetrahedrite, and galena and make up 5 to 10 percent of the ore.

Production: Little data available; 1891 output was $15,000 in silver and $3400 in gold. Shipments in 1913-1914 averaged $50 per ton.

References: Lindgren, 1901:694; Swartley, 1914:180; Parks and Swartley, 1916:156; Department Bulletin 14-B, 1941:84; Allen, 1948:38.

Owl (Red Bird and Virginia) mine

Greenhorn District, A-19

Location: Grant County, secs. 10 and 11, T. 10 S., R. 35 E. about 2 miles northeast of Greenhorn.

Development: Poor record. Probably about 1000 feet, including two shafts. Small stopes.

Geology: Indefinite. Country rocks include granodiorite and older gabbro. Several narrow quartz veins exposed; one is associated with a shear zone as much as 22 feet wide in granodiorite; strikes N. 35° W., dips 75° to 85°, steepening with depth. Another vein is in crushed gabbro. Ore mined from near-surface pockets.

Production: Probably about $50,000. A $20,000 pocket was taken from the Virginia claim in the 1890's. Red Bird and Owl claims, located in 1915 and 1921 respectively, operated intermittently until World War II. Thirty tons milled in 1940 ran $18 per ton in free gold, 700 fine.

References: Parks and Swartley, 1916:187, 229; Department Bulletin 14-B, 1941:86-87.

Phoenix mine
Greenhorn District, A-20

Location: Baker County, sec. 15, T. 10 S., R. 35 E.

Development: Three adits; total about 1500 feet.

Geology: Vein in serpentine. Ore minerals are quartz and chalcopyrite.

Production: Small, no mill record.

References: Hewett, 1931:20, 36; Department Bulletin 14-A, 1939:56.

Portland Consolidated mine
Greenhorn District, A-21

Location: Grant County, SE$\frac{1}{4}$ sec. 33, T. 9 S., R. 34 E., at the head of the South Fork of Desolation Creek.

Development: Seventy-five-foot shaft and open cuts.

Geology: Quartz vein in argillite; strikes N. 28° E.; contains pyrite and galena.

Production: No record.

References: Department Bulletin 14-B, 1941:105.

Pyx mine
Greenhorn District, A-22

Location: Grant County, secs. 1 and 2, T. 10 S., R. 35 E., between forks of McWillis Gulch.

Development: Includes shaft 150 feet deep and several short adits; one is 600 feet long.

Geology: Vein in argillite; vein minerals are quartz and pyrite.

Production: Small output prior to 1900 and during 1907-1911. A 25-ton mill erected in 1954, but little used.

References: Hewett, 1931:20; Department Bulletin 14-B, 1941:88.

Quick Action (Ornament) mine
Greenhorn District, A-23

Location: Grant County, N$\frac{1}{2}$ sec. 10, T. 10 S., R. 34 E., on Granite Boulder Creek.

Development: Three drifts.

Geology: Two northeast-trending veins near granodiorite-argillite contact. Vein width averages about 24 inches.

Production: Small shipments of low-grade ore made prior to 1916.

References: Parks and Swartley, 1916:176; Department Bulletin 14-B, 1941:88.

Rabbit mine

Greenhorn District, A-24

Location: Grant County, SE¼ sec. 10, T. 10 S., R. 35 E., two miles northeast of Greenhorn.

Development: 1000 feet of tunnels and a 160-foot shaft with short drift levels.

Geology: Vein about 2 feet wide in granodiorite; strikes N. 10° E., dips 70° E.

Production: Discovered in 1925; said to have produced $40,000 prior to 1940 with five-stamp mill. Gold 90 percent free; 750 to 760 fine.

References: Department Bulletin 14-B, 1941:89.

Red Boy mine

Greenhorn District, A-25

Location: Grant County, SE¼ sec. 10, T. 9 S., R. 35 E., near head of Congo Gulch, 5 miles west of Granite by way of Olive Lake road.

Development: Three adits and 300-foot shaft, total about 5000 feet; attains 500 feet below outcrop.

Geology: Country rock is argillite cut by premineral felsite dikes. Five veins explored; Red Boy and Monarch most important. Red Boy vein strikes N. and dips 80° W. Monarch vein strikes N. 30° E., dips 50° to 55° W. Vein filling is 3 to 15 feet wide and consists of crushed argillite cemented by veins and stringers of quartz. Stope widths average about 28 inches.

Production: Operated 1890 to 1914. Total output estimated at $1,000,000. Returns on 83,373 tons was $666,322.10, or $8.00 per ton. Over-all average grade of ore said to be $12.00 per ton. Gold 75 to 85 percent free; sulfides low grade. Bullion 520 gold:450 silver.

References: Lindgren, 1901:681; Pardee and Hewett, 1914:110; Swartley, 1914:189; Parks and Swartley, 1916:188; Lorain, 1938:15; Department Bulletin 14-B, 1941:60.

Roberts mine

Greenhorn District, A-26

Location: Grant County, secs. 16 and 21, T. 10 S., R. 35 E., about 2 miles southwest of Greenhorn.

Development: Four short adits and open cuts.

Geology: Crushed zone in serpentine and gabbro; strikes N. 70° W. and dips 40° NE.; said to be more than 40 feet wide; contains occasional small kidneys of quartz breccia, some of which have yielded high-grade oxidized ore.

Production: Total unknown; $10,000 produced from hydraulicking surface pocket. Discovered in 1899. Operated 1912 to 1917, with intermittent development from 1925 to 1942.

References: Parks and Swartley, 1916:194; Department Bulletin 14-B, 1941:90.

Royal White mine

Greenhorn District, A-27

Location: Grant County, sec. 3, T. 10 S., R. 35 E., about one mile north of Greenhorn.

Development: Adit containing 600 feet of drift and crosscuts attains depth of 95 feet.

Geology: Vein in argillite; strike N. 40 E., dip steeply west. Offset by cross fault. Vein filling is argillite breccia cemented by chalcedony and quartz. Ore oxidized. Two small stopes; ore 1 to 3 feet wide. Manganese oxides present.

Production: Several hundred tons of sorted ore yielded $25 to $28 per ton during 1904 to 1910 period. Small output in 1930's.

References: Pardee and Hewett, 1914:115; Swartley, 1914:187; Parks and Swartley, 1916:195; Hewett, 1931:20; Department Bulletin 14-B, 1941:91.

Ruby Creek mine

Greenhorn District, A-28

Location: Grant County, secs. 21 and 22, T. 9 S., R. 35 E., on Ruby Creek.

Development: Two adits, longest said to be 400 feet.

Geology: Country rock is cherty argillite. One adit crosscuts four veins within 160 feet of portal. Veins strike N. 10° E. and dip 80° E. Small stopes on each. No data on lower adit.

Production: Discovered in 1924. Output 1932 to 1936 was $7,430 from five-stamp mill. Previous production unknown.

References: Department Bulletin 14-B, 1941:63.

Snow Creek mine

Greenhorn District, A-29

Location: Baker County, sec. 16, T. 10 S., R. 35 E., near head of Snow Creek, 2 miles southwest of Greenhorn.

Development: Numerous adits; includes 1600-foot crosscut, a shaft more than 200 feet deep, and more than 1300 feet of drifts.

Geology: Country rocks argillite, serpentine, and greenstone; cut by quartz-monzonite dikes with which mineralized quartz seams 1 to 6 inches thick are associated. Main vein, mostly barren quartz, is 2 to 10 feet wide, strikes N. 60° W. to W., and dips 50° to 75° S. Ore shoots rich in galena, chalcopyrite, and pyrite. One stope 80 feet by 200 feet by 4 feet thick on crosscut level.

Production: $52,000 between 1902 and 1905. Ten-stamp mill on property in 1916; idle. Ore shipped during 1925-1927 and 1939.

References: Parks and Swartley, 1916:208; Hewett, 1931:19; Allen, 1948:41; Department Bulletin 14-A, 1939:56.

Stalter (Heppner) mine — Greenhorn District, A-30

Location: Grant County, secs. 8 and 17, T. 10 S., R. 34 E.

Development: 3,900 feet.

Geology: Granodiorite cut by porphyry dikes. A zone 1000 feet wide contains a dozen or more quartz veins which are 1 to 20 feet wide, strike N. 40 E., dip 50° to 75° E. Some are traceable for several hundred feet. Pyrite present locally. Gold free.

Production: Small. In 1937 to 1938, about 200 tons treated by flotation produced concentrates assaying $32.00 per ton. Concentrate ratio 30:1. Ore previously worked in two-stamp mill.

References: Swartley, 1914:174; Parks and Swartley, 1916:119; Department Bulletin 14-B, 1941:92.

Tempest mine — Greenhorn District, A-31

Location: Grant County, secs. 3 and 10, T. 10 S., R. 34 E., near the Chloride mine.

Development: Several short adits.

Geology: The country rock is granodiorite. There are said to be five veins consisting of crushed and sericitized granodiorite in which are small lenses and stringers of quartz with arsenopyrite, pyrite, and sphalerite. The chief values are in silver rather than gold. One vein is as much as 4 feet wide, strikes N. 35° E., and is nearly vertical.

Production: Probably small, although "quite a little ore was shipped" prior to 1916.

References: Lindgren, 1901:695; Swartley, 1914:175; Parks and Swartley, 1916:221; Department Bulletin 14-B, 1941:93; Allen, 1948:36.

Tiger mine — Greenhorn District, A-32

Location: Grant County, sec. 2, T. 10 S., R. 34 E., a mile and a half west of the Bi metallic.

Development: Two adits of unknown length.

Geology: North-trending vein in granodiorite cut by diorite-porphyry dikes. Sulfides make up about 5 percent of the vein material; include arsenopyrite, pyrite, sphalerite, tetrahedrite, and chalcopyrite.

Production: No record.

References: Department Bulletin 14-B, 1941:94; Allen, 1948:37.

West Side mine — Greenhorn District, A-33

Location: Grant County, SE$\frac{1}{4}$ sec. 9, T. 10 S., R. 35 E., immediately west of Greenhorn.

Development: 500-foot adit drift and a shallow shaft.

Geology: Vein in the form of broken lenses and blocks in greenstones, argillite, and serpentine all badly altered and structurally disturbed. Vein difficult to follow. Ore minerals are chiefly quartz, calcite, and dolomite with pyrite, galena, gold, and silver.

Production: Small. A few carloads of ore shipped in 1914; returns were \$50 to \$75 per ton.

References: Swartley, 1914:186; Parks and Swartley, 1916:235; Hewett, 1931:19; Department Bulletin 14-B, 1941:96.

Wray mine — Greenhorn District, A-34

Location: Grant County, W½ sec. 17, T. 10 S., R. 34 E.

Development: Small.

Geology: Copper minerals and associated gold along fractures in greenstone.

Production: None recorded.

References: Swartley, 1914:173.

B. SUSANVILLE DISTRICT

Badger mine — Susanville District, B-1

Location: Grant County, S½ sec. 7, T. 10 S., R. 33 E., on the south side of Elk Creek, about 2 miles above its junction with the Middle Fork of the John Day River.

Development: A 900-foot shaft with several drift levels, raises, and stopes. A 1600-foot crosscut adit connects with the 500-foot level of the shaft.

Geology: The country rock is mostly slate with some argillite, shale, and quartzite. The vein strikes a little north of east and dips 60° to 70° S. In parts of the mine; two veins about 6 feet apart. The principal ore shoot was 190 feet long and from 1 to 20 feet wide, consisting of quartz and partly replaced country rock containing pyrite, arsenopyrite, sphalerite, galena, chalcopyrite, and tetrahedrite containing high values in silver and gold. Sorted ore was kept above \$150 per ton. The grade is said to decrease below the 500 level, and little ore has been removed.

Production: Probably between \$250,000 and \$500,000, mostly during 1899-1905. Mill recovery averaged about 60 percent of assay values. The vein discovered in 1878.

References: Lindgren, 1901:706; Swartley, 1914:170; Parks and Swartley, 1916:19; Gilluly, Reed, and Park, 1933:111; Department Bulletin 14-B, 1941:133.

Chattanooga mine

Susanville District, B-2

Location: Grant County, secs. 5 and 6, T. 10 S., R. 33 E., southwest of the Reserve Group.

Development: Poor record; includes shaft 210 feet deep.

Geology: Vein in talc schist; strikes N. 65° E., and dips 60° E. The predominant gangue is quartz but ankerite, mariposite, and sericite are present. Sulfides include pyrite, chalcopyrite, sphalerite, galena, and arsenopyrite. Vein said to range from 1 to 8 feet in width.

Production: A little ore was shipped in early days; amount unknown.

References: Swartley, 1914:171; Parks and Swartley, 1916:165; Gilluly, Reed, and Park, 1933:113; Department Bulletin 14-B, 1941:139.

Daisy mine

Susanville District, B-3

Location: Grant County, NW$\frac{1}{4}$ sec. 5, T. 10 S., R. 33 E.

Development: Open cuts.

Geology: Rusty quartz vein in talc schist; strikes N. 20° E., and dips 65° E. Quartz said to contain a little pyrrhotite, pyrite, chalcopyrite, and free gold.

Production: No record.

References: Gilluly, Reed, and Park, 1933:113; Department Bulletin 14-B, 1941:140.

Gem mine

Susanville District, B-4

Location: Grant County, N$\frac{1}{2}$ sec. 5, T. 10 S., R. 33 E., between the forks of Elk Creek.

Development: Shaft 350 feet deep, with four levels.

Geology: Country rock is talc schist and peridotite. Vein, mostly quartz with some ankerite, strikes N. 45° E., dips 60° SE., and is about 3 feet wide. Free gold is present along with chalcopyrite, pyrite, and pyrrhotite.

Production: Small.

References: Parks and Swartley, 1916:164; Gilluly, Reed, and Park, 1933:112; Department Bulletin 14-B, 1941:140.

Golden Gate (Poorman) mine

Susanville District, B-5

Location: Grant County, secs. 7 and 8, T. 10 S., R. 33 E., east of the Badger and north of Elk Creek.

Development: Poor record; includes several adits and cuts, one adit being 833 feet long.

Geology: The country rock is talc schist with some slate. The series is cut by an aplite dike about 50 feet thick which is locally mineralized. Numerous small veins and lenses of quartz are found in the aplite and enclosing schist. Some reportedly contain free gold. Sulfides present are sphalerite, pyrite, galena, chalcopyrite, and stibnite. A shoot on the Beaver vein said to be 70 feet long, $2\frac{1}{2}$ feet thick, and 100 feet deep was worked prior to 1870.

Production: Small (?).

References: Gilluly, Reed, and Park, 1933:114; Department Bulletin 14-B, 1941:141.

Homestake (Bull of the Woods, Mockingbird) mine — Susanville District, B-6

Location: Grant County, secs. 7 and 8, T. 10 S., R. 33 E., on the north side of Elk Creek opposite the Badger.

Development: Several adits, two of which are connected by a shaft.

Geology: The country rock is serpentine and talc schist. The workings explore several mineralized fracture zones which locally contain a little quartz, pyrite, chalcopyrite, pyrrhotite, and galena.

Production: Small output reported in 1905.

References: Parks and Swartley, 1916:124; Gilluly, Reed, and Park, 1933:113, 116.

Princess mine — Susanville District, B-7

Location: Grant County, SW$\frac{1}{4}$ sec. 32, T. 9 S., R. 33 E.

Development: Includes a 460-foot adit and 60-foot shaft.

Geology: Nearly vertical pipe-like quartz replacement body in schist near contact with quartz diorite. Stringers extend from the mass in several directions. Pyrite, chalcopyrite, pyrrhotite, sphalerite, and free gold have been reported.

Production: No record.

References: Swartley, 1914:172; Parks and Swartley, 1916:182; Gilluly, Reed, and Park, 1933:114; Department Bulletin 14-B, 1941:146.

Rescue mine — Susanville District, B-8

Location: Grant County, NW$\frac{1}{4}$ sec. 5, T. 10 S., R. 33 E., adjoining Gem group on the west.

Development: Shaft 140 feet deep and an adit of unknown length.

Geology: Vein in schist strikes N. 10° E., and dips 60° E. The vein is mainly quartz, with ankerite, pyrite, sphalerite, pyrrhotite, chalcopyrite, galena, and arsenopyrite.

Production: No record.

References: Gilluly, Reed, and Park, 1933:113; Department Bulletin 14-B, 1941:146.

Side Issue (Black Hawk) mine — Susanville District, B-9

Location: Grant County, secs. 7 and 8, T. 10 S., R. 33 E., on the south side of Elk Creek.

Development: 130-foot inclined shaft and short adit.

Geology: Vein consisting of quartz stringers and gouge 4 inches to 20 inches wide in talc schist; strikes N. 65° E., and dips 60° at the surface, flattening to 40° with depth. Pyrite, galena, and sphalerite present in small amounts.

Production: No record.

References: Gilluly, Reed, and Park, 1933:116; Department Bulletin 14-B, 1941:138.

Stockton mine — Susanville District, B-10

Location: Grant County, S½ sec. 7, T. 10 S., R. 33 E., southeast of the Badger.

Development: Poor record; includes a 200-foot shaft.

Geology: Vein in slate; details of mineralization unknown.

Production: No record.

References: Lindgren, 1901:707; Swartley, 1914:171; Parks and Swartley, 1916:214; Gilluly, Reed, and Park, 1933:117; Department Bulletin 14-B, 1941:147.

LOOKOUT MOUNTAIN-PEDRO MOUNTAIN AREA

Location and Geography

The Lookout Mountain-Pedro Mountain area encompasses the Connor Creek, Mormon Basin, and Weatherby lode mining districts and the old placer-mining areas of Rye Valley, Malheur, Clarks Creek, and Eldorado in southeastern Baker and northern Malheur Counties (figure 28).

The area is about 25 miles long, extending from Snake River southwesterly across Mormon Basin. U.S. Highway 30 cuts diagonally across the region, following Burnt River, and good gravel roads lead into the mining districts. The topography is in general rugged, with elevations ranging from 2077 feet at Brownlee Reservoir on Snake River to 7120 feet on Lookout Mountain and 6453 feet on Pedro Mountain. Parts of the upper flanks of Pedro and Lookout Mountains are sparsely timbered, whereas the remainder of the country supports little more than brush and desert grasses.

Geology

The gold deposits in the area appear to be genetically related to a northeasterly aligned group of exposures of Jurassic-Cretaceous intrusive rock, mainly granodiorite of medium grain size. The two largest exposures underlie Lookout Mountain and Pedro Mountain. Older bedded rocks in most of the area are part of a thick sequence of phyllite, slate, massive to schistose greenstones, and limestone of indefinite age which Gilluly (1937) named the Burnt River Schist for exposures in Burnt River Canyon. In the southern part of the area thick sections of metavolcanic rocks and graywacke sandstones and shales of Upper Triassic-Jurassic age are exposed. Small- to moderate-sized masses of altered gabbro are scattered throughout the area underlain by the Burnt River Schist. Gold deposits occur in the granitic rocks and in most of the older rocks.

Production

The Lookout Mountain-Pedro Mountain area contains both lode and placer mines of past importance. No estimates of total placer output can be given because few authentic records are available. Some of the placers, notably those in the Malheur, Eldorado, Rye Valley, and Mormon Basin areas, are generally believed to have produced considerable gold.

Total lode production is probably close to $4,500,000, more than half of which was produced by the Rainbow mine in the Mormon Basin district. Much of the remainder came from the Connor Creek mine in the Connor Creek district. Other mines of note are the Gold Ridge mine in the Weatherby district, the Humboldt and Sunday Hill in the Mormon Basin district, and the Bay Horse silver mine in the Connor Creek district.

Placer Mines

Mormon Basin and adjacent placers

Placer mines were in operation in the Mormon Basin district as early as 1863. Gravels cover a large part of the basin floor to varying depths. Despite a scarcity of water, most of these gravels have been worked; however, it seems likely that exploitation has not been as thorough here as in areas where water is more abundant. Little is known about total production. Lindgren (1901, p. 772) states: "In 1882 two

American and one Chinese company were operating, with a total yield of $40,000. In 1883 a yield of $35,200 was reported. Since that time the production has greatly diminished."

Several areas adjacent to Mormon Basin have produced large amounts of placer gold. To the north, on the lower reaches of Clarks Creek and on Burnt River below that stream, a small dredge was operated periodically during the years 1917 to 1936. To the east of Mormon Basin, on the south fork of Dixie Creek, are the Rye Valley placers where work began in 1862. These placers produced more than $1,000,000 (Swartley, 1914, p. 228), much of which was won from hydraulic mining of high gravels. Southwest of Mormon Basin on the Willow Creek slope are the famed early-day placer camps of Malheur, Eldorado, and Amelia. The 100-mile-long Eldorado Ditch, completed in 1873, carried water to these diggings all the way from the head of Burnt River.

Lower Burnt River and Weatherby placers

The gravel bars of Burnt River from Durkee to Huntington are said to have been placer mined in early days (Swartley, 1914, p. 217). Gravels along several tributary streams have also been worked, notably those on Sisley, Chicken, and Shirttail Creeks, which drain the Weatherby district. For a short time around 1900 the Pomeroy Dredging Co. operated a small dredge on Burnt River below the mouth of Sisley Creek. In Burnt River Canyon upstream from Durkee there is abundant evidence of small-scale placer activity.

Connor Creek placers

Connor Creek was one of the first to be mined for placer gold in the early days; it had been worked over twice when seen by Lindgren in 1901. Parts of the creek are still being mined intermittently in a small way. Probably most of the placer gold was eroded from the Connor Creek vein; little placering has been done in the creek above it. Total production of placer gold to 1914 was said by Swartley (1914, p. 213) to be about $125,000. According to local historians, this figure represents only a small part of the actual production. Several other gulches that drain into Snake River in this area have been worked for placer gold.

Connor Creek District

General information

The Connor Creek district includes all of the Snake River Canyon area between the mouths of Burnt River near Huntington and Powder River near Richland (figure 28). The relief is high and topography is rugged.

The district encompasses a wide variety of greenstones, slates, phyllites, low-grade schists, and old plutonic rocks of Permian (?) and Late Triassic age. These rocks have been intruded by Jurassic-Cretaceous granodiorite. Lookout Mountain is the largest exposure of the intrusive rock.

Most of the gold produced in the district has come from the Connor Creek mine, which has a total output of about $1,250,000. Nearly 150,000 ounces of silver have been produced by the Bay Horse mine. Little or no output has been reported for other deposits in the district. Placer production from Connor Creek has been discussed above.

Principal lode mines

Connor Creek mine: The Connor Creek mine, located 3 miles up Connor Creek from Snake River, is developed by six adits over a vertical interval of more than a thousand feet. This, together with the raises, amounts to well over 8000 feet of development work. The vein is enclosed in dark-colored slate and phyllite with a generally N. 40° W. strike and a 70° to 75° SW. dip. Above the outcrops to the northwest is a large exposure of limestone which, however, is not cut by the vein or by any of the drifts

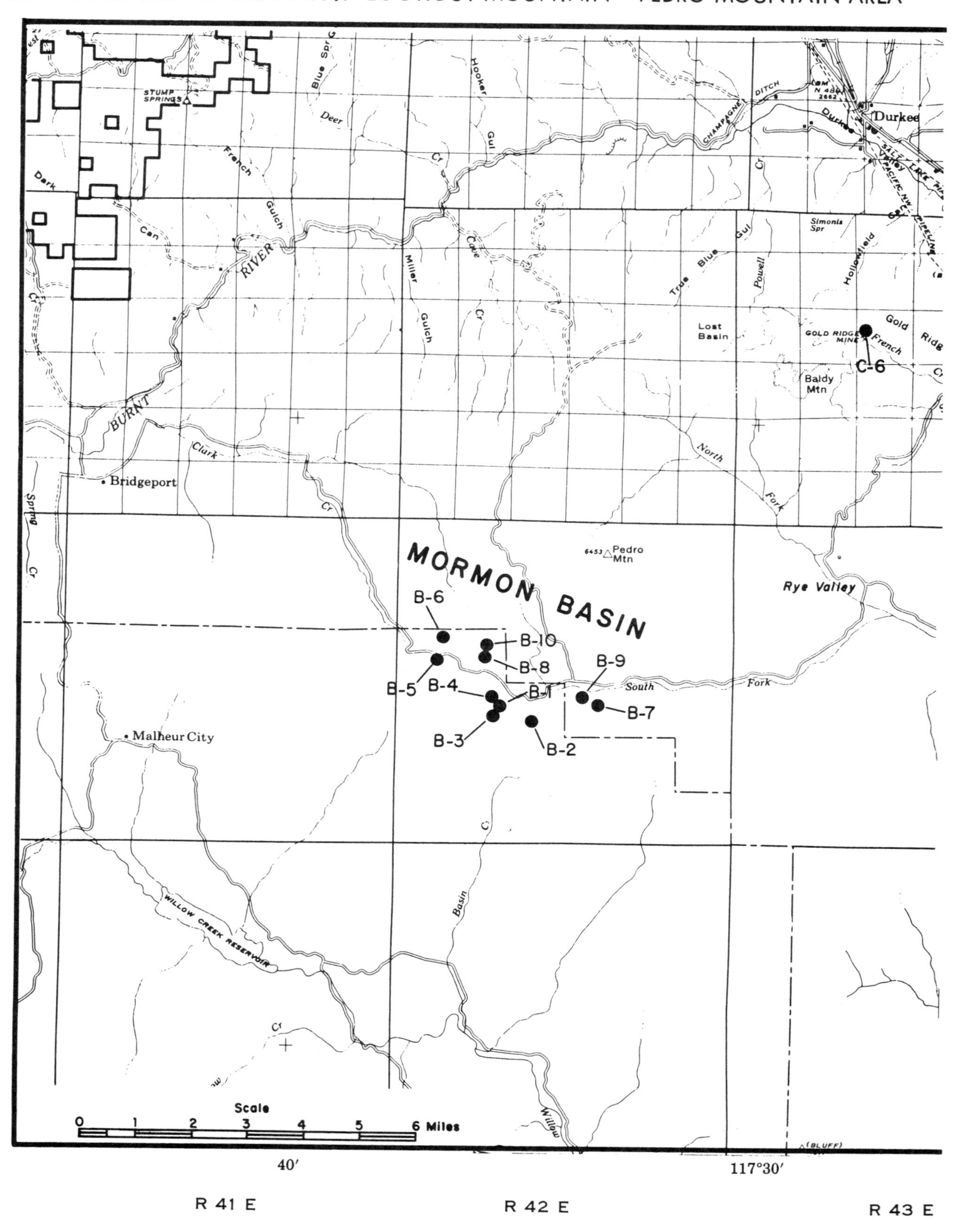

Figure 28. Index map of the Lookout Mountain - Pedro Mountain area.

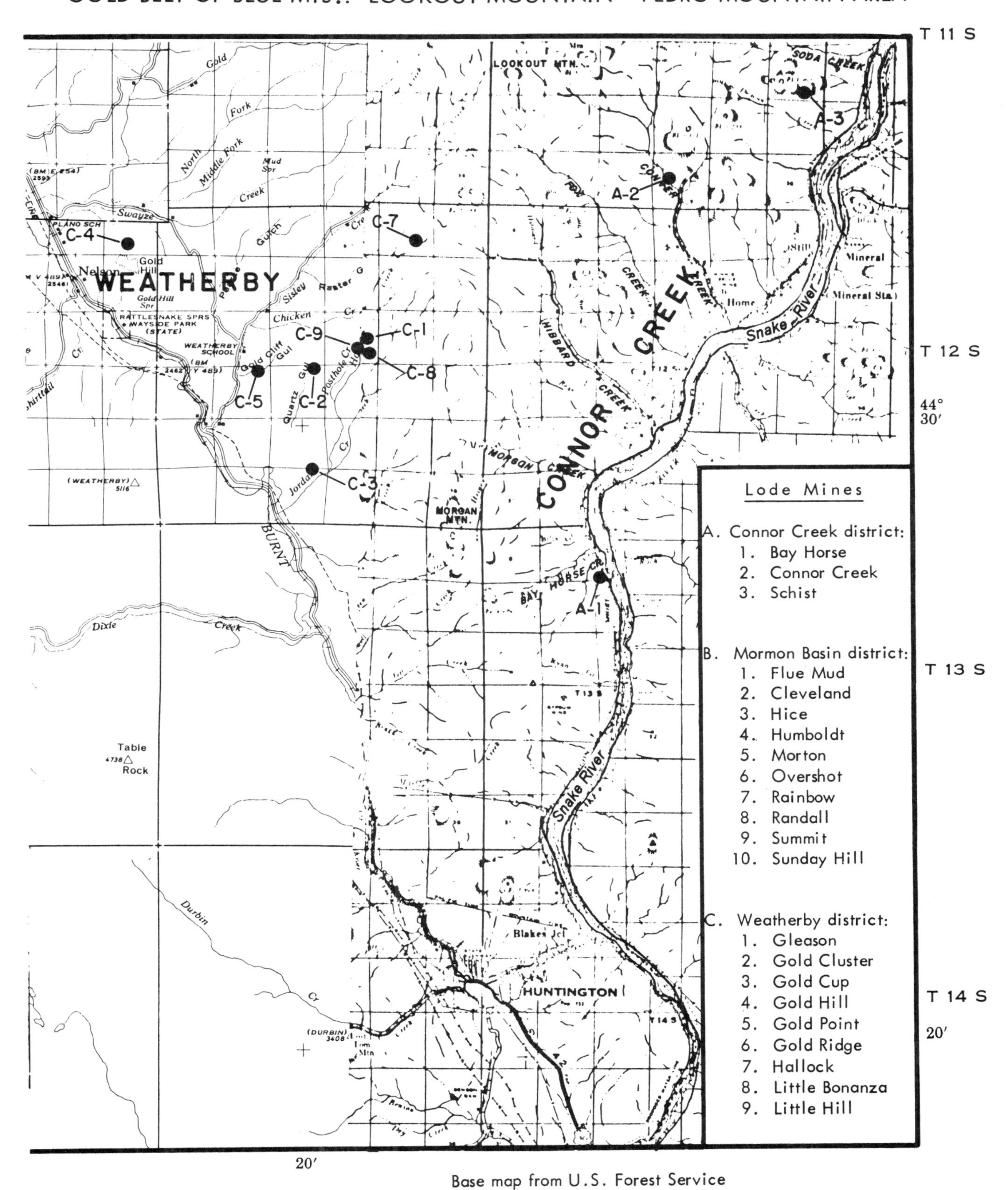

Base map from U.S. Forest Service

or crosscuts. The vein has been followed and stoped on all levels in a westerly direction up to a zone of sheared and chloritized slate about 130 feet wide that strikes N. 31° E. and dips 45° to 60° SE. The vein has been found west of this "final cut-off" only on the lowest level of the mine.

The vein ranges in width from a narrow seam to 8 feet, but the average is between $1\frac{1}{2}$ and 4 feet. The filling is coarse white quartz that contains coarse gold, almost entirely native, with some argentite and pyrite. The gold is unusually fine. The main ore shoot had a maximum length of about 1400 feet on the lower levels and extended to the surface. The mill ore contained 0.15 to 0.5 ounce per ton of gold. Small rich pockets of coarse gold were scattered through the shoot.

Swartley (1914, p. 216) estimated the total production of the Connor Creek mine at about $1,250,-000. Although work was started in 1871, the greatest period of activity was between 1880 and 1890 (Lindgren, 1901, p. 757). Except for a small production of probably not more than $20,000 in 1915-1918 (Gilluly, Reed, and Park, 1933, p. 50), little mining has been done since that period.

Bay Horse mine: The Bay Horse mine is located on a steep slope a few hundred feet above Snake River 7 miles below Huntington. According to Livingston (1925, p. 17-18), silver-bearing tennantite is the sole ore mineral, associated in only a few places with small amounts of other sulfides. The country rocks are andesite and rhyolite flows of Late Triassic(?) age. The ore body, made up of fractured and silicified rock containing seams and impregnations of tennantite, is irregular in outline but reportedly had an over-all strike of N. 70° W. to west with a dip of 10 feet in a hundred into the mountain (Livingston, 1925). The mine is developed by two adits 130 feet apart vertically.

The Bay Horse mine is credited with production as early as 1891 (Lindgren, 1901, p. 753-754), but the main period of activity was during the years 1920-1925 when 145,459.54 troy ounces of silver were recovered from 4,895.053 tons of ore shipped to smelters at Tacoma, Wash., and Kellogg, Ida.

Mormon Basin District

General information

The Mormon Basin district straddles the Baker-Malheur County line (figure 28). It encompasses an oval-shaped depression about 3 by 2 miles in extent at the heads of Basin Creek and the South Fork of Dixie Creek. Basin Creek flows south to Willow Creek, a tributary of the Malheur River, and the South Fork of Dixie Creek flows east and south to Burnt River. Elevations range from 4500 to 6000 feet.

Rocks of the district comprise quartz-rich schists, slates, and greenstones cut by numerous intensely sheared and altered igneous bodies of basic and ultrabasic composition. Intrusive into these rocks are several small offshoots of the Jurassic-Cretaceous granodiorite stock that underlies Pedro Mountain at the north edge of the district. Intrusion of the stock resulted in considerable deformation of the enclosing rocks. Minor offsetts and abrupt changes in the attitude of the bedded rocks are common, although on the whole the planes of schistocity are parallel to the edges of the intrusive. Wolff (1965, p. 56) states: "The direction of intrusion of the Pedro Mountain pluton has been upward and southwesterly and....the Burnt River Schist which trends easterly, has been deflected around the intrusion and has been broken by a series of faults." He suggests that the Burnt River Schist in the Mormon Basin area is a fault block thrust upward and southwestward ahead of the Pedro Mountain intrusive and that "The ultramafic rocks around the periphery of the Burnt River Schist provided the surface upon which the plate was thrust."

Mine and prospect workings are abundantly scattered throughout the district and numerous veins have been exposed. Few of these veins are persistent for any great length and in general they are irregular in both width and attitude. Quartz with lesser ankerite and fuchsite are the chief gangue minerals. Pyrite and arsenopyrite are the most abundant metallic minerals; galena, sphalerite, polybasite, hessite, and tetrahedrite are present locally.

The quartz mines of the district have produced more than $2,650,000 in gold, most of it from the Rainbow mine. Other properties in the district where development was fairly extensive include the Humboldt and Sunday Hill mines. Mills were erected on both properties. There is little record of production from the Sunday Hill.

Principal lode mines

Rainbow mine: The Rainbow mine was first worked in 1901. The following figures, presented in a lengthy appraisal of the property by W. Elmer and G. Hogg dated January 26, 1923, were reportedly obtained from company records:

Operator	Period	Gross value	Tons milled	Gross value per ton	Recovery
Commercial Mining Co.	1901-1910	$ 242,000.00	?	?	?
United States Smelting, Mining & Refining Co.	1911-1915	1,111,796.76	103,547.5	$11.24	$10.737
Commercial Mining Co.	1916-1919	969,295.70	113,064	8.79	8.57
	Total	$ 2,323,092.46			

A 15-stamp mill and cyanide plant of 100-ton-per-day capacity was operated more or less continuously from 1912 to 1919. Since then production has been small and sporadic. In 1923 the surface plant was destroyed by fire. Condor Gold Mines produced 1222 ounces gold and 361 ounces silver in 1934 but closed in October of that year because of financial difficulties. At the Rainbow mine, a fault breccia zone as much as 50 feet wide occurs in slate and greenstone, the latter derived in part from old intrusives. Into portions of this zone a diorite dike was injected and later considerably altered. Quartz veins occur on both walls. The best and most persistent values of the mine are in the footwall vein, which strikes roughly N. 60° E. and dips 55° to 65° NW. In places the breccia and dike are cemented and veined with quartz, which is locally ore bearing. The ore shoots occur as lenses that are continuous for lengths of as much as 350 feet and are recurrent both on strike and dip. Ore has been mined at intervals through a maximum distance of 1500 feet along the strike and 500 feet down the dip. The gold was largely free milling; sulfide minerals are generally sparse even on the lowest levels.

Development includes about 7000 feet of drift on four levels turned from a 400-foot vertical shaft and upper-level adit. A winze was sunk to the 500-foot level and a little drifting was done but no mining.

Humboldt mine: The Humboldt mine (figure 29) operated rather steadily between 1909 and 1915 inclusive, producing roughly 35,000 tons of ore which yielded about $225,000 in gold and silver. The mine was equipped with a 20-stamp amalgamating and concentrating mill and in 1915 was developed by a 500-foot vertical shaft and 3500 feet of drifts.

Sunday Hill mine: A mill was built at the Sunday Hill mine in 1868 and the mine reputedly produced $80,000 in early days. Operations during the 1930's brought the total output of the mine up to around $100,000. The country rock is quartz-mica schist. One vein, the Phalen, strikes N. 40° to 50° W. and dips 50° N. Average mining width is about 2½ feet. Several other narrower veins are known on the property. The vein matter is quartz and gouge with pyrite and lesser arsenopyrite, galena, and sphalerite. In 1934, 400 tons of ore treated by tabling and amalgamation yielded 120 ounces gold and 35 ounces silver. The ore was crushed in a 5-stamp mill.

Weatherby District

The Weatherby district (part of the old Lower Burnt River district) lies about 35 miles southeast of Baker (figure 28). It straddles U. S. Highway 30, which follows Burnt River. The district is in the Durkee quadrangle, which has recently been mapped by Prostka (1967). In this area a heterogeneous assemblage of phyllite, slate, chert, greenschist, graywacke, marble and metagabbro has been intruded by granodiorite. The largest exposure of the granodiorite underlies Lookout Mountain and an area of several square miles to the southeast, including the heads of Sisley and Chicken Creeks in the northeastern part of the district.

The Weatherby district is noted for the old placers along Chicken and Sisley Creeks north of the highway and for several small but high-grade auriferous quartz veins. The veins are not persistent; none can be traced for long distances, nor are directions of strike and dip constant. The veins occur mainly in the granodiorite, but some are in adjacent older schists and greenstones.

The more prominent quartz mines in the district are the Gold Hill, about 4 miles southeast of Durkee; the Gold Ridge 4 miles due south of Durkee; and the Little Bonanza, Little Hill, Gleason, and Hallock in the vicinity of Chicken Creek. The Gold Ridge mine, which worked several small, east-trending veins in granodiorite, is said to have produced about $210,000, practically all of it extracted between 1881 and 1886 (Lindgren, 1901, p. 765).

Figure 29. Humboldt mine, Mormon Basin district, Malheur County, prior to 1914.

Lode Mines of the Lookout Mountain - Pedro Mountain Area

CONNOR CREEK DISTRICT

Bay Horse mine — Connor Creek District, A-1

Location: Baker County, sec. 9, T. 13 S., R. 45 E., on the east bank of Snake River.

Development: About 1800 feet of drifts and crosscuts, mainly from two adit levels 130 feet apart and an intermediate shaft level.

Geology: The country rocks are altered andesite and rhyolite of probable Upper Triassic age. The ore occurs as small replacement masses of irregular shape scattered through a poorly defined zone of faulting. The ore zone is about 50 feet in maximum width and has a westerly trend.

Production: Output during 1920-1925 amounted to 145,459.54 troy ounces silver from 4,895 tons of ore shipped to smelters. Previous output unknown.

References: Lindgren, 1901:753; Livingston, 1925; Gilluly, Reed, and Park, 1933:52.

Connor Creek mine — Connor Creek District, A-2

Location: Baker County, sec. 34, T. 11 S., R. 45 E., about 3 miles up Connor Creek from Snake River.

Development: Six adits over vertical interval of 1000 feet. Total with raises more than 8000 feet.

Geology: Quartz vein in slate and greenstone; strikes N. 40° W., and dips 70°-75° SW. Width of vein 1½ to 4 feet. The pay shoot was 1400 feet long on the Dry Creek level and extended to the surface. Ore minerals are coarse gold with a little argentite and pyrite.

Production: Estimated at $1,250,000. Main periods of operation were 1870-1910 and 1915-1918. Mill ore ran between $3 and $10 per ton. The gold averaged about 900 fine.

References: Lindgren, 1901:756; Swartley, 1914:216; Parks and Swartley, 1916:68; Gilluly, Reed, and Park, 1933:50; Department Bulletin 14-A, 1939:22, 23.

Schist (Snake River) mine — Connor Creek District, A-3

Location: Baker County, SE¼ sec. 24, T. 11 S., R. 45 E.

Development: Several hundred feet of crosscut and drift from adit levels.

Geology: The country rocks are phyllite and marble. The phyllite is impregnated with reticulated veins and small, irregularly shaped masses of gold-bearing quartz. Channel sampling is reported to have yielded about $3.50 per ton over a width of 90 feet.

Production: No record. There was a 75-tons-per-day mill on the property in 1916.

References: Swartley, 1914:215; Gilluly, Reed, and Park, 1933:52; Parks and Swartley, 1916: 207; Department Bulletin 14-A, 1939:23-24.

MORMON BASIN DISTRICT

Blue Mud prospect

Mormon Basin District, B-1

Location: Malheur County, E½ sec. 20, T. 13 S., R. 42 E., about 1500 feet south of the Humboldt mine.

Development: Several tunnels and 200-foot shaft.

Geology: Workings said to expose vein in greenstone, chlorite schist, and metagabbro. Width and altitude unknown. Rich float found in Tertiary gravels nearby.

Production: No record.

References: Gilluly, Reed, and Park, 1933:47; Department Bulletin 14-A, 1939:76.

Cleveland Development Co. mine

Mormon Basin District, B-2

Location: Malheur County, SW¼ sec. 21, T. 13 S., R. 42 E., on the south side of California Gulch just above confluence with Basin Creek.

Development: Two adits and several pits.

Geology: Quartz seams in porphyry dike cutting schist and greenstone. Dike is a few feet wide, strikes N. 75° E., dips steeply south.

Production: No record.

References: Parks and Swartley, 1916:56; Gilluly, Reed, and Park, 1933:47; Department Bulletin 14-A, 1939:76.

Hice mine

Mormon Basin District, B-3

Location: Malheur County, E½ sec. 20, T. 13 S., R. 42 E., about 2000 feet south of the Humboldt mine.

Development: 250-foot adit with branches; total 400 feet.

Geology: Many quartz seams less than 4 inches thick in quartz diorite. Diorite locally silicified and impregnated with sulfides.

Production: No record.

References: Gilluly, Reed, and Park, 1933:46; Department Bulletin 14-A, 1939: 78.

Humboldt mine — Mormon Basin District, B-4

Location: Malheur County, N$\frac{1}{2}$ sec. 20, T. 13 S., R. 42 E., in the southwestern part of Mormon Basin.

Development: 500-foot shaft with 3500 feet of drifts distributed among several levels.

Geology: Quartz vein in slate, diorite, and trachyte. Strike E-W., dip 75° N. in upper levels, steeper below. Vein zone, including altered country rock, is locally 40 feet wide, contains arsenopyrite, pyrite, galena, and sphalerite, some calcite. Free gold in upper levels.

Production: Output 1909-1915 was $225,000 from 35,000 tons of ore. No mill record.

References: Swartley, 1914:224; Gilluly, Reed, and Park, 1933:45; Department Bulletin 14-A, 1939: 78.

Morton mine — Mormon Basin District, B-5

Location: Malheur County, SE$\frac{1}{4}$ sec. 18, T. 13 S., R. 42 E., on the divide between Glengarry and French Gulches.

Development: Shaft 96 feet deep.

Geology: Rich quartz float in decomposed diorite debris, which the shaft failed to penetrate.

Production: No record.

References: Gilluly, Reed, and Park, 1933:48; Department Bulletin 14-A, 1939:80.

Overshot mine — Mormon Basin District, B-6

Location: Malheur County, NE$\frac{1}{4}$ sec. 18, T. 13 S., R. 42 E., near the head of the north fork of Glengarry Gulch.

Development: Adits; the longest is 300 feet.

Geology: Slate and quartz-mica schist cut by thin sheets and sills of altered quartz diorite. Gold said to have been found in the diorite and near its contact with the schist.

Production: No record.

References: Gilluly, Reed, and Park, 1933:43; Department Bulletin 14-A, 1939:80.

Rainbow mine

Mormon Basin District, B-7

Location: Baker County, sec. 22, T. 13 S., R. 42 E.

Development: 500-foot shaft with several levels and an upper-level adit; about 7000 feet of drift.

Geology: Fault zone as much as 50 feet wide in slate, chloritized ultrabasic intrusive rocks and minor limestone with some granitic dikes. Strike N. 60° E.; dip 55° to 65° N. Quartz veins on both walls contain free gold and very small amounts of pyrite and arsenopyrite.

Production: Output during 1901-1919 totaled $2,323,092.46 from ore averaging about $10 per ton treated in a 100-tons-per-day stamp mill and cyanide plant.

References: Swartley, 1914:220; Gilluly, Reed, and Park, 1933:37; Department Bulletin 14-A, 1939:81.

Randall mine

Mormon Basin District, B-8

Location: Malheur County, SE$\frac{1}{4}$ sec. 17, T. 13 S., R. 42 E., on the south end of Sunday Hill.

Development: Two adits 24 feet apart vertically, total 1200 feet.

Geology: Three quartz veins in quartz-mica schist and diorite. Small amounts of pyrite, arsenopyrite, hessite, galena, and sphalerite are present. Gold largely free, since workings are in the oxidized zone. Two of the veins intersect in the lower level; one strikes N. 60° W., dips 32° SW., and is 2 to 5 feet wide; the other strikes N., dips 32° E., and is more than 8 feet thick.

Production: No record.

References: Gilluly, Reed, and Park, 1933:41; Department Bulletin 14-A, 1939:82.

Summit mine

Mormon Basin District, B-9

Location: Baker County, NW$\frac{1}{4}$ sec. 22, T. 13 S., R. 42 E., on the hillside south of the South Fork of Dixie Creek west of Rainbow Gulch.

Development: Several adits; one is 1400 feet long, another 750 feet long.

Geology: The country rocks are slate, phyllite, and granodiorite. Vein strikes NE. and is 1 to 5 feet wide, cut off in the upper levels by an E.-W. fault. Sparse pyrite and galena. Gold mostly free.

Production: No record.

References: Gilluly, Reed, and Park, 1933:44; Department Bulletin 14-A, 1939:83.

Sunday Hill mine — Mormon Basin District, B-10

Location: Malheur County, E$\frac{1}{2}$ sec. 17, T. 13 S., R. 42 E.

Development: Many hundred feet of drifts and crosscuts from adit levels and a shaft.

Geology: One strong vein and several narrower ones in quartz-mica schist which is cut by quartz-diorite dikes. Main vein strikes N. 40° to 50° W. and dips 50° N. It is 2$\frac{1}{2}$ feet thick in mined area. Vein is faulted. Vein matter is chiefly quartz and gouge with pyrite, arsenopyrite, galena, and sphalerite. Some free gold.

Production: About $100,000. Located in 1867; mill erected in 1868. Some production also in 1920's and 1930's. Gold-to-silver ratio about 3.5 to 1.

References: Gilluly, Reed, and Park, 1933:39; Department Bulletin 14-A, 1939:83.

WEATHERBY DISTRICT

Gleason mine — Weatherby District, C-1

Location: Baker County, SE$\frac{1}{4}$ sec. 15, T. 12 S., R. 44 E., on Hogback Creek north of the Little Bonanza mine.

Development: Two thousand feet of workings from two adits and a shaft.

Geology: Country rock is granodiorite cut by a porphyry dike. Quartz vein 2 to 4 feet wide along footwall of the dike strikes N. 7° E.; dips 65° W. Gold free milling.

Production: $150,000. The property was first located about 1867.

References: Parks and Swartley, 1916:99; Lorain, 1938:35; Department Bulletin 14-A, 1939:67.

Gold Cluster mine — Weatherby District, C-2

Location: Baker County, NE$\frac{1}{4}$ sec. 21, T. 12 S., R. 44 E., east of Quartz Gulch.

Development: 600 hundred feet of tunnel and shallow shafts.

Geology: Discontinuous quartz stringers in a shear zone in granodiorite. Strike N. 7° E.; dip 35° W. Step-faulted easterly.

Production: $8000 in 1928-1938; early production not recorded. Gold mostly free; 770 to 825 fine.

References: Department Bulletin 14-A, 1939: 67.

Gold Cup mine

Weatherby District, C-3

Location: Baker County, secs. 28 and 33, T. 12 S., R. 44 E.

[Four patented mining claims -- no other data available.]

Gold Hill mine

Weatherby District, C-4

Location: Baker County, $S\frac{1}{2}$ sec. 1, T. 12 S., R. 43 E., on the north slope of Gold Hill.

Development: 4200 feet of drifts, three shafts, and several cuts.

Geology: Country rock is granodiorite. Phyllite and slate are exposed nearby. Eight small quartz veins are cut by the workings; strike N. 60° to 65° W.; dip S.; average thickness less than 6 inches. Ore minerals include pyrite, sphalerite, and galena. The wall rocks are intensely sericitized adjacent to the veins.

Production: Small.

References: Swartley, 1914:218; Gilluly, Reed & Park, 1933:54; Dept. Bulletin 14-A, 1939:68.

Gold Point mine

Weatherby District, C-5

Location: Baker County, $NE\frac{1}{4}$ sec. 20, T. 12 S., R. 44 E., in Gold Cliff Gulch, about three-quarters of a mile from the Sisley Creek road.

Development: No record.

Geology: Quartz stringers in granodiorite. Average 2 to 4 inches thick. Strike N. 10° to 30° E.; dip 25° to 45° W.

Production: $40,000.

References: Department Bulletin 14-A, 1939: 69.

Gold Ridge mine

Weatherby District, C-6

Location: Baker County, $W\frac{1}{2}$ sec. 16, T. 12 S., R. 43 E., near the head of French Creek, about 4 miles due south of Durkee.

Developments: 250-foot shaft; 2000 feet of drifts and crosscuts, mostly from adit levels.

Geology: The country rock is quartz diorite. Three well-defined quartz veins, 2 to 3 feet wide, strike N. 50° to 70° W., dip 65° SW. Ore zone said to have been 780 feet long.

Production: About $210,000 between 1881 and 1886, with 10-stamp mill. Oxidized ore ran $12 to $15 per ton in gold about 870 fine.

References: Lindgren, 1901:765; Parks and Swartley, 1916:109; Gilluly, Reed, and Park, 1933:56; Department Bulletin 14-A, 1939:66.

Hallock mine — Weatherby District, C-7

Location: Baker County, SE$\frac{1}{4}$ sec. 2, T. 12 S., R. 44 E., near the head of Chicken Creek.

Development: Six adits total 1650 feet, all shallow depth.

Geology: Quartz-calcite veins in granodiorite near irregular contact with sericite schist and limestone. Veins strike N. 68° W., dip 70° N., and vary from 4 to 16 inches thick. Gold free milling.

Production: Small.

References: Department Bulletin 14-A, 1939:69.

Little Bonanza mine — Weatherby District, C-8

Location: Baker County, SE$\frac{1}{4}$ sec. 15, T. 12 S., R. 44 E., on Hogback Creek.

Development: Two thousand feet of tunnels and two winzes.

Geology: The country rocks are granodiorite cut by a basic lamprophyric dike 20 to 30 feet wide. A quartz vein as much as 4 feet wide associated with the dike contains pyrite galena, and free-milling gold.

Production: $200,000. Located about 1890.

Reference: Department Bulletin 14-A, 1939:70.

Little Hill mine — Weatherby District, C-9

Location: Baker County, SE$\frac{1}{4}$ sec. 15, T. 12 S., R. 44 E., on Hogback Creek.

Development: Three or four tunnels 200 to 300 feet long and 60-foot shaft.

Geology: Narrow quartz veins in granodiorite strike N. 65° W., dip south; sulfides, including pyrrhotite, in lower levels.

Production: $200,000. Located in 1882. Gold free milling.

Reference: Parks and Swartley, 1916:142; Department Bulletin 14-A, 1939:70.

VIRTUE AREA

Location

The Virtue area includes the mines and prospects in the vicinity of Virtue Flat, about 10 miles east of Baker (figure 30). Lesser deposits occur in the Farley Hills (not shown on map) near Haines and North Powder, about 15 miles north of Baker.

The principal gold deposits of the area are scattered among the low hills bordering Virtue Flat, a gently rolling depression about 8 miles long and 3 miles wide drained by Ruckles Creek. Elevations range from 3400 to 5000 feet. The general region surrounding Virtue Flat has been known as the Virtue mining district.

Geology and Mineralization

Much of the area is covered by Cenozoic basalt flows and lacustrine and fluviatile sediments. Pre-Tertiary rocks are exposed in the hills bordering Virtue Flat and in the Farley Hills north of the map area. The pre-Tertiary units include Elkhorn Ridge Argillite, Clover Creek Greenstone, and gabbro and related rocks which intrude both formations. South of Virtue Flat the argillite series predominates, while to the north greenstones and gabbro are most abundant. All these rocks are structurally contorted and have undergone considerable shearing and dynamic metamorphism. A younger group of intrusive rocks consists largely of quartz diorite, probably of Jurassic-Cretaceous age. Exposures of the quartz diorite are small and scattered, but it is probable that the intrusive body occupies a considerably larger area beneath the younger basalts and alluvium.

The veins of the Virtue district lie along the edges of an intrusion of diorite and gabbro into argillite and greenstone. At different times during the intrusion, the intrusive as well as the intruded rocks were fractured. Into these fractures came many basic to acidic dikes and, later, quartz veins. The quartz veins strike in many directions and none are traceable for long distances. Several of the veins, notably those of the Virtue mine, strike northwest in marked contrast to most of the important veins elsewhere in eastern Oregon. Most of the ore deposits are simple quartz veins containing very small amounts of sulfides. The gold occurs largely in the free state, is coarse, and contains little silver.

The several gold prospects in the Farley Hills north of the map area are in greenstones and altered gabbro; some of the deposits contain copper minerals.

Production

Most of the production in the Virtue district has come from the Virtue mine, which was one of the largest gold producers in eastern Oregon. Its total output was more than $2,000,000. The White Swan mine has a reported production of about $700,000. Several other mines in the district have produced lesser amounts, notably the Flagstaff, Emma, Hidden Treasure, Friday, Rachel, and Mable mines. Production records for these mines are scarce, but none is believed to have produced more than $200,000.

The gulches leading up to the Virtue and White Swan are said to have contained rich placers in the early days.

Principal Lode Mines

Virtue mine: The Virtue was operated from 1862 to 1884 and from 1893 to 1898. A small output was made in 1906-1907. Since that time the mine has been closed. According to Lindgren, the production up

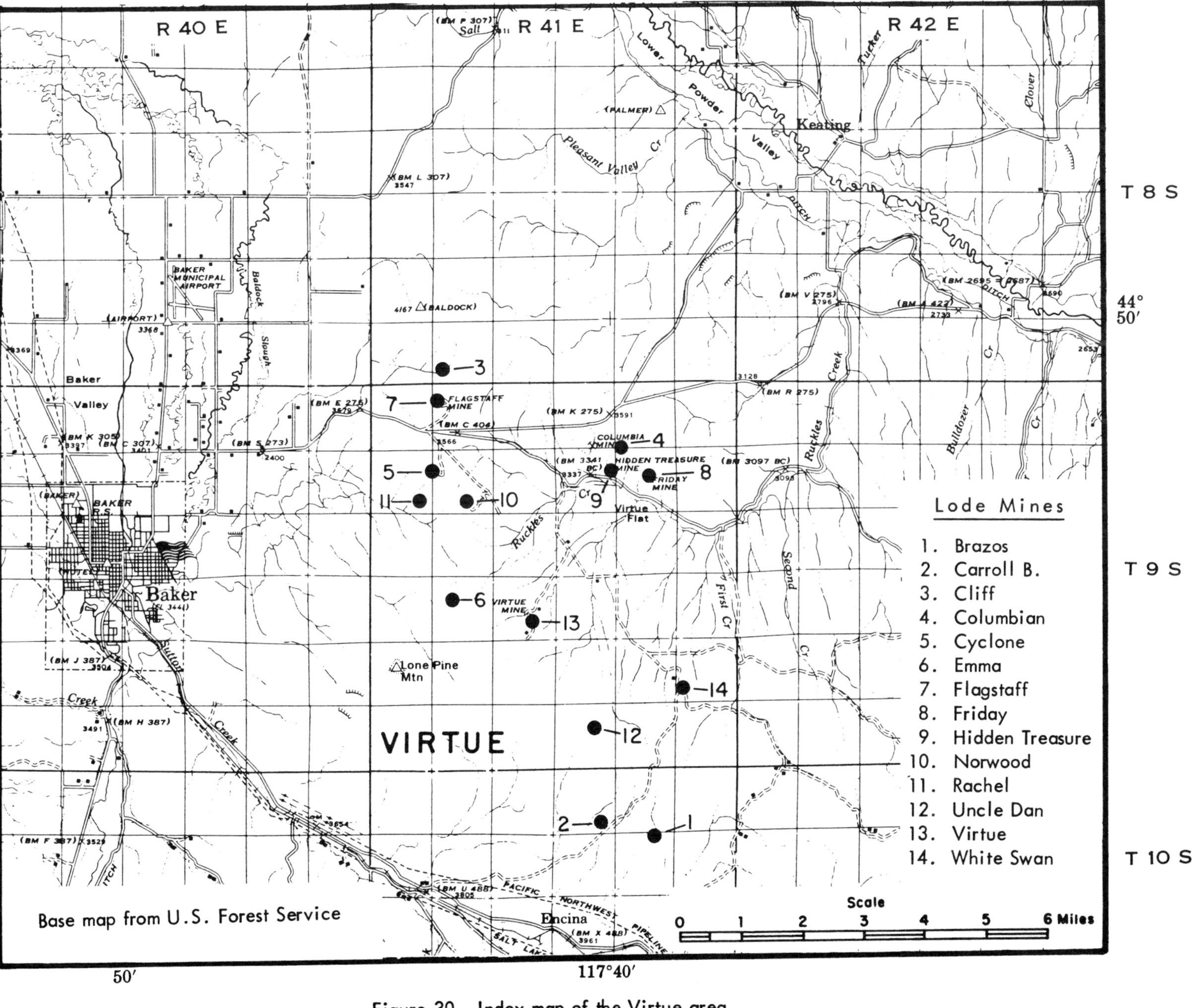

Figure 30. Index map of the Virtue area.

to 1878 was $1,250,000. From 1878 to 1884, the estimated amount is $200,000. From 1893 to 1898 the production was $739,000, the maximum being reached in 1896 with $256,000, and the minimum in 1898 with $13,100. The total production is thus $2,189,000.

The mine is developed by at least 10,000 feet of workings from three adits and an 800-foot shaft. Eight veins, of which the Virtue has been most productive, have been cut in the mine workings. They are subparallel and strike N. 20° to 45° W. They dip northeast above the mill level, but steepen downward and dip southwest in the lower workings of the old mine (Gilluly, Reed, and Park, 1933, p. 72-74).

According to Lindgren (1901, p. 722-723), an ore shoot with a maximum strike length of 1200 feet was mined to a depth of about 1000 feet below the outcrop. A lean broken zone was then encountered and operations ceased. The ore was a quartz vein 6 inches to 12 feet wide. The quartz averaged about 14 inches wide and contained from 0.5 to 1.0 ounces of gold per ton. During short periods values were often considerably higher. The quartz was sparsely mineralized with pyrite and chalcopyrite. The gold was largely free and very high grade, the bullion averaging more than 925 fine. Operations at depth were handicapped by a heavy flow of warm water.

White Swan mine: The White Swan mine is said to have produced at least $600,000 from 1890 to 1897, $84,000 in 1916, and about $40,000 during 1935-1937. The mine develops several subparallel quartz veins in a broad east-trending shear zone in argillite. Only one of the veins has been an important producer. This vein dips 70° S. and the workings include a shaft 350 feet deep. The only minerals observed in the veins were quartz, sericite, and calcite, with limonite indicating the former presence of sulfides which, however, must have been small in amount (Gilluly, Reed, and Park, 1933, p. 77).

Lode Mines of the Virtue Area.

VIRTUE DISTRICT.

Brazos mine

Virtue District, 1

Location: Baker County, secs. 2 and 11, T. 10 S., R. 41 E.

Development: 600-foot shaft and several drifts.

Geology: Quartz stringers and nodules along fault zone in argillite. Strike NW.; dip gently SW. Width 3 to 4 feet. Pay shoot 400 feet long; low grade.

Production: Small.

References: Lindgren, 1901:726; Parks and Swartley, 1916:43; Gilluly, Reed, and Park, 1933:79; Gilluly, 1937:100.

Carroll B mine

Virtue District, 2

Location: Baker County, SE$\frac{1}{4}$ sec. 3, T. 10 S., R. 41 E.

Development: Shaft and two adits several hundred feet long.

Geology: Quartz vein 1 to 2 feet wide in argillite and greenstone.

Production: No record.

References: Gilluly, Reed, and Park, 1933:78; Gilluly, 1937:100; Department Bulletin 14-A, 1939:104.

Cliff mine

Virtue District, 3

Location: Baker County, SW¼ sec. 32, T. 8 S., R. 41 E.

Development: 225-foot shaft with short drifts on 4 levels.

Geology: Brecciated quartz vein in altered diorite; width 3 feet. Scheelite present in small quantity.

Production: 135 tons milled netted $13.55 per ton in early days. Gold free milling.

References: Lindgren, 1901:725; Parks and Swartley, 1916:56; Gilluly, Reed, and Park, 1933:77; Gilluly, 1937:98; Department Bulletin 14-A, 1939: 104.

Columbian mine

Virtue District, 4

Location: Baker County, secs. 2 and 11, T. 9 S., R. 41 E.

[Information on development and production not available.]

Cyclone mine

Virtue District, 5

Location: Baker County, secs. 7 and 8, T. 9 S., R. 41 E.

Geology: Vein in greenstone. Strike NW. Width 2 feet.

Production: No records.

Reference: Department Bulletin 14-A, 1939: 104.

Emma mine

Virtue District, 6

Location: Baker County, sec. 20, T. 9 S., R. 41 E.

Geology: Vein 2 to 5 feet wide in argillite strikes northeast.

Production: $250,000(?). Equipped with 20-stamp mill in 1905.

Reference: Department Bulletin 14-A, 1939:104-105.

Flagstaff mine

Virtue District, 7

Location: Baker County, NW¼ sec. 5, T. 9 S., R. 41 E.

Geology: Several brecciated zones arranged in a "horsetail" pattern in sheared gabbro and diorite. Strike N. 45° E. to north. Dip SE. Contain masses of gouge and quartz lenses as much as 1½ feet thick. Sulfides sparse.

Production: About $100,000.

References: Lindgren, 1901:724; Swartley, 1914:130; Parks and Swartley, 1916:93; Grant and Cady, 1914:152; Gilluly, Reed, and Park, 1933:74; Gilluly, 1937:96; and Department Bulletin 14-A, 1939:105-106.

Friday mine — Virtue District, 8

Location: Baker County, sec. 11, T. 9 S., R. 41 E.

Development: 200-foot inclined shaft with short drift levels.

Geology: East extension of Hidden Treasure vein.

Production: Small.

References: Lindgren, 1901:725; Gilluly, 1937:100.

Hidden Treasure mine — Virtue District, 9

Location: Baker County, NE¼ sec. 10, T. 9 S., R. 41 E.

Development: 138-foot inclined shaft with 3 short drift levels.

Geology: Vein in sheared, highly altered greenstone. Strike NW. Dip 60° S. Vein minerals include manganese oxides and stibnite. Gold mineralization spotty.

Production: $24,000 from crude ore shipped during 1933-1938; previous output unknown.

References: Gilluly, 1937:100; Department Bulletin 14-A, 1939:107.

Norwood mine — Virtue District, 10

Location: Baker County, S½ sec. 8, T. 9 S., R. 41 E.

Development: Unknown.

Geology: 2-foot quartz vein in greenstone. Strike E. to W.; dips steeply.

Production: No record. Mill installed in 1913.

References: Swartley, 1914:131; Grant and Cady 1914:152; Parks and Swartley, 1916:174; Gilluly, Reed, and Park, 1933:76; Gilluly, 1937:98; Department Bulletin 14-A, 1939:108.

Rachel mine

Virtue District, 11

Location: Baker County, SE$\frac{1}{4}$ sec. 7, T. 9 S., R. 41 E.

Development: 800-foot incline with 3,000 feet of lateral workings.

Geology: One- to three-foot wide vein in argillite and greenstone.

Production: $150,000 (?).

References: Department Bulletin 14-A, 1939:108.

Uncle Dan mine

Virtue District, 12

Location: Baker County, NE$\frac{1}{4}$ sec. 34, T. 9 S., R. 41 E.

[Information on development and production not available.]

Virtue mine

Virtue District, 13

Location: Baker County, sec. 21, T. 9 S., R. 41 E.

Development: Three adits and 800-foot shaft with several levels. Total 10,000 feet.

Geology: Country rock is strongly sheared greenstone derived mainly from gabbro. Eight sub-parallel veins cut by mine workings. Strike N. 20° to 45° W.; dip NE. near surface and SW. in lowest levels. Virtue vein most productive; average about 14 inches thick. Quartz, subordinate calcite, very minor pyrite, and chalcopyrite. Ore averaged 0.5 to 1.0 ounce gold per ton. Some high grade.

Production: $2,200,000 during 1862-1884; 1893-1899; 1906-1907. Gold free and coarse; average more than 920 fine.

References: Lindgren, 1901:722; Grant and Cady, 1914:150; Parks and Swartley, 1916:229; Gilluly, Reed, and Park, 1933:72; Gilluly, 1937:94; Department Bulletin 14-A, 1939:108-109.

White Swan mine

Virtue District, 14

Location: Baker County, SW$\frac{1}{4}$ sec. 25, T. 9 S., R. 41 E.

Development: 300-foot shaft with 4 levels; total 2000 feet.

Geology: Several quartz veins, a few inches to 1$\frac{1}{2}$ feet thick, in broad shear zone in argillite. Most of production from one vein. Strike west; dip steeply south.

Production: $724,000 during 1890-1897, 1916, and 1935-1937. 1936 pilot mill flow-sheet given by Lorain. No other records available.

References: Lindgren, 1901:725; Swartley, 1914:131; Gilluly, Reed, and Park, 1933:77; Gilluly, 1937:98; Lorain, 1938:26; Department Bulletin 14-A, 1939:110.

CANYON AREA

Location

The gold deposits of the Canyon district in Grant County lie within a few miles of John Day, a small town situated at the junction of Canyon Creek and the John Day River (figure 31). Canyon City, one of the earliest mining camps in eastern Oregon, lies about one mile south up Canyon Creek from this point. Through this area the John Day River traverses a broad, fertile valley bordered on the south by the high and rugged Strawberry Range. Canyon Creek flows through a steep-walled canyon cutting across the range. Elevations range from about 3100 feet at John Day to 8007 feet at the top of Canyon Mountain.

Geology

The Canyon District is underlain chiefly by gabbro, peridotite, and serpentinite of the Canyon Mountain Complex of post-Permian-pre-Upper Triassic age (Thayer, 1956). Paleozoic and Upper Triassic sedimentary rocks are exposed on Miller Mountain and in the vicinity of Prairie Diggings.

Production

Records of early-day gold production for the Canyon district are sketchy but indicate that the total output was several millions of dollars. Most of the gold produced prior to 1916 came from the gravels of Canyon Creek and its tributary streams and gulches near Canyon City. In later years dredges were employed on Canyon Creek and the John Day River. Production from lode mines in the Canyon district has been small.

Placer Mines

Gold was discovered in the Canyon district in early 1862 and by mid-summer of that year more than a thousand miners were at work on the gravel bars of Canyon Creek and in the gulches of the surrounding hills. In his report on the district, Lindgren (1901, p. 712-720) states: "During the first few years production was very great, but exact figures will probably never be known. Estimates are made varying from $3,000,000 to $5,000,000 a year. In 1865 the product was estimated at $22,000 a week (Raymond's report, 1870), or about $1,000,000 a year. By 1870 it had already fallen to $300,000 a year. In the following year the production was still further reduced, but remained for a long time about $100,000. The Mint reports for 1883 and 1884 estimate $87,000 and $80,000; for 1890 $72,000, and for 1891, $100,000. While the figures are incomplete and untrustworthy, it is scarcely probable that the total production much exceeds $15,000,000."

From 1908 on, production gradually decreased to only a few thousand dollars each year until 1916. In that year a dredge was installed by the Empire Dredge Co. near the town of John Day. It operated almost continuously until it was dismantled and moved to Prairie City in 1929. The dredge reportedly produced about $1,750,000 in gold and silver, according to the Engineering and Mining Journal (1929, p. 736-737).

A large dragline dredge owned by Ferris and Marchbank began work in the John Day River near John Day in 1935, and in 1937 a connected bucket dredge was installed by Western Dredging Co. Both operations ceased in 1942. Recorded production from the Canyon district for the years 1935-1942 was $2,539,214, most of it from the dredges. Combined recorded output from dredges operating in Canyon Creek and in the John Day Valley near the town of John Day during 1916-1942 was 123,911 ounces gold and 13,066 ounces silver.

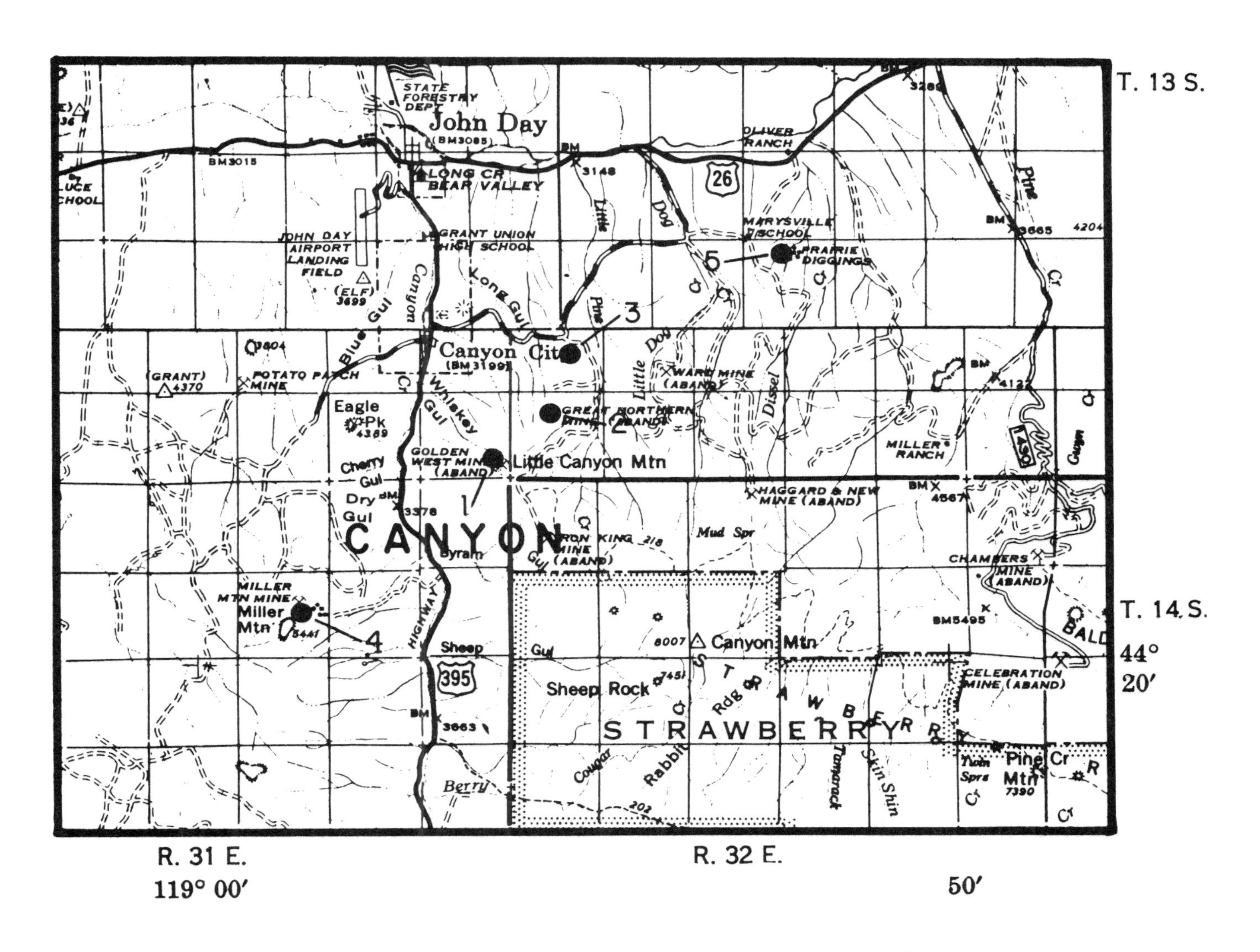

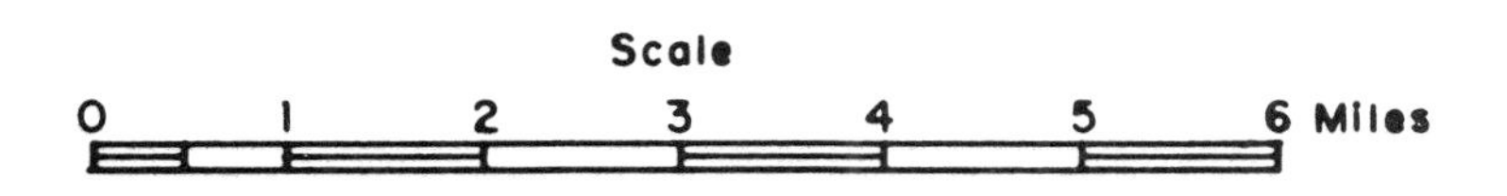

Base map from U.S. Forest Service

Lode Mines

1. Golden West
2. Great Northern
3. Haight
4. Miller Mountain
5. Prairie Diggings

Figure 31. Index map of the Canyon area.

Principal Lode Mines

Numerous small veins and irregular masses of quartz have been found on Canyon Mountain east of Canyon Creek and on Miller Mountain to the west. From several of them small pockets of coarse gold have been extracted. In 1898, gold worth $30,000 was recovered from a quartz seam opened by a surface cut at the Great Northern mine.

At Prairie Diggings, 3 miles east of John Day, placers containing rough quartz gold were worked during early days. In the same vicinity is a broad quartz-impregnated shear zone in cherty metavolcanic and sedimentary rocks. In Raymond's report of 1870 it is stated that the zone is 400 feet wide, strikes northeast and dips 60° SE. In 1872 a mill had been erected and $10,000 extracted. Subsequently, little has been produced, since the values are reportedly scattered and low grade.

Lode Mines of the Canyon Area

CANYON DISTRICT

Golden West mine — Canyon District, 1

Location: Grant County, SE$\frac{1}{4}$ sec. 12, T. 14 S., R. 31 E.

[No other data available.]

Great Northern mine — Canyon District, 2

Location: Grant County, N$\frac{1}{2}$ sec. 7, T. 14 S., R. 32 E. on Canyon Mountain 2 miles southeast of Canyon City and about 1500 feet above the town.

Development: Numerous prospect pits and short adits.

Geology: The country rock is mostly altered gabbro. From some of the numerous quartz-calcite seams cutting the gabbro small pockets of high-grade gold ore have been extracted. Quartz veins 1 to 2 feet thick are also present. These show a scattering of pyrite, but reportedly contain little or no gold.

Production: $30,000 from a pocket in 1898. Lesser pockets have since been extracted.

References: Lindgren, 1901:712-720; Parks and Swartley, 1916:111-112; Department Bulletin 14-B, 1941:24.

Haight mine — Canyon District, 3

Location: Grant County, NE$\frac{1}{4}$ sec. 6, T. 14 S., R. 32 E. on Canyon Mountain about 1000 feet below Great Northern mine.

Development: Tunnel and several pits.

Geology: Numerous quartz seams 6 inches to 2 feet apart; strike N. 35° W. and dip 60° E. The quartz has been brecciated and recemented by hematite and calcite. Gold found in scattered pockets, usually where veins intersect or change in attitude. Elsewhere the veins are barren or nearly so.

Production: Said to have been "considerable" in early days.

References: Department Bulletin 14-B, 1941:26.

Miller Mountain mine

Canyon District, 4

Location: Grant County, NE$\frac{1}{4}$ sec. 22, T. 14 S., R. 31 E. on the northeast slope of Miller Mountain.

Development: Totals about 4000 feet, including raises from several adits and shallow shafts.

Geology: Rocks exposed in the area include dark-colored meta-argillites and subordinate greenstones of Permian age overlain by upper Triassic graywackes and shales. Workings develop two quartz veins: one strikes N. 45° to 65° W. and dips 75° to 85° NE.; the other strikes N. 65° W. and dips 45° to 50° NE. Powell tunnel drifts 800 feet westerly on steep vein; about half this distance has been stoped to the surface. Stope width averages 3 to 4 feet. High-grade stringers penetrate hanging wall. The ore is mostly quartz, containing veinlets and patches of chalcopyrite, pyrite, and malachite in thin films, together with considerable manganese-oxide stains.

Production: Unknown, probably not less than $30,000 nor more than $100,000. Ten-stamp mill operated intermittently up to World War II.

References: Department Bulletin 14-B, 1941: 30-32.

Prairie Diggings

Canyon District, 5

Location: Grant County, N$\frac{1}{2}$ sec. 33, T. 13 S., R. 32 E.

Development: Surface cuts and shallow shafts.

Geology: Broad system of quartz veins and mineralized fractures in dark-colored slaty rocks. Mineralized zone said to be 400 feet wide and half a mile long in northeasterly direction.

Production: Little data available. The property was evidently placered extensively in early days. By 1872 a mill had been built and $10,000 in quartz gold extracted.

References: Lindgren, 1901:712-720; Swartley, 1914:205; Department Bulletin 14-B, 1941:34.

QUARTZBURG AREA

Location

The Quartzburg area, which includes the Quartzburg district, lies in Grant County mainly in the drainage of Dixie Creek, a southward-flowing tributary which enters the John Day River at Prairie City (figure 32). The area also extends northward over the Dixie Creek divide into the headwaters of Ruby Creek, a tributary of the Middle Fork of the John Day River. Elevations range from 3500 feet at Prairie City to 7592 feet at the top of Dixie Butte.

Geology and Mineralization

The rocks of the Quartzburg area are chiefly metamorphosed andesitic to basaltic lavas and tuffs, with subordinate argillite, gabbro, diabase, and serpentine into which numerous relatively fresh stocks and dikes of intermediate to silicic composition have been intruded.

The older rocks have been considerably sheared and altered. The volcanic rocks are in general quite chloritic. Uralite, biotite, and talc are common alteration products of the intrusive rocks.

The district contains quartz-carbonate-sulfide veins, sulfidized fault and shear zones, and quartz-tourmaline-sulfide replacement masses. The principal sulfide minerals are pyrite and chalcopyrite. The cobalt minerals glaucodot, cobaltite, and erythrite are sparingly present in several deposits. Other metallic minerals found locally include arsenopyrite, native bismuth, bismuthinite, tetrahedrite, pyrrhotite, sphalerite, and galena. Gilluly, Reed, and Park (1933, p. 85-105) describe 18 small mines and prospects.

Most of the work has been directed toward development of the veins which are valuable mainly for gold, although small amounts of copper and cobalt have been recovered from some of them. The values are contained in discontinuous lenses, which are rarely as much as 4 feet wide, and are composed chiefly of crushed and altered country rock with stringers of quartz, calcite, and sulfide minerals. In places considerable thicknesses of sheared and altered country rock have been impregnated with sulfide minerals. Little systematic prospecting and sampling of these zones has been done.

Production

No reliable estimate of total gold output can be made for the Quartzburg district. Although there is much evidence of early-day mining activity, authentic records of production are not available.

In his report of 1914 Swartley (p. 198) stated: "Locally the gross production from the Dixie placers is reported from $600,000 to $6,000,000. Probably the lesser amount approximates the truth." Statements by Lindgren (1901, p. 712) and Swartley (1914, p. 196) suggest that the combined production of all lode mines to 1914 was less than $200,000. On the other hand, production from the Equity mine alone was estimated by other authorities (Oregon Dept. of Geology and Mineral Industries, 1941, p. 124) at $400,000 to $600,000 prior to 1910. Dredges operating on the John Day River just below Prairie City during 1930-1936 and on Dixie Creek during 1938-1941 produced more than 22,500 ounces of gold.

Placer Mines

The most important placers of the district were on Dixie Creek, but many of the streams and gulches heading in the vicinity of Dixie Butte contained auriferous gravels. In 1901 (p. 712) Lindgren reported:

"The Dixie Creek placer mines were discovered about 1862, and were reported rich, though no data

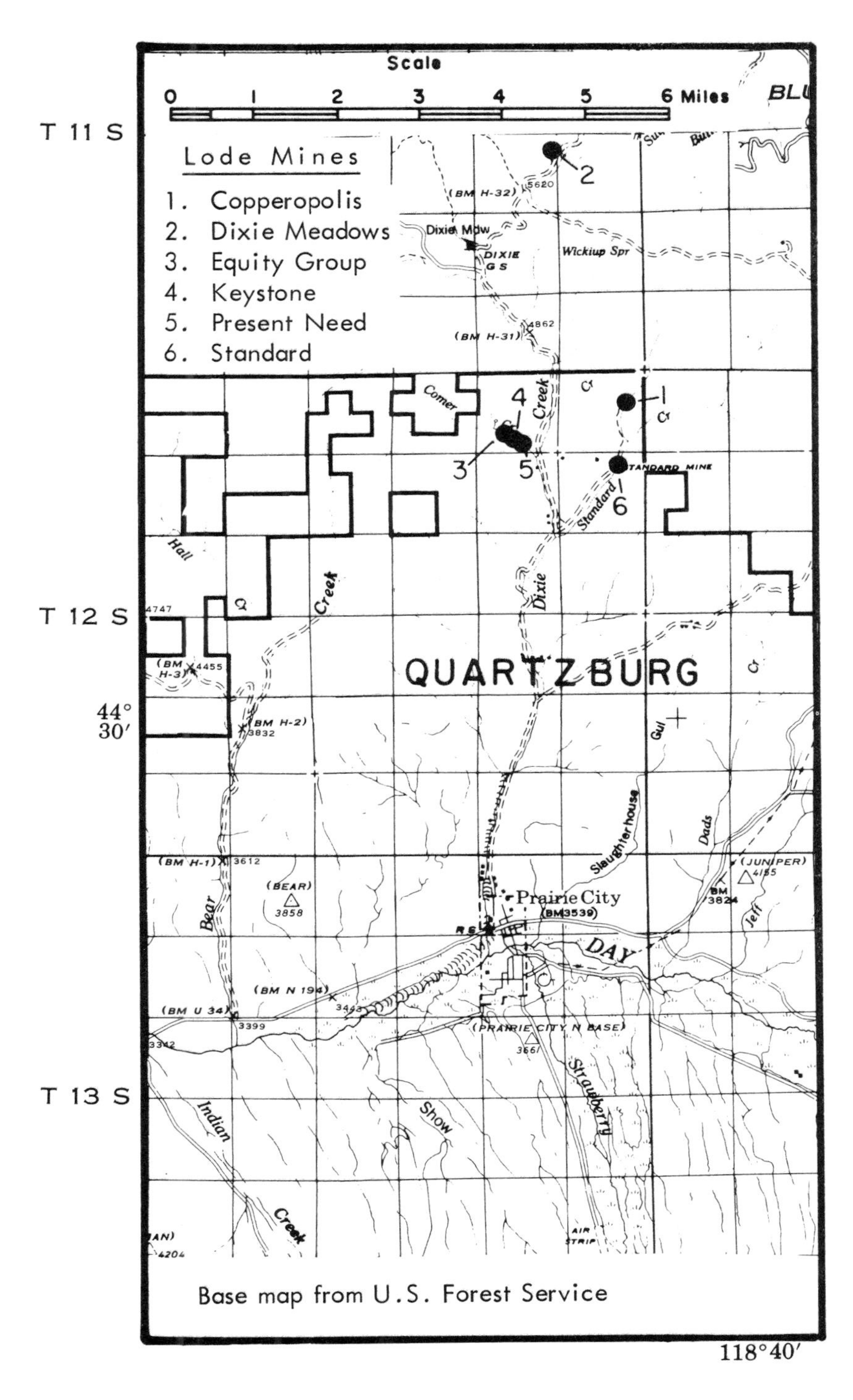

Figure 32. Index map of the Quartzburg area.

as to production are at hand. Raymond's report for 1870 contains the statement that at that time there were 100 white men and 200 Chinamen employed, and that the fine, scaly gold was 860 fine. In 1873 the creek is reported as turned over to Chinese labor. In 1882 two small hydraulic plants were in operation, producing $30,000 (Mint report). At the present very little placer mining is done.

"The placers consist of the gravels accumulated in the present creek to a depth of 10 or 15 feet. The workings extend upstream from Prairie City for 5 miles, or to the entrance of the diorite canyon, where the grade becomes very steep. The width of the gravel-covered river bottom is from 300 to 800 feet, the whole of which has been worked."

During 1930-1936 a bucket-line dredge owned by the Empire Dredging Co. was employed on the John Day River just below Prairie City. The H. D. England Co. dragline dredge worked in Dixie Creek above Prairie City almost continuously between October 1938 and April 1941.

Principal Lode Mines

Equity mine: The Equity group is on the West Fork of Dixie Creek 7 miles from Prairie City. According to the Oregon Department of Geology and Mineral Industries (1941, p. 124), "This property was discovered and located in 1878 and from that date up to 1910 was worked almost continuously.... The exact production is not known, but has been estimated from four to six hundred thousand dollars. Judging from the stoped areas these estimates would seem to be very nearly correct. From 1910 until 1933 very little work was done on the property, for it had been worked out down to the lowest point which could be attained by drifting or crosscutting." During 1933-1941 winzes were sunk and a small amount of drifting was done about 75 feet below the lowest level.

The country rock is nearly all sheared gabbro. The main vein strikes N. 35° and dips 80° SE. It has been opened by three adit levels, the top adit 95 feet above the lowest. Three distinct shoots have been found on this vein. Two of these have been mined out from the lowest level to the surface, a distance of 275 feet. The combined length of the three ore shoots is approximately 350 feet. The vein, which is about a foot wide in the stopes, is composed largely of quartz, dolomite, and massive sulfides--pyrite, chalcopyrite, galena, and sphalerite. The ore is said to contain as much as $500 a ton in gold.

Standard mine: The Standard mine is on the East Fork of Dixie Creek about one mile upstream from the forks. The Standard vein is said to have been discovered in the 1860's. According to Lindgren (1901, p. 701-712) a few tons of copper-gold ore had been shipped and about 600 feet of development work done prior to his visit in 1900. The principal period of development was during the years 1900-1907. A small smelter was erected on the property in 1901, but was shut down after only a few months' use. Crude ore and concentrates were railed to the Sumpter smelter. Ore containing cobalt and gold was reportedly shipped to Europe and to the Edison Laboratories in New York. In 1906 and 1907 shipments of 415 tons of concentrates netted about $18,200. Since 1907 attempts to work the mine have been periodic and short lived. In recent years several truckloads of hand-sorted copper ore mined from upper levels have been delivered to the Tacoma smelter. Three shipments totaling 57 tons made during 1964 averaged 0.5 ounces gold, 1.2 silver, and 20 percent copper.

On the Standard mine property several veins have been explored, including the Standard, Grover Cleveland, Juniper, and Smuggler. The Standard vein is the most extensively developed, with about 4000 feet of drift from four adits and a shaft. The levels are about 80 feet apart vertically and interconnected by numerous raises and stopes. The lowest and longest is just above the creek level. It contains 1700 feet of drift, above which are several large stopes. The average width of material mined from these stopes was probably about 18 inches.

The Standard vein is mainly in andesite, strikes N. 70° to 75° E., dips steeply south, and ranges in width from a few inches to 4 feet. It is not a persistent vein, but rather a group of mineralized stringers with considerable replaced wall rock. The main gangue mineral is quartz, but some ferriferous dolomite and calcite are also present. The ore minerals are pyrite, chalcopyrite, arsenopyrite, cobaltite, glaucodot, bismuthinite, native bismuth, galena, and sphalerite (Gilluly, Reed, and Park, 1933, p. 101). The cobalt minerals are said to be found mainly in the lower levels.

Dixie Meadows mine: The Dixie Meadows mine is at the head of Ruby Creek 12 miles north of Prairie City. During 1900-1910 between 8000 and 9000 tons of ore, averaging $8 to the ton in gold, were milled and about 350 tons of concentrates averaging $50 per ton shipped. Two adit levels 100 feet apart develop the deposit, the upper containing about 6000 feet of workings and the lower about 1200 feet. The workings explore a fault zone as much as 60 feet wide in a complex of sheared and altered andesite, tuff, diorite, and serpentine cut by dikes of granodiorite. The fault zone strikes N. 30° to 35° E. and dips 65° to 70° E. The rocks in the fault zone have been impregnated with quartz and sulfides, including pyrite, arsenopyrite, chalcopyrite, pyrrhotite, galena, marcasite, and sphalerite. The ore is but slightly oxidized; probably 75 percent of the gold is contained in the sulfides, mainly pyrite with lesser amounts of marcasite and pyrrhotite.

Lode Mines of the Quartzburg Area

QUARTZBURG DISTRICT

Copperopolis mine — Quartzburg District, 1

Location: Grant County, NE$\frac{1}{4}$ sec. 1, T. 12 S., R. 33 E.

Development: Several adits, including one about 1500 feet long that cuts the deposit at a depth of 300 feet.

Geology: Steeply dipping quartz replacement body in meta-andesite, metadiorite, and diabase. Strikes N. 60° E. and is more than 1000 feet long and 40 to 75 feet wide, including unreplaced blocks of country rock and gouge layers. Walls well defined. Contains tourmaline, pyrite, chalcopyrite, and lesser amounts of magnetite, hematite, cobaltite, tetrahedrite, bornite, galena, and sphalerite. Content of copper and gold appears to be very low.

Production: About 250 tons milled in small concentrator prior to 1906.

References: Gilluly, Reed, and Park, 1933:100-101; Department Bulletin 14-B, 1941:113.

Dixie Meadows mine — Quartzburg District, 2

Location: Grant County, secs. 23 and 24, T. 11 S., R. 33 E., near the head of Ruby Creek just north of the Dixie Creek divide.

Development: Includes two adits from which about 7200 feet of drifts, crosscuts, and raises have been driven. A new development campaign has been under way in recent years.

Geology: Low-grade quartz-sulfides replacement body up to 60 feet wide localized along a fault in a complex of greenstone, meta-andesite, tuff, metadiorite, and serpentine. The zone strikes N. 30° to 35° E. and dips 65° to 75° SE. Contains much decomposed country rock. Sulfides present are pyrite, arsenopyrite, chalcopyrite, pyrrhotite, galena, marcasite, and sphalerite. Ore only slightly oxidized at the surface. Little free gold present.

Production: Total said to be less than $100,000. Development began about 1900. Between 1903 and 1910, about 8000 or 9000 tons of ore averaging $8 to the ton in goldwas milled and about 350 tons of concentrates averaging about $50 to the ton were shipped.

References: Swartley, 1914:196; Parks and Swartley, 1916:86; Gilluly, Reed, and Park, 1933:88; Department Bulletin 14-B, 1941:114.

Equity Group

Quartzburg District, 3

Location: Grant County, secs. 2 and 11, T. 12 S., R. 33 E., on the West Fork of Dixie Creek, a few hundred feet northwest of the Keystone.

Development: Three adit levels. The upper one is 95 feet above the lowest, which in 1930 was 750 feet long. Some work has been done from shafts sunk below the lowest level.

Geology: Narrow vein in sheared gabbro strikes N. 35° to 45° E. and dips 75° to 80° SE. It is about one foot wide in the stopes and is composed largely of quartz, dolomite, and massive sulfides including pyrite, chalcopyrite, galena, and sphalerite. Numerous small cross faults cut the vein. Three distinct shoots with total length of 350 feet have been mined, two of which extend from the lowest level to the surface, a distance of 275 feet. Some of the ore contained as much as $500.00 in gold per ton.

Production: Estimated at $400,000 to $600,000. The mine is said to have been almost continuously active from 1878 to 1910.

References: Lindgren, 1901:711; Swartley, 1914:196; Parks and Swartley, 1916:91; Gilluly, Reed, and Park, 1933:91-93; Department Bulletin 14-B, 1941:124-125.

Keystone mine

Quartzburg District, 4

Location: Grant County, secs. 2 and 11, T. 12 S., R. 33 E., in a gulch west of the West Fork of Dixie Creek, 7 miles from Prairie City.

Development: Poor record. The pay shoots are said to be worked out down to creek level.

Geology: Similar to the Present Need. The vein strikes NE., dips SW., and is 4 feet wide. The pay forms narrow streaks on the hanging wall, on the footwall, or on both. The gangue contains much calcite. In 1933 a shaft was sunk on the vein from the surface, exposing a 9-foot width with 4 feet of low-grade ore assaying $8.50 for the base ore.

Production: Small output in 1882 and 1889; no other record available.

References: Lindgren, 1901:710; Parks and Swartley, 1916:137; Gilluly, Reed, and Park, 1933:93; Department Bulletin 14-B, 1941:119.

Present Need

Quartzburg District, 5

Location: Grant County, secs. 2 and 11, T. 12 S., R. 33 E., on the West Fork of Dixie Creek about a mile above the forks and a few hundred feet below the Keystone.

Development: Poor record; the vein is said to have been exploited down to creek level.

Geology: Narrow and discontinuous quartz vein in diabase. Strikes N. 20 E., dips 70° SE. Contains pyrite, marcasite, and a little chalcopyrite, sphalerite, and galena. Two shoots 4 inches to 2 feet wide and about 70 feet long yielded ore averaging 6 to 25 ounces silver and 4 to 5 ounces gold per ton. The shoots are separated by 70 feet of barren ground.

Production: Small output during 1890's and later. Total unknown.

References: Lindgren, 1901:710; Gilluly, Reed, and Park, 1933:93; Department Bulletin 14-B, 1941:122.

Standard mine

Quartzburg District, 6

Location: Grant County, NE$\frac{1}{4}$ sec. 12, T. 12 S., R. 33 E., on the east side of the East Fork of Dixie Creek.

Development: Totals about 4000 feet from three adits at 80-foot intervals and an inclined shaft about 70 feet deep. Considerable stoping has been done and the levels are interconnected by raises. The lowest and longest adit level is just above creek level.

Geology: The country rocks are chiefly porphyitic andesite cut by dikes of granodiorite porphyry and a few of diabase. The Standard vein strikes N. 70° E., dips steeply south, and ranges from a few inches to 4 feet in width. The ore minerals are pyrite, chalcopyrite, malachite, arsenopyrite, cobaltite, glaucodot, erythrite, bismuthinite, native bismuth, galena, and sphalerite. The cobaltiferous ores are found mainly in the lower levels and are said to contain much gold. Copper-gold ores occur in the intermediate and upper levels. In recent years, rich copper ores have been shipped periodically from surface cuts and short adits above the upper level. In addition to the Standard vein, at least three other veins have been explored.

Production: Unknown. Considerable stoping has been done on the Standard vein.

References: Lindgren, 1901:711; Swartley, 1914:197; Parks and Swartley, 1916:210; Gilluly, Reed, and Park, 1933:101; Department Bulletin 14-B, 1941: 125-127.

UNITY AREA

Location

The Unity area includes the Unity mining district (sometimes called the Bullrun district), which lies 5 to 9 miles southwest of Unity in Baker County (figure 33). The gold mines are situated on the northern flanks of Mine Ridge and Bullrun Mountain--two north-trending mountain spurs. Bullrun Creek, a tributary of the South Fork of the Burnt River, flows northward between the two mountain ridges and out onto an alluviated plain that slopes northward toward Unity.

Geology

Jurassic-Cretaceous diorite and quartz diorite intrude limy tuffs of the Rastus Formation of Late Triassic-Middle Jurassic age and older ultramafic rocks and schists. The geology of this region is described by Lowry (in press).

Production

Quartz mines in the district known to have been productive are the Record, Orion, and Bull Run. Production from the Record is said to be about $103,000, more than half between 1933 and 1937. Production from the others is unknown but, judging from the amount of stoping, it must have been quite small; the Orion was most active between 1903 and 1917.

The Ferris Mining Co. operated a dragline dredge with a 3-cubic-yard excavator on Bullrun Creek from October 1940 to July 1941. Output totaled 2161 ounces gold and 253 ounces silver from 61,000 yards of gravel.

Principal Lode Mines

The principal ore shoot of the Record mine is composed of a number of closely spaced parallel stringers of high-grade quartz in an irregular felsite-porphyry dike along the contact between granodiorite and serpentinite. Most of the ore was oxidized and contained a high percentage of free gold. The zone is 5 to 10 feet wide, is nearly vertical, and strikes about N. 60° E. The main ore shoot was 100 feet long and 260 feet high. Gold is also found in thin seams of hornblende associated with the felsite dike. A 150-ton amalgamation mill was erected on the property in 1929.

The ore of the Bull Run mine occurred as irregular veins, small lenses, and disseminations in argillite. Ore mineralization at the Orion mine consisted of veinlets of pyrite and arsenopyrite in a shear zone in argillite (Lorain, 1938, p. 43). The ore mined was highly oxidized.

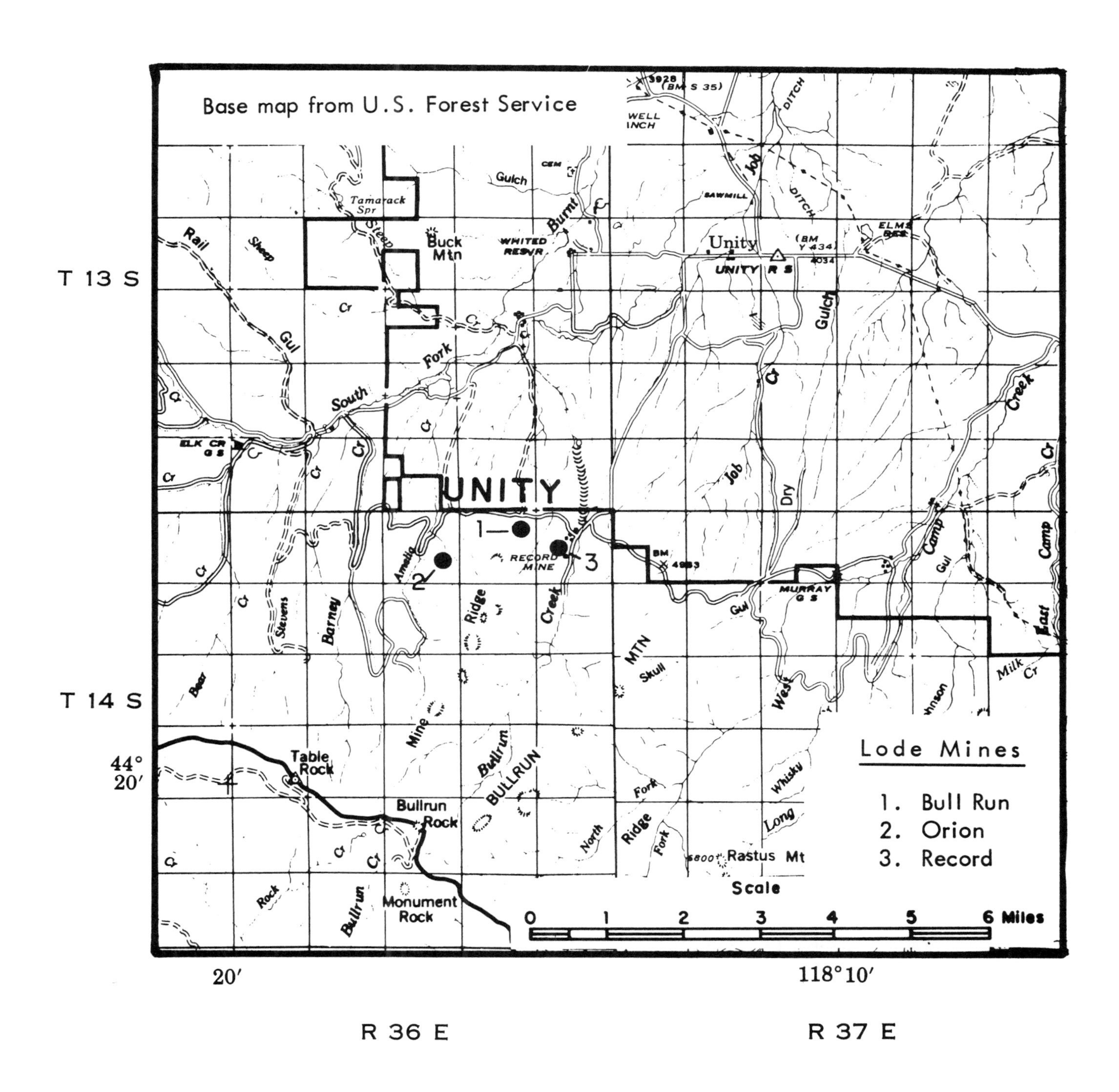

Figure 33. Index map of the Unity area.

Lode Mines of the Unity Area.

UNITY DISTRICT

Bull Run mine

Unity District, 1

Location: Baker County, sec. 2, T. 14 S., R. 36 E.

Development: 1200 feet of tunnel, a glory-hole raise, and 3 shallow shafts.

Geology: Small quartz veins in argillite.

Production: $10,000 (?)

References: Lorain, 1938:42-43; Department Bulletin 14-A, 1939: 98.

Orion mine

Unity District, 2

Location: Baker County, sec. 3, T. 14 S., R. 36 E.

Development: 280-foot adit drift and 425-foot crosscut adit.

Geology: Altered shear zone in argillite. Contains veinlets of pyrite and arsenopyrite.

Production: Small.

Reference: Lorain, 1938:43.

Record mine

Unity District, 3

Location: Baker County, sec. 1, T. 14 S., R. 36 E.

Development: Five adits total about 5000 feet.

Geology: Mineralization along irregular granodiorite-serpentine contact, which is locally intruded by felsite porphyry dike. Strike averages about N. 45° W. and dip is steep. Chalcopyrite, pyrite, and molybdenite. Free gold in amphibole veinlets and in quartz stringers. Ore zone locally attains width of 5 feet.

Production: $103,000 estimated. Gold 65 to 68 percent free, with 92 percent minus 100-mesh grind. Extraction 95 percent from 0.2 ounce heads in gravity mill. Gold 900 fine.

References: Lorain, 1938:41-43; Department Bulletin 14-A, 1939:101-102.

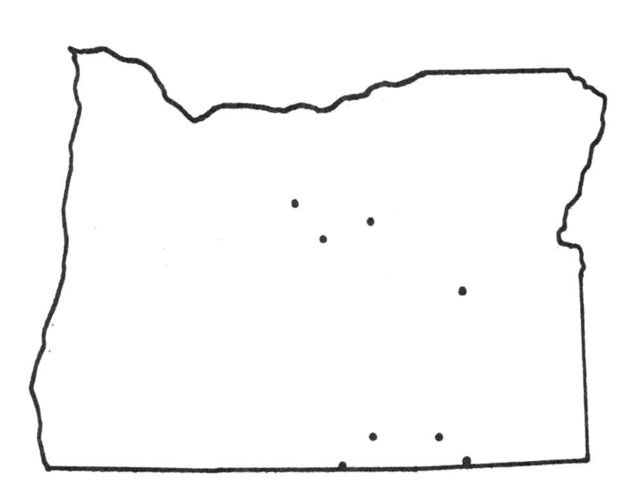

Part II-B Deposits in Eastern Oregon Isolated Gold Mining Disrricts

Gold scales for field use "for weighing gold nuggets used by Charlie Brown at Canyon City." Exhibited in Grant County Museum, Canyon City.
—Courtesy of Grace Williams, Grant County Museum : Photo by Bert Webber.

Bert Webber examines rocks among tailings in the Old Humboldt Diggins, where dredges once worked, on the side of the hill west of Canyon City.
—Bert Webber photo

ISOLATED GOLD MINING DISTRICTS

Table of Contents

Wing dam on the Rogue River between Agness and Illahe at Old Diggin's Riffle *ca* 1912. White gravel on left is tailings from Gold Bar Placer Mine.
—Curry County Historical Society

ISOLATED GOLD MINING DISTRICTS

A number of small mining districts occur as isolated areas in eastern Oregon. Their location is shown in Figure 7, page 29. Of the seven districts described below, five are in Tertiary rocks and two in pre-Tertiary. Recorded production has been small.

ASHWOOD DISTRICT

The Ashwood district is situated in eastern Jefferson County, and it encompasses the village of Ashwood 26 miles east of Madras. It lies within the arbitrary bounds of the Horse Heaven quicksilver mining district. Rugged, rolling hills dotted with patches of sagebrush, desert grasses, and juniper characterize the landscape. The area is underlain by a complex assemblage of Tertiary volcanics and subordinate sedimentary rocks, mainly of the Eocene-Oligocene Clarno Formation. Topographically prominent volcanic plugs are common.

Gold-silver production in the region has been mainly, if not entirely, from the Oregon King mine 3 miles by road northeast of Ashwood. Other prospects occur in the vicinity of Axhandle Butte, a few miles farther to the east. Development of the Oregon King vein began in 1898, when some sheepherders dug into the outcrop and found rich silver ore. J. G. Edwards, then half owner of the nearby Hay Creek ranch, financed the first shaft in 1899.

Libbey and Corcoran (1962, p. 5-6) stated:

"The ore zone at the Oregon King mine is associated with a fault zone trending N. 75° W. The dip on the fault averages about 75° SW., but it is steeper in the lower levels of the mine, according to mapping by the Alaska Juneau Co. The andesitic rocks along the fault have been brecciated, silicified, and impregnated with quartz and pyrite, together with smaller amounts of chalcopyrite, galena, and sphalerite. Cerargyrite, silver chloride, and native silver have been reported in upper levels. Bunches of massive sulfides, largely pyrite, are occasionally found.

"Reportedly, the ore occurs in lenticular pipelike masses of variable size ranging in width from a few feet to as much as 20 feet. Generally, walls of the shoots must be determined by assay. It is said that some of the shipping-grade ore was difficult to distinguish visually from low grade.

"Values are mainly in silver with a smaller amount of gold. Copper, lead, and zinc sulfides are found in relatively minor amounts except locally. A very little arsenic was reported in the concentrates, and stibnite was reported among the sulfides."

Development of the Oregon King now totals about 4000 lineal feet, including a 700-foot shaft. Production was recorded for the years 1899, 1901, 1904, 1935, 1940-1947, 1950, and 1963-1965. Smelter value of the 7334 dry tons of ore and concentrates shipped through 1950 was $233,693. The 1935-1950 shipments contained 2419 ounces gold, 232,402 ounces silver, 59,076 pounds copper, and 110,071 pounds lead. In September 1950 a fire in the shaft, which also destroyed much of the machinery and equipment on the surface, caused a stoppage of all mining activity.

The main shaft was reopened in 1962 by Oregon King Consolidated Mines, Inc., and new ore developed from the workings gave a significant boost to Oregon gold and silver production the following year. In 1963 the state produced 58,234 ounces of silver and 1281 ounces of lode gold, and nearly all of it came from this one mine. The bulk of the ore was taken from exploratory work on the east 400-foot level and west 600-foot level with lesser amounts obtained from between the 200-foot and 400-foot levels. Production dwindled in 1964 while the company conducted an exploration drilling program in some of the lower drifts. The mine is presently idle and renewal of operations presumably is contingent upon discovery and development of new ore bodies.

HARNEY DISTRICT

The Harney district, also known as the Idol City-Trout Creek district, lies 20 miles northeast of Burns, Harney County, in sec. 4, T. 21 S., R. 32 E., in the vicinity of Trout Creek, a branch of Silvies River.

The country rock is a porphyritic andesite of probable late Miocene age. The andesite underlies most of the larger hills in this region and presumably is a part of the Strawberry Volcanics (Brown and Thayer, 1966). Mineralization appears to be confined to a northwest-trending shear zone along which the andesite has been altered or bleached for a distance of at least a mile.

A small amount of underground development work has been done, but most of the gold production has come from placering the Recent valley fill in one of the tributaries of Trout Creek. The placers have yielded about $50,000 since discovery in 1891 (Parks and Swartley, 1916, p. 273).

HIGH GRADE DISTRICT

The High Grade district is largely in Modoc County, California, but it extends into the southern edge of Lake County in Oregon. Here it includes several prospects in the hills a few miles east of New Pine Creek, a small community on U.S. Highway 395 about 15 miles south of Lakeview.

The veins in the area are small and occur along fractures in early (?) Tertiary volcanic rocks. Much prospecting was done in the early days and a few properties were equipped with small plants. No records of production are available for deposits in the Oregon part of the district.

HOWARD DISTRICT

The Howard district lies in Crook County, 26 miles east of Prineville in the timbered hills bordering Ochoco Creek. The district includes some small placer and lode deposits.

The area is underlain by Tertiary andesites. Mineralization has taken place along rather broad fracture zones in which the andesite is intensely altered and cut by carbonate-quartz-sulfide veinlets. The veinlets are commonly less than an inch thick but locally enlarge into ore shoots, especially at junctions or intersections.

The principal placers, all of them small, worked deposits along the lower part of Scissors Creek, which crosses the mineralized area and enters Ochoco Creek from the southwest.

Lode mine production from the district has come mainly from the Ophir-Mayflower or Ochoco mine. Its lower adit, a crosscut 1435 feet long, enters the west bank of Ochoco Creek just above creek level. A second crosscut adit enters the slope about 200 feet higher. Many other short adits and opencuts dot the adjacent slopes.

According to Parks and Swartley (1916, p. 167-168), one ore shoot in the Ophir-Mayflower mine measured 70 feet long, 250 feet high, and 1 to 6 feet wide. Another measured 20 feet long, 40 feet high, and as much as 4 feet wide.

Available records (Gilluly, 1933, p. 125) show that the district produced $79,885 in gold up to 1923 from intermittent operations. Of this, $17,560 is based on estimates prior to 1902 and $62,325 is based on records since 1902. The earliest record of production in Mint reports shows an output of $10,000 for 1885. Recorded output from 1903 through 1923 amounted to 536.80 fine ounces of gold and 79 fine ounces of silver from placers, and 2225 tons of crude ore from which 2,478.35 fine ounces of gold, 442 fine ounces of silver, and 2662 pounds of lead were recovered. The greatest output was $26,623 in 1918, of which $24,092 was in gold. Subsequent production has been very small. A few ounces of gold were produced in 1923 and 1933.

LOST CABIN DISTRICT

The Lost Cabin district (also known as the Coyote Hill, or Camp Loftus district), is in T. 35 S., R. 23 E., about 10 miles north of Plush in Lake County. Gold was reportedly discovered here in 1906 by the Loftus Brothers. There was a small rush to the area and much shallow prospecting done but no records of production exist.

Rocks in the district include andesitic to rhyolitic flows, tuffs, breccias, and small intrusives of early (?) Tertiary age. In places these rocks are cut by small irregular fractures filled with seams of clay, limonite, and, locally, quartz. Small amounts of gold and copper-oxide minerals are said to have been found in some of these seams. A few flasks of quicksilver have been produced from the Gray prospect in sections 14 and 15, T. 35 S., R. 23 E.

SPANISH GULCH DISTRICT

The Spanish Gulch district is in southeastern Wheeler County just east of Antone. Rock and Birch Creeks, which cross the district and flow north to the John Day River, were worked for placer gold during the earliest days of placer mining in eastern Oregon. Development of quartz veins has been small, although some of the veins are reportedly wide and persistent (Parks and Swartley, 1916, p. 296). Records of output are scarce.

The district is located in a small exposure of pre-Tertiary rocks in an area otherwise covered by Tertiary lavas and sediments. The older rocks are greenstones, argillites, and serpentine which have been intruded by dikes and irregular masses of granodiorite. Many of the veins are almost entirely massive quartz, while in others the vein material is highly altered and silicified country rock. The other minerals are pyrite, chalcopyrite, and galena containing gold and silver.

One quartz vein described by Parks and Swartley (1916, p. 232) is a distinct quartz vein which strikes N. 60° E. and dips 45° S., and has an average width of 2 feet. Another strikes N. 60° E., dips 45° N., has a width of 12 feet, and can be traced on the surface for about 1000 feet.

STEENS-PUEBLO DISTRICT

A few gold prospects occur in extreme southern Harney County in the southern part of the Steens Mountain-Pueblo Mountains range. Tertiary and pre-Tertiary rocks exposed along the eastern front of the mountains are cut by a multitude of small faults related to the uplift of the range during late Tertiary time. Zones of silicification along these faults locally contain cinnabar and copper minerals. A very few contain gold. Small shipments of gold are reported to have been made from the Farnham and Pueblo prospects in secs. 8 and 17, T. 40 S., R. 35 E. (Williams and Compton, 1953, p. 47).

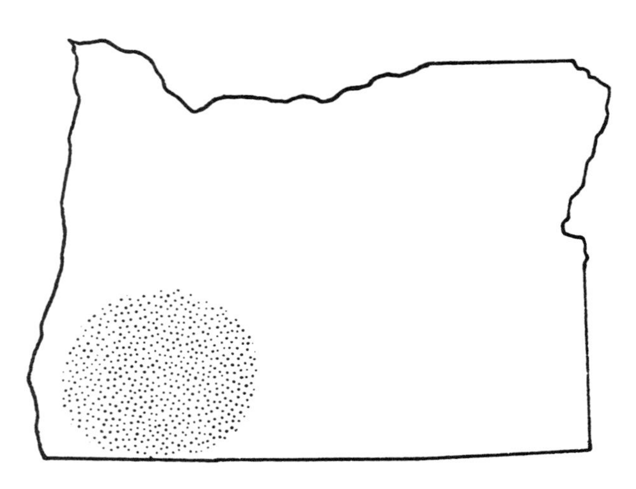

Part III-A Deposits In Western Oregon The Klamath Mountains

Flume (left) **running to Inman Mine. Hydraulic mining** (below) **either Pelaka or Russian Mike Mine on Sixes River. Both sites in Curry County.**
—Curry County Historical Society

Editor's Note: In an earlier printing, Page No. 162 was blank. Blank pages have been eliminated in the present book thus that page number does not appear.

KLAMATH MOUNTAINS

Table of Contents

THE KLAMATH MOUNTAINS

INTRODUCTION

Early settlement of the state was sparked by discovery of gold in southwestern Oregon. At least three-fourths of the gold produced in southwestern Oregon since 1850 was derived from placer deposits. The bulk of the lode production has been from six mines. Only two of these, the Greenback and the Ashland, have produced more than $1,000,000. Output from the other four has been between $500,000 and $1,000,000. Fourteen other lode mines in the region have produced between $100,000 and $500,000. Essentially all of the richer placer ground has been mined over, but one can safely say the lode-gold mining in southwestern Oregon has barely "scratched the surface."

Most of the gold deposits in the Klamath Mountains, like those in the Blue Mountains, are related to Late Jurassic diorite intrusives. Their areal distribution is shown in figure 34. Mine groupings away from large igneous outcrops may indicate the presence of these bodies at depth.

GEOGRAPHY

The Klamath Mountains geomorphic province in southwestern Oregon occupies the area between the Pacific Ocean and the Western Cascades, lying south of the Coast Range and north of the California border. Elevations range from sea level to 7500 feet. The topography of most of the region is one of deep canyons, narrow ridges, and craggy peaks. The ridge tops in general are of fairly uniform height, about 4000 feet above sea level, and are believed to represent an old erosion surface (Diller, 1902).

The region is drained primarily by the vast Rogue River system, which includes the Illinois and Applegate Rivers and Bear, Evans, and Grave Creeks. Other important streams are the Chetco and Coquille Rivers on the coast and the South Umpqua and its tributaries which drain the northern part of the Klamath Mountains. Broad, flat valleys have developed in a few places such as along Bear Creek in the vicinity of Medford and the Rogue River near the mouth of the Applegate, and in the Illinois Valley around Cave Junction. A few small lakes occur in the area; they are of glacial origin and are situated above 3500 feet elevation.

Rainfall ranges from about 25 to 50 inches annually, and temperatures are generally mild. Vegetation is abundant, except in areas underlain by ultramafic rocks (peridotite and serpentinite), where it is typically sparse, and on bold, rocky bluffs where it is almost entirely lacking. Both conifers and deciduous trees are found in great variety in the Klamath Mountains province. Wild flowers and flowering shrubs such as rhododendrons, azalias, and many others beautify the mountain slopes. Poison oak grows nearly everywhere.

The principal areas of population are in the central valleys around Medford, Grants Pass, and Ashland. Population centers along the coast include Brookings, Gold Beach, and Port Orford. Industries in the region include lumbering, agriculture, and some mining.

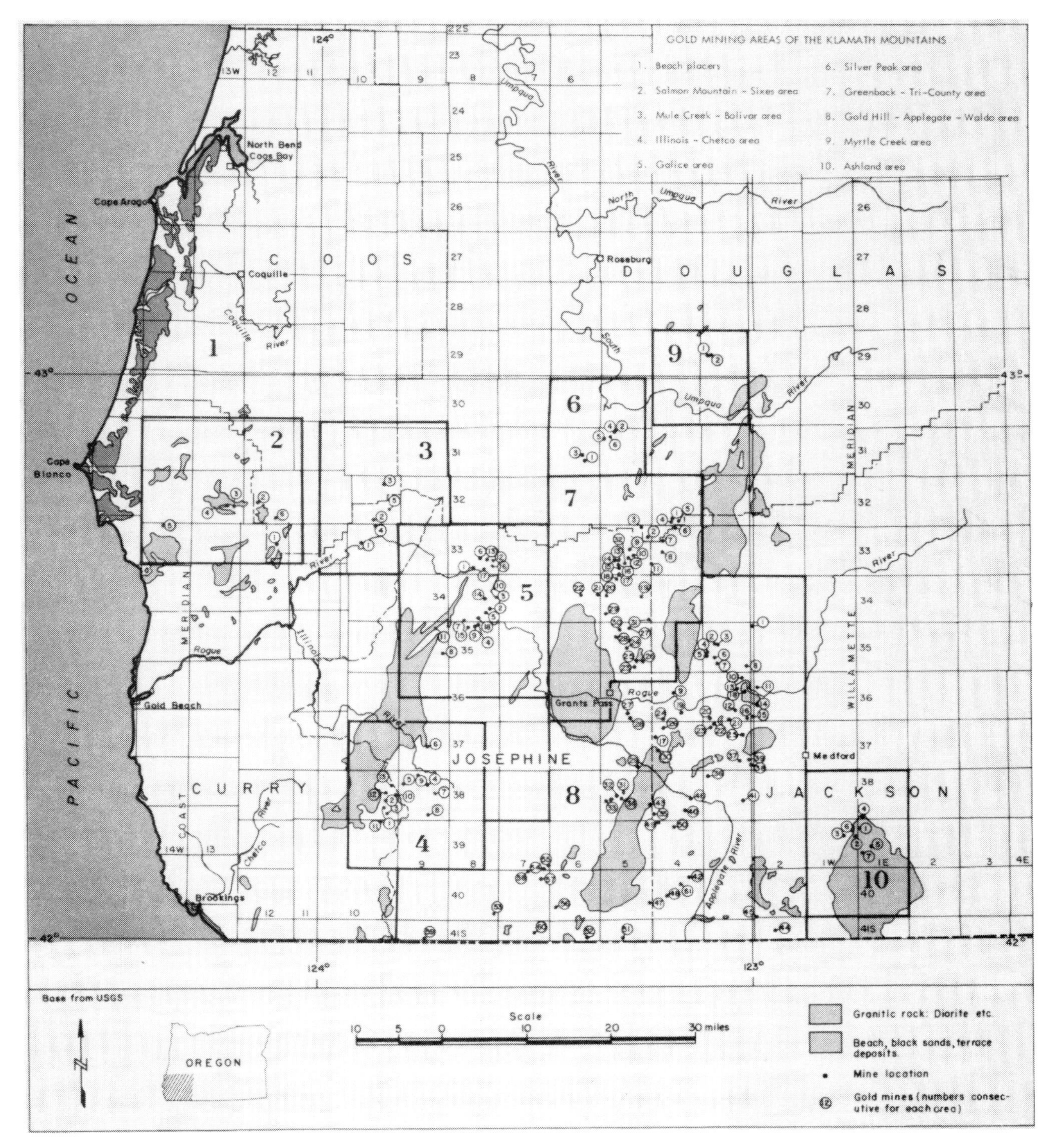

Figure 34. Map of the Klamath Mountains showing distribution of the gold mining areas and intrusive bodies.

PREVIOUS WORKERS

The earliest work published on gold mines of southwestern Oregon was by J. S. Diller and G. F. Kay (1909) and covered the mineral resources of the Grants Pass quadrangle and bordering districts. In a later report, Diller (1914) has additional information on southwestern Oregon mineral resources. A.N. Winchell (1914) described the mineral resources of Jackson and Josephine Counties. Parks and Swartley (1916) published an alphabetical list of mines in Oregon. P. J. Shenon (1933a, b, and c) reported on copper mineralization in the Squaw Creek, Silver Peak, Almeda and other areas; the Robertson, Humdinger, and Robert E. mines; and ore deposits of the Takilma-Waldo area. F. G. Wells' 1933 report on the Chieftain and Continental mines in southern Douglas County accompanied Shenon's report (1933a). J. T. Pardee (1934), R. R. Hornor (1918), and A. B. Griggs (1945) all reported on black-sand beach placers of the southern Oregon coast. Callaghan and Buddington (1938) described mineral deposits of the Cascade Range. The Oregon Metal Mines Handbook series, containing an accumulation of available published information and previously unpublished Department mine file reports, was prepared for Coos, Curry, and Douglas Counties (1940); Josephine County (1942); and Jackson County (1943). E. A. Youngberg (1947) reported on mines and prospects of the Mount Reuben district of Josephine County. F. W. Libbey (1963) summarized the history of gold mining in southwestern Oregon and (1967) described the history of production and development at the Almeda mine. In addition to these published reports, some unpublished mine reports prepared by various Department personnel since publication of the Metal Mines Handbook series supply information for this paper. Geologic maps showing mine and prospect locations, accompanied by short geologic reports, include: J. S. Diller (1898) Roseburg quadrangle; Diller (1903) Port Orford quadrangle; Diller and Kay (1924) Riddle quadrangle; Wells and others (1940) Grants Pass quadrangle; Wells, Hotz, and Cater (1949) Kerby quadrangle; Wells and Walker (1953) Galice quadrangle; Wells and others (1956, revised) Medford quadrangle; and Wells and Peck, geology of Western Oregon (1961).

HISTORY AND PRODUCTION

Placer Mining

Gold was first discovered in Oregon in the summer of 1850 by a party of miners, most of whom were originally from Illinois. They worked a placer on what is now called the Illinois River near the mouth of Josephine Creek (Spreen, 1939). The discovery that was responsible for triggering the first real goldrush into southwestern Oregon occurred in December of 1851 near what is now Jacksonville and which was formerly a camp site on the overland supply route to the California gold fields. Literally thousands of men came into this bustling mining camp to stake claims and make their fortunes.

Word of rich placer deposits spread rapidly and the rush to Jackson County soon made it the most populous county in the state. Mining camps sprang up over night in such places as Jacksonville; Buncom at the mouth of Sterling Creek on the Little Applegate River; Sailors Diggings (Waldo), discovered by a group of sailors who abandoned their ship at Crescent City and were headed for the Jacksonville diggings; Kerbyville (now Kerby) on the Illinois River; Willow Springs on Willow Creek west of Central Point; Phoenix on Bear Creek; Allentown near Takilma; Browntown on Althouse Creek; and later at Golden on Coyote Creek and at Placer on Grave Creek (figure 35, in pocket).

Some of the richer placers were on Sterling Creek, which has produced more than $3,000,000; Althouse Creek; and Sailors Diggings (where a ditch costing $75,000 paid for itself in one year). Rich Gulch at Jacksonville as well as Rich Gulch at Galice were important producers. Placers were located on Sucker Creek, Josephine Creek, Briggs Creek, Galice Creek, Grave Creek and its tributaries, Foots Creek, Sardine Creek, Galls Creek, Forest Creek, Poorman Creek, Humbug Creek, Ferris Gulch, Powell Creek, and Palmer Creek. The first official mining districts in the Oregon Territory were organized in 1852 at Jacksonville, at Sailors Diggings, and on Althouse Creek.

Most of the placer mines were worked by hydraulic methods. Water was conveyed around the hillsides in ditches, many of which were constructed by Chinese coolie labor. The famous 23-mile Sterling

Figure 36. Hydraulic elevator at the old Esterly (Llano de Oro) mine near Waldo about 1939.

mine ditch, which was fed from the Little Applegate River, was built in 1877. Remnants of old placer ditches are still common sights.

Some of the ditches had extensive wooden flumes suspended on superstructures around rocky bluffs. The 12-mile ditch from upper Grave Creek to the Columbia placer on Tom East Creek had flumes such as this; and the Red River Gold Mining Co.'s $80,000 flume in the Mule Creek area was suspended on high trestles. Four levels of ditches which served the placer mines around Waldo and Takilma included several flumes and ditch tunnels, making up a network that totaled about 32 miles in length. Most of these structures were built in the 1870's.

Hydraulic elevators were used at the old Esterly, or Llano de Oro, mine to elevate water and tailings out of the pits which were worked to bedrock 3 to 50 feet below ground level (figure 36).

The Old Channel placer mine near Galice has been referred to as one of the largest hydraulic operations in the United States (Parks and Swartley, 1916). Operation began in the late 1850's when the first high ditch was built. During the peak of its activity in 1935, a crew of 75 men was employed at rehabilitation work. The average operating crew was 20.

One of the first successful dredges in Oregon was built and operated on Foots Creek in 1903. It was powdered by steam, but in 1905 it was converted to electric power from the Ray Gold hydroelectric plant near Gold Hill. In addition, dredging has been done on Forest Creek, Poorman Creek, Thompson Creek, Oscar Creek, Althouse Creek, Pleasant Creek, Grave Creek, Wolf Creek, Sardine Creek, Jackson Creek, the Illinois River near Eight Dollar Mountain, Rogue River near Gold Hill and the town of

Rogue River, and along the Applegate River near Ruch (fig. 35).

Lode Mining

By 1880 lode-gold mining had made a good start in southwest Oregon. The Greenback mine was discovered in 1897. By 1939 it had produced $\$3\frac{1}{2}$ million. In 1916, the Greenback operated with a crew of 30 men, and at the peak of its activity (1900-1910) the mine employed 100 men.

The Benton mine was discovered in 1893 and worked intermittently to 1942. The peak of its production was reached between 1937 and 1942, when a crew of from 25 to 45 men was employed. During this five-year period, production varied between $5000 and a reported maximum of $12,000 per month.

The Almeda mine on the Rogue River below Galice produced in excess of $100,000 of gold, silver, and copper between 1908 and 1916. A small matte smelter was operated at the property for a short time. The mine was reopened in 1940, but operations were discontinued in 1942 (Libbey, 1967).

Another important producer in the area, the Ashland mine, was discovered about 1890 and its gold production totaled about $\$1\frac{1}{2}$ million to 1942. The Opp mine near Jacksonville produced about $100,000 during its major period of operation during the early 1900's. A 20-stamp mill was used to grind the ore.

During the late 1800's and through the 1930's the clatter of stamp mills was a fairly common sound. In addition to the 20-stamp at the Opp mine, a partial list of stamp mills includes the Greenback with a 40-stamp; the Granite Hill, 20-stamp; the Bone of Contention, 15-stamp; and the Oreole, Braden, Lone Eagle, Bunker Hill, and Ashland each with 10-stamp mills. A number of the smaller mines were equipped with 5-stamp units, and even smaller 2- and 3-stamp units were used.

Value of Production

Gold production for southwestern Oregon is difficult to appraise accurately. Not all of the gold mined in the area reached the Mint. Furthermore, it is likely that some of it was used as a medium of exchange and passed out of the area before it could be sold to the Mint. Some of the gold was undoubtedly taken out of the country by Chinese. Diller (1914) reports that U.S. Mint receipts for the entire state during the period 1851 to 1882 was $16,816,275.39. Nearly half of this amount is believed to have come from southwestern Oregon alone, most of it in the 10-year period prior to discovery of gold in northeastern Oregon.

From 1882 to 1899, southwestern Oregon is credited with a gold production of $5,808,831.11, and from 1900 to 1912, $5,488,941. During this latter period, value of placer production was nearly twice that from lode mines. Silver production was valued at $63,385, while placer production of platinum metals mounted to $15,293. More complete records are given in Part I of this bulletin.

GEOLOGY

Most of the Klamath Mountains province is underlain by pre-Tertiary sedimentary, igneous, and metamorphic rocks. Tertiary continental sedimentary and volcanic rocks of the Western Cascades form the eastern boundary of the province. To the north and northwest are Tertiary marine sediments of the southern Coast Range (figure 37).

The Paleozoic and Mesozoic formations which underlie the Klamath Mountains are subdivided into two groups: those that predate the Late Jurassic Nevadan orogeny, and those that were deposited after this event. The Nevadan episode is of special economic importance in southwestern Oregon, because it was during this time interval that most of the ultramafic to granitic plutons were emplaced. The mineralization that accompanied the igneous activity probably was the source for most of the gold deposits found within this province.

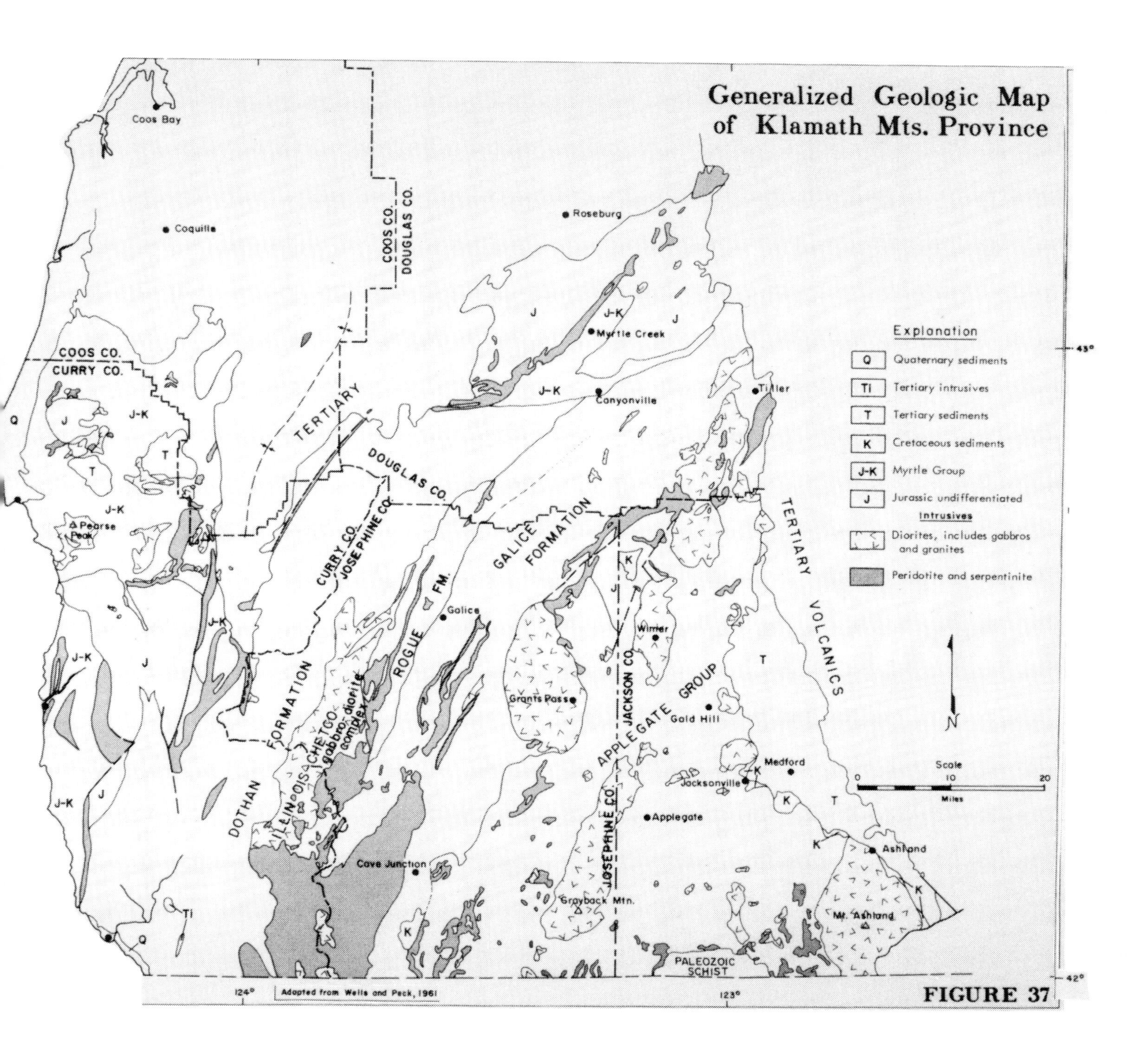
Generalized Geologic Map
of Klamath Mts. Province
Explanation
Q Quaternary sediments
Ti Tertiary intrusives
T Tertiary sediments
K Cretaceous sediments
J-K Myrtle Group
J Jurassic undifferentiated
Intrusives
Diorites, includes gabbros and granites
Peridotite and serpentinite
Coos Bay
Coquille
Roseburg
Myrtle Creek
Canyonville
Tiller
COOS CO.
DOUGLAS CO.
CURRY CO.
JOSEPHINE CO.
JACKSON CO.
Pearse Peak
TERTIARY
GALICE FORMATION
ROGUE FM.
DOTHAN FORMATION
TERTIARY VOLCANICS
APPLEGATE GROUP
Golice
Wimer
Grants Pass
Gold Hill
Medford
Jacksonville
Applegate
Ashland
Mt. Ashland
Cave Junction
Grayback Mtn.
PALEOZOIC SCHIST
Scale
0 10 20
Miles
N
43°
42°
124°
123°
Adapted from Wells and Peck, 1961

FIGURE 37

Pre-Nevadan Formations

Paleozoic schist

The oldest rocks exposed in the Klamath Mountains province of Oregon are highly metamorphosed schists of probable Paleozoic age exposed in the upper Applegate drainage about 20 miles south of Medford. Baldwin (1964) suggests a correlation with the Ordovician (?) Salmon and Abrams Formations of northern California. The Oregon rocks are of three types, which grade into each other. As described by Wells (1956), the most abundant variety is a medium- to dark-green plagioclase hornblende schist with variable amounts of epidote; another common variety is a dark-gray graphitic schist; and the third type consists of a few masses of silvery white quartz sericite schist with pyrite cubes.

Triassic rocks

Unconformably overlying the Paleozoic schist is an apparently great thickness of altered volcanic and sedimentary rock belonging to the Triassic Applegate Group. This group of tightly folded rocks is exposed over a large area of the Klamath Mountains province. It extends from the Ashland batholith on the east to the Grants Pass diorite batholith on the west, and from the Tiller area on the north southward into California.

These rocks have been subjected to regional tectonic metamorphism, as well as contact alteration around the margins of numerous granitoid intrusives where schists and gneisses are developed. Altered sedimentary rocks in the group include argillite, quartzite, marble, chert, and some highly indurated conglomerate with stretched pebbles. The volcanic phase, which makes up possibly two-thirds of the section, includes altered lavas (basalt, andesite, and some rhyolite) and altered tuff and tuffaceous sediments.

Jurassic formations

The Dothan, Rogue, and Galice Formations, taken as a group, are somewhat similar in lithology and origin to the Applegate Group but are in most places less metamorphosed. These formations are older than the Nevadan intrusive rocks in the area. They are considered to be Late Jurassic in age, but the question as to which is the oldest has never been settled. The formations crop out in the area west of Grants Pass for a distance of about 30 miles as measured across the strike to a point near Marial on the lower Rogue River. The Galice and Dothan Formations extend northward from northern California to a point east of Roseburg where they are overlain by volcanic rocks of the Western Cascades. The Rogue Formation, sandwiched between the Dothan and Galice Formations near Galice (Wells and Walker, 1953), covers a smaller area and its extent outside the Galice quadrangle has not been determined accurately.

The Dothan Formation is the most extensive of the three Upper Jurassic units. It is a thick series of massive graywacke sandstone with some shale, conglomerate, chert, and pillow basalt. It lies west of the Rogue Formation from a point near Whiskey Creek to near Marial on the Rogue River. The distance measured across its strike in this area is about 15 miles. Type locality for the Dothan Formation is along Cow Creek near West Fork Station, formerly Dothan Post Office (Diller and Kay, 1924). This broad, north- to northeast-trending band of sedimentary rock extends through the province from northern California to a point 10 miles east of Roseburg, a distance of about 100 miles. There is only one break in continuity in the Canyonville and Riddle area, where it is interrupted by a down-faulted segment of Cretaceous sediments and is also intruded by a band of ultramafic igneous rocks. The Dothan has sustained considerable deformation and is in places known to be overturned.

The Rogue Formation lies east of the Dothan Formation and is best exposed along the Rogue River from the Almeda mine north of Galice to a point down stream from Whiskey Creek, a distance of about 7 miles along the river. It is composed of fine- to coarse-grained tuff, tuffaceous sediments, agglomerate, flow breccia, and andesitic lavas. These volcanic rocks are tightly folded and metamorphosed in varying degree from siliceous greenstone in the northern part of the area to an amphibole gneiss south of Galice. The Rogue Formation appears to be the more favored for mineral deposits of the three Jurassic units. Although no important deposits are known, scattered mineralization in the form of gold-bearing fissure veins and related sulfides are found in the formation from the Canyonville area south to the Illinois River. The

GENERALIZED STRATIGRAPHIC CHART FOR KLAMATH MOUNTAINS

ERA	PERIOD	FORMATION AND DESCRIPTION
CENOZOIC	Quaternary	Beach sands and marine terraces along coast. Bench gravels and alluvium along streams and glacial moraine and till. Auriferous gravels (in former stream channels).
CENOZOIC	Tertiary	Old gravels - on Klamath peneplain. Small intrusions of dacite porphyry and nepheline syenite. Nonmarine sedimentary rocks of late Eocene age in Medford area. Tyee Formation, Umpqua Formation } Eocene marine sedimentary rocks extend into Klamath Mountains from southern Coast Range.
MESOZOIC	Late Cretaceous	Marine beds in Gold Beach area. Hornbrook Formation - Marine beds in Medford-Ashland area and upper Grave Creek.
MESOZOIC	Early Cretaceous & Late Jurassic ?	Myrtle Group coast area { Rocky Point Fm., Humbug Mt. Congl., Otter Point Fm. Myrtle Group inland area { Days Creek Fm., Riddle Fm.
MESOZOIC	Late Jurassic	Nevadan orogeny - intrusive rocks: peridotite, serpentinite, gabbro, diorite, granite, pegmatite.
MESOZOIC	Late Jurassic	Galice Formation, Rogue Formation, Dothan Formation } May be overturned.
MESOZOIC	Triassic	Applegate Group - Metasediments, metavolcanics, and intrusives. Age of lower part uncertain.
PALEO-ZOIC		Pre-Triassic schist - age uncertain.

Almeda mine lies along the eastern contact of the Rogue Formation, and the Benton mine in diorite that intrudes it near the western margin.

The Galice Formation lies east of the Rogue Formation and is in reverse-fault contact with the Applegate Group. It is composed mainly of slaty siltstone, sandstone with occasional conglomerate lenses, and thick interbeds of lavas, tuffs, and tuffaceous sediments. All of these rocks have been subjected to low-grade regional metamorphism and the volcanics are generally altered to hard, silicified, chloritized, and saussuritized greenstones. The Galice Formation is dated as Early Late Jurassic (Oxfordian-Kimmeridgian) on the basis of fossils collected from 18 localities (Imlay, 1961).

Isoclinal folding and longitudinal faulting of the entire Dothan-Rogue-Galice sequence has resulted in confusion as to which of the formations is the youngest. Diller (1914), who first studied these rocks, believed that the Dothan Formation was on top. Taliaferro (1942) examined the section critically and concluded that the Dothan belonged at the bottom of the series, as did Wells and Walker (1953). Dott (1965) also presents evidence that the Dothan is the oldest of the series. On the other hand Irwin (1964), working in the Klamath Mountains of northern California, believes the Galice and Rogue are older and that the Dothan is post-Nevadan and has been moved into its present position by thrust faulting. Field investigations of this problem are continuing, but as yet no diagnostic fossils have been found in the Dothan Formation. A Late Jurassic K-Ar date of 149 ± 4 m.y. has been obtained from an essentially unaltered Dothan rhyolite flow (Dott, 1965).

Intrusive Rocks of the Nevadan Orogeny

Igneous rocks, ranging from ultramafic to granitic, and a few related pegmatites were emplaced mainly during the Nevadan orogeny in Late Jurassic time. Koch (1966) specifies an intra-Late Jurassic age for the Pearce Peak diorite near Port Orford by both stratigraphic and radiometric (K-Ar) methods (141 ± 7 m.y. and 146 ± 4 m.y.). It is generally accepted that the more mafic intrusives, peridotite and gabbro, were intruded earlier and the diorites and granites followed. A large part of the ultramafic rocks have been altered to serpentinite.

Not all of the intrusive rocks in the area are necessarily confined to the Nevadan orogeny. It is believed likely that some of the ultramafic rocks are significantly older. A few altered hypabyssal intrusives have been found in the Applegate Group as well as in the Galice-Rogue-Dothan sequence.

Distribution of the batholiths, stocks, and some of the smaller intrusive bodies of the Nevadan orogeny is shown in figures 34 and 37. Aside from the Ashland stock (or batholith), no permanent names for these igneous masses have been adopted in the literature. For the sake of convenience in describing the geology of the various mineralized areas, the intrusive bodies are named for some prominent geographic feature in the vicinity.

Post-Nevadan Formations

Late Jurassic-Early Cretaceous rocks

Rocks of Late Jurassic-Early Cretaceous age in southwestern Oregon fall into the "Myrtle formation" originally described by Diller (1898). This formation was raised to group status by Imlay and others (1959) and Koch (1966) to include several post-Nevadan Mesozoic units (see accompanying stratigraphic chart). The Myrtle Group is exposed along the western fringe and in various down-faulted blocks within the Klamath Mountains province. It consists chiefly of siltstone, sandstone, and conglomerate. Units within the group are similar lithologically, and differentiation is based largely on evolution of a mussel-like clam of the genus Buchia. Subdivisions of the Myrtle Group along the coast from oldest to youngest include the Otter Point Formation, Humbug Mountain Conglomerate, and Rocky Point Formation (Koch, 1966). Similar rocks exposed inland consist of the Riddle and Days Creek Formations (Imlay and others, 1959).

The Otter Point Formation, which is exposed in the Gold Beach area, consists of about 10,000 feet of mudstone, sandstone, bedded chert, tuff, and pillow lava. The rocks are steeply folded, but much less altered than the underlying pre-Nevadan Jurassic formations. Where they are exposed in the Port Orford area, the Humbug Mountain Conglomerate and the overlying Rocky Point Formation (dominantly mudstone and sandstone) have a combined thickness of about 9000 feet.

The Riddle Formation unconformably overlies pre-Nevadan Jurassic rocks along the South Umpqua River near the town of Days Creek and along Cow Creek near Riddle. The formation consists of dark gray siltstone with limestone lenses, chert-pebble conglomerate, and sandstone; at the type locality near Days Creek it is overturned and measures about 1088 feet in thickness.

The Days Creek Formation overlies the Riddle Formation concordantly at the type section along the South Umpqua River near Days Creek. The formation is composed of alternating dark gray sandstone and siltstone and measures about 800 feet thick in this area. It occurs in all areas where the Riddle is found, as well as in the Takilma-Waldo region east of O'Brien where it unconformably overlies the Galice Formation.

Late Cretaceous units

Near-shore marine deposits of sandstone, siltstone, and conglomerate with fairly abundant marine fossils of Late Cretaceous age are found as scattered remnants in the Jacksonville-Medford-Ashland area and as a downfaulted block on upper Grave Creek. These rocks belong to the Hornbrook Formation (Peck, Imlay, and Popenoe, 1956). The rocks are gently folded, but dip as much as 40° where the formation laps on the east flanks of the uplifted Ashland batholith. Some of the conglomerates of the Hornbrook Formation have been mined for their gold content, as at the Forty-nine Diggings placer on Bear Creek near Ashland.

Cretaceous rocks younger than the Hornbrook Formation were mapped and described by Howard and Dott (1961) between Cape Sebastian and Pistol River. A lower section of massive sandstone, with minor conglomerate at the base, is about 800 feet thick. The upper section of alternating shale and sandstone with a few distinctive fossils is about 600 feet thick.

Cenozoic Rocks

Tertiary formations

Tertiary rocks of the southern Coast Range province extend about half way into the Klamath Mountains as a large synclinal trough. On the east edge of the Klamaths near Medford, Tertiary rocks lap onto pre-Tertiary formations.

At Eden Ridge in southern Coos County and northern Curry County these rocks are Eocene marine sediments of the Umpqua and Tyee Formations. The underlying Umpqua Formation extends as far south as a point on the Illinois River near the mouth of Silver Creek, where it lies unconformably on the Dothan Formation and gabbroic rocks. The Umpqua Formation in this area is composed chiefly of well-bedded sandstone and siltstone with some massive conglomerate. Placer gold, derived by weathering of coarse Umpqua Formation conglomerate, has been mined in the Olalla Creek drainage east of Camas Valley. The overlying Tyee Formation, which extends as far south as Bald Knob on the north side of the Rogue River, consists of coal-bearing shales, sandstone, and conglomerate.

Tertiary sediments in the Medford-Ashland area consist of nonmarine, light-colored micaceous sandstone with subordinate shale, conglomerate, and coal. These rocks were formerly assigned to the Umpqua Formation of early Eocene age by Wells (1956). More recent mapping by Wells and Peck (1961) and plant identification by Brown (1956) raise the stratigraphic position of these rocks to the upper part of the Eocene. The sediments are overlapped on the east by the volcanic rocks of the Western Cascades.

Tertiary intrusive rocks

Tertiary intrusives are also found within the Klamath Mountains province. They consist of a few dacite porphyry dikes and the nepheline syenite of Mount Emily in southern Curry County. Some evidence of mineralization is found related to these Tertiary intrusives, but most of the gold and silver ore bodies appear to be associated with the igneous activity of the Nevadan orogeny.

Late Tertiary-Quaternary deposits

Remnant deposits of sand and gravel lie at various elevations above present-day streams in the Klamath Mountains and on marine terraces along the coast. These deposits give evidence of stillstand periods during uplift of the region, which probably began in Miocene time and has continued to the present. More recent sands and gravels occur on valley floors and along ocean beaches.

Stream deposits: Four categories of stream gravels, based on age and elevation, have been mapped by Wells and others (1949). Many of the deposits listed below have been the source of placer gold:

1. "old gravels" of Miocene or Pliocene age. These deposits lie on erosion surfaces at about 4000 feet elevation. The poorly sorted cobbles and pebbles are partially decomposed and firmly cemented in a sandy matrix. Examples are Gold Basin and York Butte.

2. "auriferous gravels of the second cycle of erosion" of late Pliocene or Pleistocene age. These are poorly sorted, partly cemented gravels that occur as detached remnants of former stream channels on divides of the present drainage system.

3. "Llano de Oro formation," or high-bench gravels of Pleistocene age. These include terrace deposits lying as much as 400 feet above the channels of the present-day streams that deposited them. Lower bench gravels occur in some areas as mapped near Galice by Wells and Walker (1953) and are evidence of continuing recent uplift of the province.

4. Recent alluvial deposits, in part fluvio-glacial in origin. The deposits consist of stratified gravel, sand, and silt on lower flood terraces, banks, and bars of the streams.

Marine deposits: Baldwin (1964, p. 34) and Griggs (1945) describe six major terraces ranging from sea level to as high as 1600 feet that were cut during sea-level changes from late Pliocene through Pleistocene time. The lower and more extensive terraces contain black-sand deposits that have been mined in the past for chromite, gold, and platinum. Marine terrace deposits are discussed more fully in this report under the heading of "Beach Placers."

ORE DEPOSITS

Host Rocks

According to Diller (1914, p. 24), the majority of lode-gold deposits of the Klamath Mountains occurs in greenstones, but some occur in diorites, some in metasedimentary rocks (argillites, quartzites, etc.), and a few in serpentinite*. Of the 20 mines that recorded the largest production in the area, 7 are in greenstone, 6 in metasediments, 3 in diorite, 3 in metagabbro, and 1 in serpentinite.

Attempts to specify the preferred type of host rocks in this area can lead to error, because there are generally several different rock types in the immediate vicinity of a deposit, and in a few cases the veins cross greenstone-metasedimentary rock contacts or lie in sheared contacts of these or other rocks.

Diller (1914, p. 24) notes that a great many mines are relatively near serpentinite bodies, but that the veins are usually faulted off and rarely extend into this rock.

Types of Mineralization

Small quartz fissure veins with mixed sulfides and some free gold are the most common type of deposit. Sulfides in these veins usually include pyrite, chalcopyrite, and arsenopyrite with less common galena, pyrrhotite, and sphalerite. Gold is often mechanically intermixed with these sulfides, and in places sulfide concentrates are exceptionally rich in gold. In a few deposits, arsenopyrite has been mistakenly identified as a gold-telluride because of an abnormally high gold content. Calcite often occurs with quartz as a subsidiary gangue mineral. Fissure-vein deposits are variable in size, but generally small. Most veins prospected and mined in southwestern Oregon are less than a foot thick. Some form a network of multiple quartz veinlets. A few veins are more than 10 feet wide where they have been stoped. Some of these wider veins generally contained segments of wall rock and/or brecciated areas not completely replaced by vein material. Average thickness of the Greenback vein, the best producer in the area, was about 20 inches, but it ranges from less than 6 inches to more than 4 feet. This vein was mined to about

* The terms "serpentinite" and "serpentine" are used interchangeably in the mine descriptions.

1000 feet down dip and about 700 feet horizontal distance. Ore zone widths of 20 to 30 feet of sheared mineralized diorite with multiple, sulfide-bearing quartz veins have been worked in the Benton mine. Development at the Benton mine explores the vein system for 1500 feet along the strike and to about 500 feet of depth. The Ashland mine fissure vein, which has been worked to a depth of 1200 feet down dip and for about 2200 feet horizontal distance, ranges in thickness from 2 to 12 feet.

Two other types of mineralization deserve mention because of their greater size. These are: 1) mineralized shear zones in which sulfides are disseminated in talcose serpentinite, and 2) broad zones of altered, usually silicified, rock which has been replaced in varying amounts by disseminated to massive sulfides. An example of a mineralized shear zone was exposed by exploration work on the Wild Rose claim of the Ida Group on Louse Creek in the Greenback-Tri-County area, where iron-stained, sheared talcose serpentinite of a little more than 10 feet wide assayed 0.14 oz. per ton gold.

Examples of broad zones of rock alteration and sulfide-impregnation include the Forget-Me-Not mine in the Greenback-Tri-County area of southern Douglas County, where low-grade mineralization is found in altered greenstone adjacent to serpentinite over a width of about 100 yards, and the Almeda mine on the Rogue River in the Galice area, where an altered mineralized zone 200 feet wide occurs in a contact zone of greenstone and slate. The alteration is associated with a sill-like body of dacite porphyry.

A number of the mines have been small, high-grade, near-surface, pocket-type deposits that were operated for only a very short time. A few of these are included with the more important deposits in order to give the reader a better insight into the true nature and distribution of gold deposition in the Klamath Mountains.

Depth of Oxidation

The depth of oxidation in southwestern Oregon deposits is relatively shallow. Most of the area has sustained recent uplift and extensive erosion, so that it is not uncommon to find fresh sulfides exposed at the surface. Diller (1914, p. 25) reports that the lower limit of oxidation is normally less than 100 feet below the surface but in a few places exceeds 200 feet. Fracturing of the ore, which allows ground water to percolate more freely, hastens the oxidation and leaching process. Some residual enrichment of gold takes place in the oxidized zone by leaching of the more soluble elements. Under special conditions discussed earlier, supergene enrichment may have taken place to form the rich pocket deposits.

Black sand mining (left) **on the beach back from the high-tide line near Gold Beach. The mill** (lower) **that processed the beach sand, was close by.**
—Curry County Historical Society

GOLD MINING AREAS OF THE KLAMATH MOUNTAINS

For convenience, the principal concentrations of gold mines have been arbitrarily grouped into 10 separate areas (see figure 34). In general, the areas have distinctive geologic features and the deposits within the areas have definite similarities.

Information given for each area includes location, geology and mineralization, history and production, and facts about the placer mines and the principal lode mines. Following the discussion of each area, the lode mines and prospects of any significance are listed alphabetically and briefly described in outline form.

The areas are listed below in the order they will be discussed. Since it was not feasible to arrange them according to their productive rank, as was done for the Blue Mountains region, they are presented here in geographic sequence from west to east, with a few salient geologic features.

1. Beach placers: Black-sand deposits in ancient elevated beach terraces, present-day active beaches, and under-water deposits along the coast of Coos and Curry Counties.

2. Salmon Mountain-Sixes area: Mineralization in altered gabbro, diorite, and serpentinite which intrude the Galice (?) Formation.

3. Mule Creek-Bolivar area: Mineralization in altered volcanics of the Dothan Formation intruded by diorite and gabbro.

4. Illinois-Chetco area: Mineralization along contact zones of serpentinite and Galice Formation in vicinity of gabbro-diorite complex.

5. Galice area: Mineralization in and near narrow quartz-diorite bodies and in veins in altered volcanics of the Rogue Formation.

6. Silver Peak area: "Big Yank lode" type of mineralization (sulfides in barite gangue and sericite schist) in altered contact zone of Dothan Formation and Galice or Rogue (?) volcanics.

7. Greenback-Tri-County area: Mineralization adjacent to faulted serpentinite-intruded zone between the Grants Pass and the Evans Creek-Cow Creek diorite bodies.

8. Gold Hill-Applegate-Waldo area: Small, rich, pocket-type deposits in altered sediments and volcanics of the Applegate Group.

9. Myrtle Creek area: Quartz veins in metagabbro-diorite.

10. Ashland area: Contact aureole of Ashland batholith and Applegate Group.

BEACH PLACERS

Location

Areas along present beaches, mainly from Coos Bay south to California, and the uplifted terraces of Pleistocene beaches extending as much as 5 miles inland have been prospected and in a few places worked for the gold and platinum content of their black-sand horizons (figure 38).

Geology

The placer deposits are wave-concentrated layers of heavier black sands composed largely of magnetite, chromite, ilmenite, garnet, zircon, and other common heavy minerals with small amounts of extremely fine particles of gold and platinum. The original source of these materials was the ultramafic rocks and mineralized veins of the Klamath Mountains. The heavy minerals were eroded by streams, carried to the sea, and deposited in most abundance near the mouths of rivers. Some of the material was probably deposited first in Tertiary beds, and has come to its present position in Pleistocene terraces or in present-day beaches through several cycles of erosion.

Pleistocene terraces have been recognized at six levels, the highest between 1500 and 1600 feet. Baldwin (1964, p. 31) has traced some of the terraces for many miles and finds a considerable variation in elevation and altitude of the deposits caused by warping during or after their deposition. The terraces are the combined result of gradual uplift of the Klamath Mountains and sea-level changes effected by the advance and retreat of Pleistocene continental glaciation.

The terraces are widest and the elevated sands most extensive in the area between Coos Bay and Port Orford. Most of the black sands occur on the three lowest terraces below 350 feet elevation. Griggs (1945), in his detailed report on chromite-bearing sands of the southern Oregon coast, mapped and named these three levels from highest to lowest: Seven Devils terrace, Pioneer terrace, and Whiskey Run terrace.

History and Production

Gold was discovered on southern Oregon beaches in 1852 and for a while a few areas were worked at a profit. About 20 years later similar gold-bearing black sand was discovered on some of the ancient elevated beaches. Spreen (1939, p. 11) reported on the beach placer mining as follows:

> "The earliest beach mining in Oregon, of which there is any record, was in 1852 at the mouth of Whiskey Run, a few miles north of the Coquille River. A few half-breed Indians discovered gold-bearing sand on the beach and worked their placers quietly for a portion of two summers, and undoubtedly saved a considerable amount of gold. In the summer of 1853 they sold out to the McNamara Brothers for twenty thousand dollars. Pans of black sand from this claim yielded from eight to ten dollars. It has been estimated that during the fifties and sixties more than one hundred thousand dollars were taken from this one claim. The town of Randolph sprang into existence over night on Whiskey Run. In the fall of 1853 more than a thousand men were there. Its lodging houses could not accommodate all the miners so Randolph became a tent city. It was a typical mining camp. Poker tables were adorned with cocked pistols; and whiskey, straight and mixed, went gurgling down the throats of those who sat around waiting for spring. Books were opened on "bedrock" credit, and all went merry as a marriage ball. But the sand, outside the one claim, did not pay."

Pardee (1934) records the production of beach placers from 1903 to 1929 as $60,615 total value of

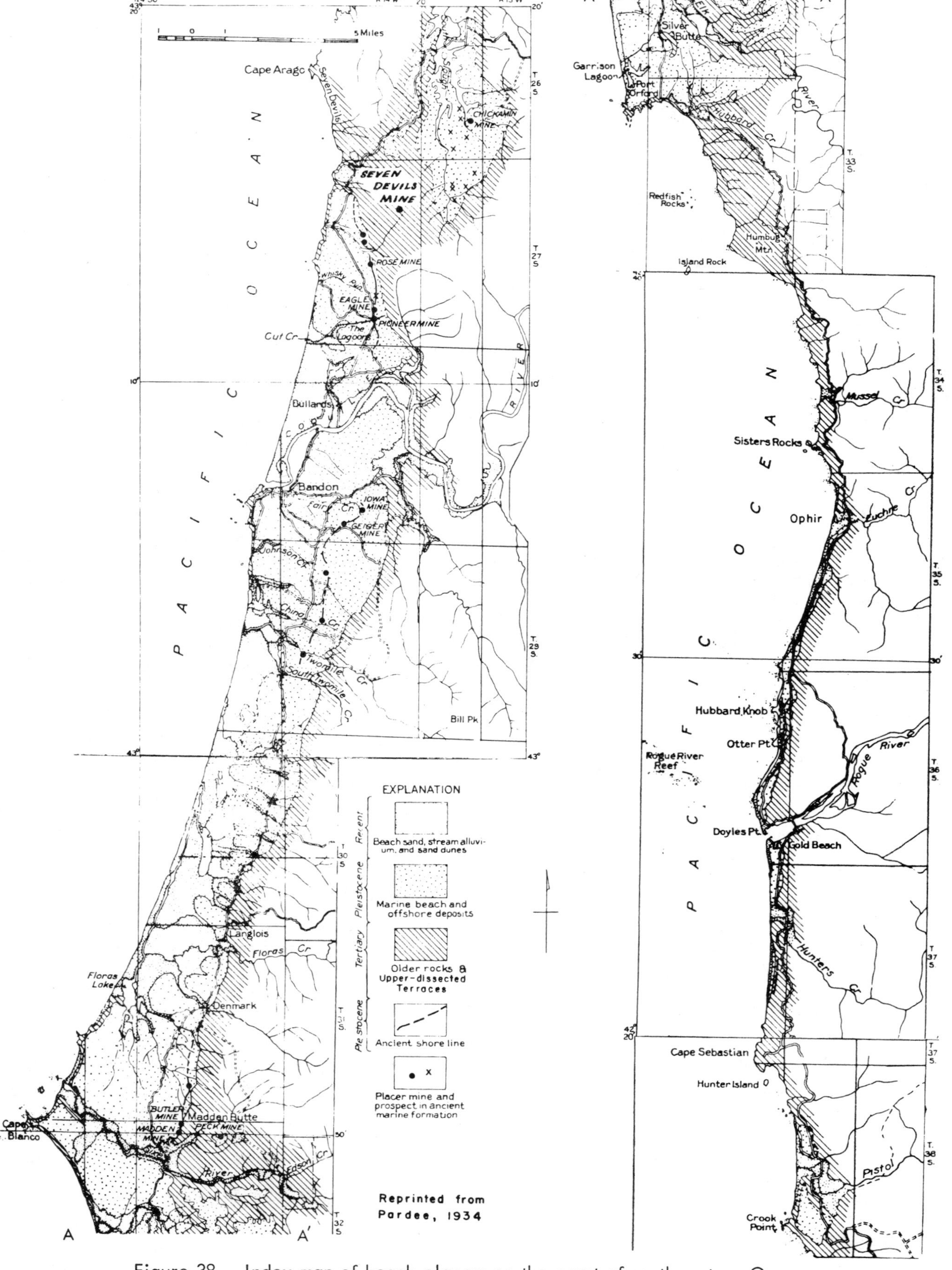

Figure 38. Index map of beach placers on the coast of southwestern Oregon.

gold and platinum from several small operators. The platinum amounted to $1,948. Both the gold and platinum were valued at about $20 per ounce so that the amount of gold produced for the 26-year period was 2848 ounces compared to about 100 ounces of platinum metals. Records of earlier production from beach placers are not available.

Beach mining

Placer operations on the beaches were itinerant because the sand deposits tend to change with the season. After a heavy storm some of the beaches may be coated with a new deposit of black sand.

Beach mining was usually done during low tide by scraping up the thin layers of black sand left by the waves. This natural concentrate was carried to a point above high tide and then washed through sluice boxes and over amalgamation plates. Selective mining of the beaches in this fashion paid the miners well in certain of the better areas. Some mechanical slushers and horse-drawn scrapers were used to speed the sand-skimming process.

Terrace mining

In addition to Whiskey Run, a few of the operations attempted on the older elevated beach terraces left evidence of fairly extensive development.

The Pioneer mine is one of the better known. It is located about 6 miles north of Bandon on Cut Creek and is reached by way of the Seven Devils Road. The mine has a history of numerous attempts to set up a paying operation. Because of extensive overburden, more than 50 feet on the average, and various other causes, only the earlier operation was profitable. Pardee (1934, p. 38) describes the deposit as follows:

> "The pay streak is a layer of black sand 3 feet or more thick, the richer part of which was mined through drifts said to have been made more than 60 years ago. Some of the mining timbers as well as an occasional huge log of drift wood are exposed by the present workings. Samples of the black sand remaining averaged about 3 percent of magnetite and 55 percent of chromite and ilmenite together. Gold and platinum alloy were being recovered by sluicing. A sample of the platinum alloy as determined by a spectrographic examination by George Steiger in the laboratory of the United States Geological Survey is composed of a relatively very large amount of platinum and smaller amounts of iridium and ruthenium. It contains, in addition, a possible trace of rhodium but no osmium or palladium....A sample from a hole 3 feet deep at one place contained 4 percent of magnetite and 60 percent of chromite and ilmenite. It is said that the tailings in the lagoons contain unrecovered gold and platinum..."

The Pioneer mine was developed by three long tunnels (now caved), the longest of which was 1340 feet.

The Eagle mine, situated just north of the Pioneer, was worked in a similar manner by drifting and sluicing. The deposit as described by Hornor (1918) contains a bed of black sand 200 to 250 feet wide and several hundred feet long, with a lenticular cross section 6 to 8 feet thick near the middle and thinning toward the edges.

The Seven Devils (Last Chance) mine in sec. 10, T. 27 S., R. 14 W., about 10 miles north of Bandon had been mined to only a limited extent in the early days for gold and platinum, but was worked for its chromite content by the Krome Corp. in 1942-43. Griggs (1945) indicated in excess of a million cubic yards of black sand in two adjacent ore bodies at the mine. Of this, about 200,000 cubic yards were mined and processed for chromite content. The ore averaged about 6 percent Cr_2O_3. Gold and platinum production was not reported but it is likely that these values were salvaged. Krome Corp. operated a 1900-ton plant with 16 concentrating tables that were later replaced by 64 Humphreys spiral concentrators. Short life of the operation was due to government suspension of domestic chrome-purchasing agreements (Mining World, January 1944, p. 7-12).

Several other operations which have had a similar history are indicated on the map (fig. 38). Natural groupings occur around the mouths of the Coquille and Sixes Rivers, indicating the probable feeding

George W. Billings in April 1932 at the mouth of his mine on Mule Mountain. He named it The Alva Vein for his friend Alva Marsters of Roseburg. Turn to page 186 for the story. See page 190 for specifications.
—Curry County Historical Society

source of the gold. When examined by Pardee in 1931, mining equipment was present in the open pit of the Chickamin mine in South Slough. Other terrace placers active at that time were the Pioneer, Eagle, and Madden.

Current and future operations

Various attempts have been made from time to time and will probably continue to be made to establish profitable operations on the intriguing black sands. A salvage operation on black-sand concentrates was in progress at the Pioneer mine in 1964, and black sands were being mined in a small way on the beach at Cape Blanco during 1966.

Offshore deposits on the continental shelf may represent a more interesting potential resource than the beach and terrace deposits. The U.S. Geological Survey and Oregon State University have been studying offshore deposits under a joint research contract. Their investigations during 1966 and 1967 disclose that concentrations of gold-bearing black sands do occur in the surface sediments on the continental shelf off the coast of southwestern Oregon. The transitory nature of the offshore sands makes them difficult to evaluate. Any future operation to process these deposits will probably have to develop markets for most of the economic minerals contained in order to realize a profit (Libbey, 1963).

SALMON MOUNTAIN-SIXES AREA

Location

The Salmon Mountain-Sixes area is in southern Coos and northern Curry Counties in Tps. 32 and 33 S., Rs. 12, 13, and 14 W. (figure 39). It may be described as a fairly narrow, west-trending belt about 18 miles long which lies between Sixes River on the north and Elk River on the south and stretches from the South Fork of the Coquille River nearly to the coastal plain. The main zone of mineralization extends from Johnson Creek across Salmon Mountain and Rusty Butte.

Geology and Mineralization

The area was originally mapped by Diller (1903) in his study of the Port Orford 30' quadrangle. Later work by Wells and Peck (1961) and by Dott (1966) has modified some of Diller's interpretations, particularly in regard to strata mapped as "Myrtle Formation," which were found to include rocks of the Galice(?) Formation.

The area is underlain by graywacke and argillite of Jurassic age intruded by serpentinite, gabbro, diorite, and dacite-porphyry dikes. The Jurassic sediments and plutonic rocks are unconformably overlain by conglomerate, sandstone, and mudstone of the Myrtle Group of Early Cretaceous age. Patches of lower Eocene marine sandstone and shale unconformably overlie the Mesozoic rocks.

The gold mineralization is believed to be related to scattered quartz veins which occur both in diorite and in the altered Jurassic sediments. The area was considered to be an important mining region by Diller (1903), who called it the "gold belt of the Port Orford quadrangle."

History and Production

Diller (1903, p. 5) states:

> "Nearly all of the gold which has thus far been obtained in the Port Orford quadrangle has come from placer mines, some of which are along beaches in marine deposits and the rest in river gravels, especially along the South Fork of the Sixes and at the heads of Salmon and Johnson Creeks, with a smaller area at the head of Boulder and Rock Creeks near the south end of Iron Mountain. There is one quartz mill in the region.
>
> "The gold belt of the Port Orford quadrangle has long been the most active mining region of the Oregon coast. It has yielded considerable gold in the past, and is yet a moderate producer. The total product from the quadrangle since 1852 is probably not far from a million....
>
> "Placer mines were once active along Johnson Creek throughout the greater part of its course, and paid moderately, but in the severe weather of the spring of 1890 landslides so filled up the stream bed that mining has since been unprofitable."

The most successful placer mines on Johnson Creek, according to Diller, were near the head of the stream close to the belt of dacite porphyry which crosses the divide toward the Salmon Mountain mine.

Significant gold placer production has also come from lower Johnson Creek. In recent years the Big Slide placer near the northeast corner of sec. 34, T. 32 S., R. 12 W. has been worked in a small way on a seasonal basis.

Numerous placer operations were active along the South Fork of Sixes River during the late 1800's. Many of these worked bench gravels from about 50 feet to as much as 130 feet above the present stream. Very little mining was done on the Sixes River above the mouth of South Fork, and activity on the South

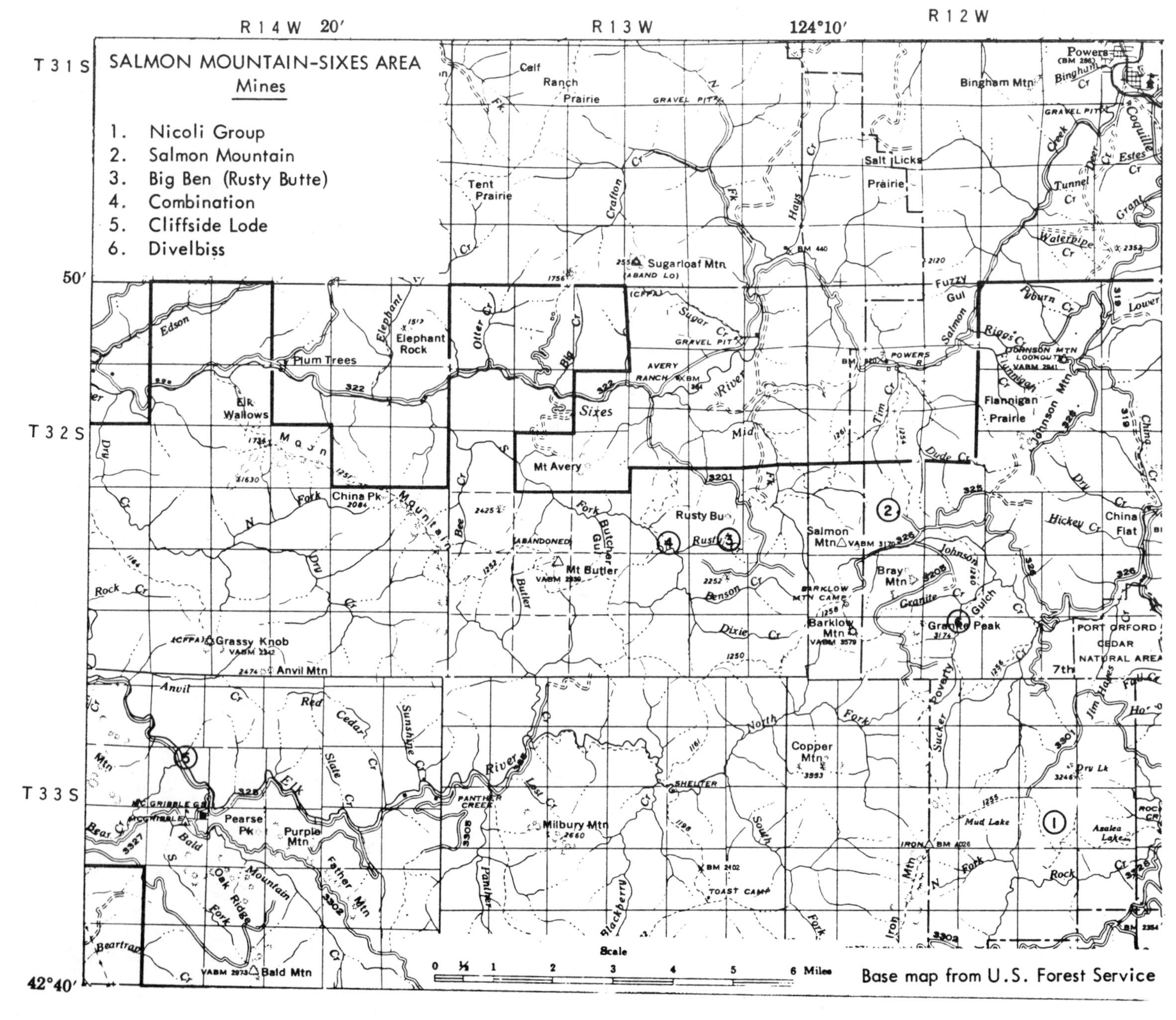

Figure 39. Index map of the Salmon Mountain - Sixes area.

Fork was limited to the area below Rusty Creek.

Production from lode mines in the area has been negligible, except from the Salmon Mountain mine which was a combined lode and placer operation. Production from this mine prior to 1936 was estimated to have been between $75,000 and $100,000.

Principal Lode Mines

Salmon Mountain mine: Judging from reports by Diller (1903) and by the Department (1940, p. 44-46), the Salmon Mountain mine (map no. 2) consisted of a combined lode and placer operation in a landslide area on the northeast slope of Salmon Mountain near the head of Salmon Creek. The workings, which included a large open pit and four tunnels, were mainly in fragmental slide material. As described by Diller (1903), "The rock is dark, often purplish or greenish, sometimes brecciated, much fractured, and easily goes to pieces. Although much altered, it retains traces of its ophitic structure which connects it with the basalts. Near the upper limit of its exposure, above the bulkhead, it is more solid and is associated with a rock rich in glaucophane, with sandstones and indurated shales bounding it on both sides."

This is apparently the highly altered basalt and sediments of the Galice(?) Formation, as mapped by Dott (1966). A dacite porphyry dike strikes north through the mine area. To the east are small bodies of serpentinite. Wells and Peck (1961) map a west-trending fault just south of the mine.

Diller (1903) reported that the gold of the mine appeared to be derived from small quartz veins such as had been prospected in the immediate vicinity. At the time the property was visited by the Department (1940), the tunnels were caved, but the west tunnel was said to have cut a number of quartz stringers. Production prior to 1936 was estimated to have been between $75,000 and $100,000.

No other lode mines in the Salmon Mountain-Sixes area are known to have had significant production. Other mines and prospects in the area which may have potential value or give insight into the characteristics of mineralization in the area are described in outline form below.

Lode Mines of the Salmon Mountain-Sixes Area

Big Ben (Rusty Butte) mine — Salmon Mt.-Sixes Area, 3

Location: Curry County, SE¼ sec. 23, T. 32 S., R. 13 W., near head of Rusty Creek. Probably same as "Harrison's" claim reported by Diller (1903).

Development: A few surface cuts and caved adits.

Geology: Northeast-striking shear zone in argillite about 12 feet wide, with small quartz and calcite veins and ore minerals including gold, silver, galena, pyrite, chalcopyrite, and arsenopyrite.

Production: Production unrecorded, but probably "a few thousand dollars in gold."

References: Diller, 1903; Department mine file report, 1950.

Cliffside Lode mine — Salmon Mt.-Sixes Area, 5

Location: Curry County, NE¼ sec. 17, T. 33 S., R. 14 W., about 300 feet elevation.

Development: Surface cuts.

Geology: Vertical quartz vein strikes N. 60° W., occurs in diorite. Pyrite and chalcopyrite are reported in the diorite. Assay values spotty.

Production: Discovered 1938. No production.

Reference: Department mine file report, 1938; Department Bulletin 14-C, v. 1, 1940:82.

Combination mine

Salmon Mt.-Sixes Area, 4

Location: Curry County, SE$\frac{1}{4}$ sec. 22, T. 32 S., R. 13 W., at about 900 feet elevation.

Development: A lower 90-foot tunnel and an upper 15-foot tunnel.

Geology: Quartz veinlets in N. 20° W. nearly vertical 10- to 20-foot-wide shear zones in argillite containing spotty pyrite, galena, and sphalerite. Low gold and silver values.

Production: Small amount of hand-sorted, high-grade sulfide pods probably packed out. No other production.

References: Department mine file report, 1950, and Department assay records.

Divelbiss (Coarse Gold) mine

Salmon Mt.-Sixes Area, 6

Location: Coos County, SE$\frac{1}{4}$ sec. 29 and NE$\frac{1}{4}$ sec. 32, T. 32 S., R. 12 W., on Poverty Gulch about 2000 feet elevation.

Development: Open cut, 200-foot tunnel, and 32-foot shaft.

Geology: Ferruginous seamy quartz mass with much manganese oxide and sulfides including pyrrhotite, pyrite, galena, and sphalerite occurs on contact of dacite porphyry dike and slates.

Production: Small production from five-stamp mill about 1900. Group of claims included a placer on Poverty Gulch.

References: Diller, 1903; Department Bulletin 14-C, vol. 1, 1940:39-40; Department assay records.

Nicoli Group

Salmon Mt.-Sixes Area, 1

Location: Coos County, E$\frac{1}{2}$ sec. 23, T. 33 S., R. 12 W., at 3000 feet elevation.

Development: In 1937 three tunnels contained a total of 425 feet, plus an open cut 50 feet deep.

Geology: Two quartz veins in "gabbro," 1 to 5 feet wide, striking N. 65° E. and N. 20° W. produce some high-grade ore at their intersection. The veins also contain calcite, pyrite, and gold.

Production: No information.

Reference: Department Bulletin 14-C, v. 1, 1940:43.

Salmon Mountain mine — Salmon Mt.-Sixes Area, 2

Location: Coos County, E½ sec. 19, T. 32 S., R. 12 W., about 2100 feet on north slope of Salmon Mountain.

Development: One hydraulic cut and four caved tunnels which totaled about 1000 feet.

Geology: Country rocks are altered basalts and sediments of Galice(?) Formation, with nearby diorite and serpentine. A probable west-trending fault and a landslide area.

Production: Beginning in 1885 was operated for 13 winters as a hydraulic placer. By 1937 combined lode and placer production was estimated at between $75,000 and $100,000.

References: Diller, 1903; Department Bulletin 14-C, v. 1, 1940; Dott, 1966:85-97.

Mule Mountain Mine Group

Curry County

—Archives, Curry County Historical Society (undated)

42°43'52"N 123°54'26"W. Summit 2,969 ft.
—MARIAL TOPO

This is a general history of the Mule Mountain District as learned from Mr. George W. Billings and others. Mr. Billings is one of the oldest settlers in the district and at the present time is postmaster at Marial on the Rogue River.

The early history of the Rogue River in the vicinity of Mule Mountain District begins with the advent of the Placers miners in the 1850's. Due to the Rogue River being largely confined in box canyons, having few gravel bars, the Placer gold mining was confined to the small creeks. About 1981, Mr. John Billings and his son George did some prospecting for gold quartz.

The real beginning of this district occurred when a railroad engineer [builder of the Astoria and Columbia River Railroad – later S.P. & S.] by the name of [Andrew B.] Hammond, of Portland, Oregon, wrote offering to grubstake his friend George W. Billings to go to Alaska to prospect for Placer gold.

Mr. Billings was undecided as he had always held the belief since he was a boy that there was gold in Mule Mountain. Before replying to Hammond's offer, he made a final trip to the top of Mule Mountain and found a piece of rich looking quartz at the foot of the stump. The accompanying picture (*see page 181*) was taken in April 1932 and shows George W. Billings standing at the mouth of the drift at the point of discovery of what he named the Alva Vein on Mule Mountain. Mr. Billings said the piece of quartz he found on the surface assayed at $350.00 per ton.

This discovery caused Mr. Billings to turn down the trip to Alaska. He remained and located the claim which he called the Alva in honor of his friend Mr. Alva Marsters, a druggist and banker on Roseburg.

In 1900, a Mr. E. B. Burns of Portland, made a trip to Mule Mountain and met Mr. George W. Billings and liked the looks of the Mule Mountain ore.

Mr. Burns and George Billings had placed a shipment of 45-tons of ore from the Alva stope to the river and took it down the Rogue by boat to Gold Beach. From there is was shipped by the steamer *Copper Queen* to the Tacoma smelter in 1900. Thus, thirty-two years ago the first ore was shipped from this district. The returns from the smelter were $45 gold per ton.

Following this, Mr. Burns interested some of his friends in Portland to join him and a deal was made with George W. Billings to buy four claims. These Portland men were, Mr. Charlie Ladd of the Ladd Bank; Fred H. Green; E. B. Burns. They purchased the Paywell and Alva, Milner and Keystone claims. The consideration was $15,000 plus a 45 percent royalty on all ore mined with an assay value of over $100 per ton. No royalty was to be paid on ore under $100 per ton.

Mr. Burns provided transportation on the Rogue River by using a stern wheel steamer. By this means, they brought the equipment for a small saw mill up to Burns Creek, named for E. B. Burns.* On one of the trips up the Rogue, they found the saw mill much in need of a piece of steel cable.

The ship's Captain robbed his boat of a cable and substituted a manila line for his return trip. As the boat went down river, it reaches bad water known as Nail Keg Riffle (rapids). The rope broke at a critical time and control of the steam boat was lost. The stern struck the rocks and tore off the paddle wheel.

In this wreck on June 4, 1903, Mr. Burns was drowned. Some time later, Mr. Fred Green died and this group passed out of the picture.

The next group to become interested was composed of Judge Van Zante, John Milner and Fred King of Portland. They organized the Alpine Mining and Development Company. John Milner was the moving figure in this firm and did dome development. Their efforts stopped with his death for whom the Milner claim is named.

The next activity, about 1911, was a lease made to Mr. Hammond of Portland. He brought in and erected 2 stamp mills at the mouth of Burns Creek, then he built an aerial tramway for the Paywell Vein. Mr. Hammond's operations were carried on under lease and it is reported that he took out between $50,000 and $75,000 in gold bullion. Then came the war and the property lay inactive for some time. □

* Burns Creek is not mentioned in Geodetic Survey –editor

MULE CREEK-BOLIVAR AREA

Location

The Mule Creek-Bolivar area lies in northeastern Curry and southeastern Coos Counties, in Tps. 32 and 33 S., R. 10 W. (figure 40). The area is very steep and rugged and has only limited access. Elevations range from 350 feet at the mouth of Mule Creek to 4319 feet on Bolivar Mountain, only 6 miles to the northeast. The principal streams are Mule Creek, which enters the Rogue River at Marial, and the West Fork of Cow Creek which flows into the Umpqua River system.

Geology and Mineralization

The area is underlain by a broad, northeast-trending band of metamorphosed volcanic and igneous rocks of probable Late Jurassic age. These rocks have been mapped (reconnaissance only) as "volcanic rocks of the Dothan Formation" by Wells and Peck (1961). Bordering them to the southeast is a thick, folded section of interbedded graywacke sandstone and siltstones of the Dothan Formation. To the west is a fault contact containing a narrow zone of highly sheared serpentinite bordered by a fossil-bearing, folded Early Cretaceous marine section of Riddle and Days Creek Formations (Myrtle Group), which are in turn overlain unconformably to the west by gently dipping Eocene marine and continental sediments of the Umpqua and Tyee Formations.

Mineralization is localized in the altered lavas (greenstones) of the Dothan Formation and associated diorites and gabbros. It consists mostly of small quartz veins containing some sulfides (pyrite, minor chalcopyrite, and galena) and free gold. Altered rock zones contain a small amount of disseminated sulfides. Some copper mineralization is present.

Butler and Mitchell (1916, p. 77) describe mineralization in the Mule Creek district of Curry County as follows:

> "In general, it may be said that two classes of lode deposits occur in this district. One of these takes the form of rather narrow, considerably faulted, high-grade quartz veins containing more or less free gold. Important examples of these are the Lucky Boy (Tina H), Paradise, and Big Devil's Stairs veins. While some of the deposits of this type have been profitably mined, their small size, faulted condition, and irregularly distributed values are factors which mitigate against their profitable development. The ore contained therein is sometimes so rich, however, that numerous attempts to mine them have been made, and doubtless will be made in the future, regardless of the difficulties and discouraging features already mentioned.
>
> "The second type of deposit found in the area under consideration consists of mineralized shear-zones of considerable width and of relatively low grade. Good examples of these are the so-called "Iron Dike" of the Red River Mining Co. and the Excelsior vein on the south side of Rogue River. These are wider than the deposits first mentioned, and have not been subjected to much faulting, so are decidedly easier to mine. The ore is comparatively low grade, and much of it is in the form of sulfides, so the shear zones have not received the attention that has been accorded to the small quartz veins. It is believed, however, that if mined on a sufficiently large scale, they may prove remunerative, and they certainly seem deserving of careful investigation."

The so-called "Iron Dike" mentioned by Butler and Mitchell (1916, p. 83) is described as a shear zone in greenstone from 10 to 50 feet wide and averaging about 20 feet. It contains numerous narrow quartz veinlets, visible chalcopyrite, and lesser amounts of pyrite. Samples are said to average $3 per ton in gold. The shear zone trends northeasterly, and is nearly vertical. It is reported to traverse the

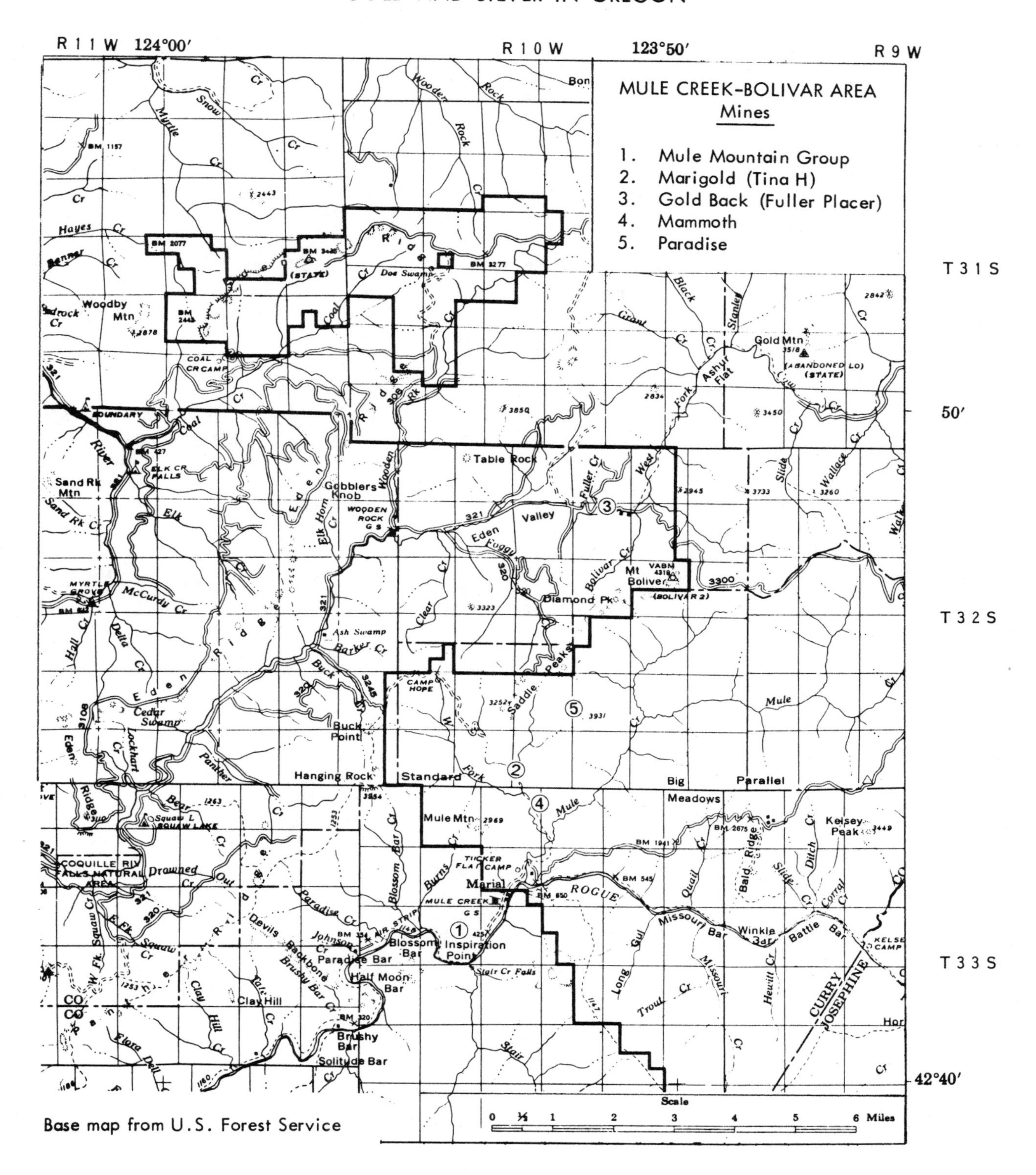

Figure 40. Index map of the Mule Creek - Bolivar area.

length of six lode claims which were a portion of the Red River Gold Mining Co. holdings on the west side of Mule Creek in sections 4 and 9, T. 33 S., R. 10 W.

No further reference to the Excelsior vein on the south side of the Rogue River was found.

History and Production

According to Butler and Mitchell (1916), the first mining in the Mule Creek area was done in 1891 by John Billings. Attention was first given to the placers along Rogue River, especially on the Red River bar, half a mile below Mule Creek. The search for sources of the placer gold soon resulted in the discovery of a number of veins.

In 1914, when the area was visited by Butler and Mitchell (1916), four lode-mining properties had been developed with some reported production. These were: the Paradise mine, the Lucky Boy (later known as Marigold and Tina H), Red River Gold Mining Co. claims, and the Mule Mountain group of mines. A two-stamp mill and cyanide plant were operating at both the Lucky Boy and Mule Mountain properties.

It is interesting to note that most of the heavy mining equipment used in the area was brought in by way of the Rogue River. Several sizes of placer pipe as large as 24 and 30 inches in diameter were brought down the river in boats from Grants Pass for use at the Red River Gold Mining Co. placer. Parts of the two-stamp mill set up at Blossom Bar to mill ore from the Mule Mountain group were brought up the river from Gold Beach on a barge, using a cable and hand winch. A five-ton mortar box for this mill, transported down the river from Grants Pass in a specially built river boat in 1910, may still be seen at Blossom Bar (R. L. Venner, oral communication, 1967).

Production from the area has been small. It has come mainly from early-day gold-placer operations on lower Mule Creek, Rogue River bench gravels below Mule Creek, and along upper West Fork of Cow Creek. Production records for the area are incomplete.

There are no important gold-lode producers in the Mule Creek area. Total production of the Mule Mountain Group, which was derived from at least three separate veins, may have been as much as $60,000. The Marigold (Lucky Boy or Tina H) produced about $50,000. Other mines in the area, although reported to have some high-grade specimen ore, had very little production.

Placer Mines

One of the larger placer operations in the area was that of the Red River Gold Mining Co. described by Diller (1914) and by Butler and Mitchell (1916, p. 121-123). The operation worked the bench gravels along the northwest side of Rogue River and west side of Mule Creek near its mouth. Water for the hydraulic operation was supplied by an $80,000 flume from the right or main fork of Mule Creek, giving a 180-foot head. Several acres were reportedly mined but total production was not reported. One operator reported the gravel ran about 9 cents per yard and material passing through his grizzly into the sluice boxes averaged 18 cents. Diller described the deposit as 30 feet of fairly coarse gravel on a bench 20 feet above the river that is capped by about 35 feet of fine, sandy material. The principal period of operation was during the late 1800's and early 1900's. Floods of later years have obliterated practically all evidence of this mining activity.

No. 3 spotted on the map in southern Coos County represents placer-mining activity that took place on upper West Fork of Cow Creek in the N½ of sec. 10, T. 32 S., R. 10 W., near Fuller Creek and on down West Fork near Sweat Creek in sec. 36, T. 31 S., R. 10 W. Some placer mining done between 1936 and 1941 at the Gold Back placer in sec. 10 was reported by Ray Treasher in 1941 (unpublished file report). At that time a low bench gravel was being hydraulicked by two giants in a pit about 200 feet in diameter. The report indicated a sizeable area of gravel from 10 to 75 feet in depth, but did not record values.

No lode deposits are known north of the drainage divide in this area that might be the source of placer deposits in upper West Fork of Cow Creek; however, small pieces of angular placer gold panned from the bend of the creek in the northwest quarter of section 10 suggest a probable nearby source.

Lode Mines of the Mule Creek-Bolivar Area

Mammoth mine

Mule Creek-Bolivar area, 4

Location: Curry County, NW$\frac{1}{4}$ sec. 3 and NE$\frac{1}{4}$ sec. 4, T. 33 S., R. 10 W., about 1600 feet elevation.

Development: There are 350 feet in two adits, plus a 12- by 25-foot stope.

Geology: A northeast-striking, southeast-dipping quartz vein 3 inches to 2 feet thick occurs in altered hornblende diorite or metagabbro country rock. Owner reported that 43 samples average assay value of $55.30 per ton.

Production: Four tons were milled in the Tina H stamp mill and 8 tons in an arrastra near Mule Creek below the mine. Recovery poor.

Reference: Department mine file report, 1946.

Marigold (Tina H) mine

Mule Creek-Bolivar area, 2

Location: Curry County, SW$\frac{1}{4}$ sec. 33, T. 32 S., R. 10 W., at about 1000 feet elevation.

Development: Total underground work is about 450 feet in three drifts.

Geology: Free gold and minor chalcopyrite occur in 1 or 2 inches to 1 foot multiple quartz vein which strikes northeast and dips steeply northwest in chloritic country rock.

Production: Formerly called Lucky Boy. Original location 1902 or 1903. Produced $48,000 prior to 1910 using water-powdered two-stamp mill. Production from later work not reported.

References: Butler and Mitchell, 1916:79-82; Department Bulletin 14-C, v. 1, 1940:76-77.

Mule Mountain group

Mule Creek-Bolivar area, 1

Location: Curry County, secs. 17 and 20, T. 33 S., R. 10 W., between 400 and 1700 feet elevation. Included 11 lode claims, 1 placer, and 1 millsite in 1916.

Development: Workings on Mule Mountain vein total about 600 feet in three adits, 83-foot vertical and 16-foot inclined shafts. Big Devil Stairs vein was opened by a 100-foot tunnel and extensive stope, and the Keystone tunnel on south side of the river was about 120 feet long in 1916. Other short adits and surface cuts are reported.

Geology: Country rocks are recrystallized greenstone in part diorite and gneiss. The Mule Mountain vein has its apex on the ridge in sec. 17 at about 1680 feet elevation. It strikes about N. 35° E. and dips 55° to 60° SE. The main vein (eastern of the two parallel veins) is 8 inches to 2 feet thick. Fine particles of free gold, probably derived from breakdown of pyrite, occur in a limonite-bearing gouge. The

Big Devils Stairs Creek workings are about 200 yards northwest and 340 feet below the Mule Mountain shaft. This lower vein is offset by numerous faults that break it into several tapering segments, ranging from 0 to as much as 4 feet thick. The strike is about N. 80° E. and the dip 54° to 64° SE. Some pyrite and chalcopyrite occur in the vein, which branches and fans out to a mineralized shear zone 15 feet or more wide. The Keystone vein consists of parallel shear zones striking about N. 25° W., and dipping 76° E. One or more quartz veins with sheared rock, gouge, and chlorite occur in zones from a few inches to more than 20 feet wide.

Production: Mule Mountain vein was discovered about 1896. Early mining on the ridge resulted in about $3000 production. A 400-foot adit was driven to the Mule Mountain vein from the east side of the ridge about 300 feet lower in 1931-1932 with promotion funds, but no production resulted. The two parallel veins on the ridge have been worked in a small way during recent years by R. L. Venner. Total production from the mill that was situated near Blossom Bar is reported to have been about $60,000, probably all prior to 1920.

References: Butler and Mitchell, 1916:83-90; Department Bulletin 14-C, v. 1, 1940:77; Department mine file field notes and map, 1956; and R. L. Venner, oral communication, 1965.

Paradise mine

Mule Creek-Bolivar area, 5

Location: Curry County, $W\frac{1}{2}SW\frac{1}{4}$ sec. 27, T. 32 S., R. 10 W., about 3200 feet elevation, near Saddle Peaks, reached by $4\frac{1}{2}$ miles of steep trail.

Development: Three tunnels separated by about 100 feet elevation. The lower is about 230 feet long, the middle 150 feet, and the upper, now caved, about 45 feet long.

Geology: The vein is faulted with discontinuous tapering segments widening to as much as 4 feet in places, but generally narrower. It strikes about east and dips steeply south. Only ore mineral noted is free gold. Values are occasionally rich but spotty.

Production: History is mostly lacking. One thousand pounds of selected ore was reportedly shipped from the property. Dates were not reported, but may have been about 1900.

References: Butler and Mitchell, 1916:78-79; Department Bulletin 14-C, v. 1, 1940:77-78; Department mine file field notes, 1961.

Original hand-made sign for KLAMATH MINE. It is displayed at the Curry County Historical Society Museum in Gold Beach. —Bert Webber photo

ILLINOIS-CHETCO AREA

Location

The Illinois-Chetco area is situated in southwestern Josephine County and southeastern Curry County (figure 41). It includes lode and placer mines in the central portion of the Illinois River drainage in Josephine County and in the headwaters of tributaries of the Chetco River in Curry County.

The area is accessible by way of the Illinois River road that trends west from U.S. 199 (Redwood Highway) at Selma, and the Eight Dollar-Onion Camp road that leaves the highway about 3½ miles south of Selma. The area is rugged, with many over-steepened slopes. Elevations range from 1000 feet on the Illinois River to 5098 feet at the top of Pearsoll Peak on the Josephine-Curry County line. The county line follows the drainage divide of the Illinois and Chetco Rivers. Other high points along the divide in the area are Gold Basin Butte (4870), Eagle Mountain (4399), Whetstone Butte (4414), and Canyon Peak (4903).

Geology and Mineralization

The Illinois-Chetco area is underlain by isoclinally folded sediments and volcanics of the Galice Formation. These rocks have been intruded by peridotite, largely altered to serpentinite and gabbro, as well as later dikes and stocks of dioritic composition mapped as hornblende diorite, granodiorite, dacite porphyry, and dolerite (Wells and others, 1949). To the west is an extensive body of rock referred to in this report as the Illinois-Chetco gabbro-diorite complex, which is composed of amphibole gneiss, gabbro, pyroxinite, and medium- to very coarse-grained hornblende diorite. The area is bounded on the south and east by the large Josephine peridotite sheet and to the north and west by the smaller central Illinois River peridotite sheet.

Interesting geologic features of probable late Tertiary age are the residual patches of old stream gravels situated west of the area on Gold Basin and York Butte at about 4000 feet elevation. These are described by Diller (1902) as remnants of the Klamath peneplain.

Gold mineralization appears to be localized along sheared contacts of the metasediments or metavolcanics (greenstone) with serpentinite. The gold generally occurs associated with sulfides in quartz veins penetrating the sheared contact zones or the older rocks nearby. The principal ore minerals are gold, pyrite, arsenopyrite, and chalcopyrite. Gangue minerals are quartz, calcite, talc, and epidote. Sericite is a common alteration product along the veins at the Peck mine.

Many of the early gold mines in the area were hillside placer operations where residual gold freed by decomposition of rocks in place along the contacts was washed into sluice boxes (Diller, 1914). The Higgins mine (Golden Dream) at the head of Slide Creek and the Hustis and Anderson on the divide between Slide and Miller Creeks were mainly this type of deposit. Diller suggests that the gold occurs in flakes between the folia of the talcose minerals in shear zones along serpentinite contacts.

Several small limonite gossan outcroppings, with associated malachite, are found between Chetco Pass and Eagle Creek. The host rocks are altered volcanics and sediments of the Galice Formation or serpentinite. One such gossan, which appears to be about 100 feet wide and 200 yards long, is exposed on Miller Creek at about 2800 feet elevation. It contains limonite, abundant pyrite, and a minor amount of chalcopyrite.

History and Production

The mining history of the area is mainly in its placer activity. The earliest mining in Oregon was done on the Illinois River near the mouth of Josephine Creek in the summer of 1850. From 1852 through

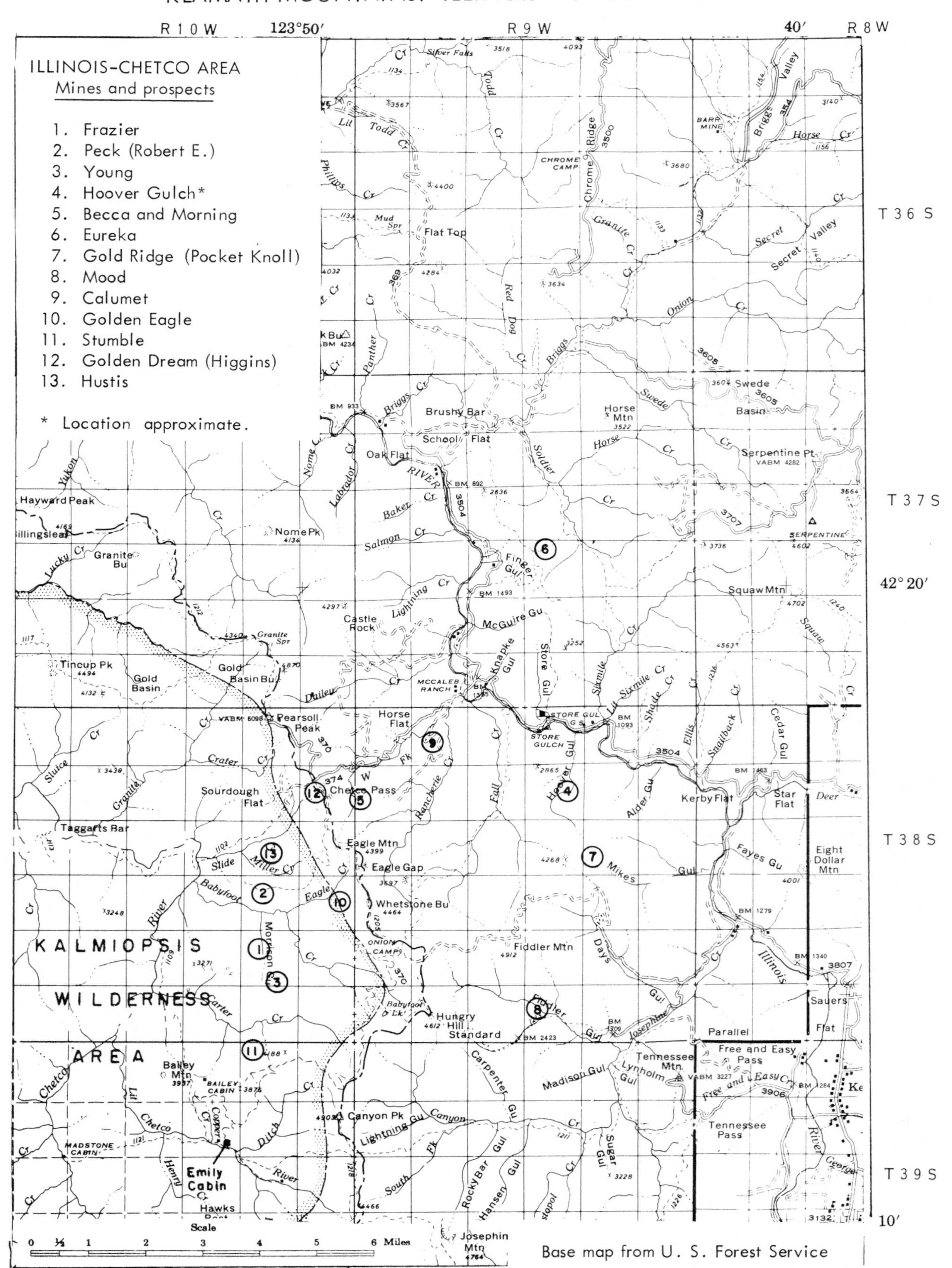

Figure 41. Index map of the Illinois - Chetco area.

the early 1900's, placer mining was especially active on Josephine Creek and along the Illinois River west of Eight Dollar Mountain. The water right on the Independence placer group near the mouth of Josephine Creek dates back to 1873.

Diller's examinations (1914, p. 64-66) were made between 1883 and 1911. He describes lode mining activity in the area between Chetco Pass and Babyfoot Creek, including the Higgins, Hustis and Anderson, and Miller and Bacon (Robert E. and Peck) which were quite active about the turn of the century. While visiting gold-quartz mines of the area, Diller found the ruins of an old arrastra near the Hustis mine on Miller Creek.

Little information is recorded in the literature on which to base estimates of production for the area. The only statistics given are partial or estimated output for a few lode mines, among which only the Eureka and Peck are known to have had significant production. It is safe to say that, except for placers on Josephine and Briggs Creeks, production for the area as a whole is not great.

Placer Mines

Josephine Creek placers

Josephine Creek and its tributaries, Canyon Creek and Fiddler Gulch, have been among the more important producers in the area. Gold in the placers is apparently derived from numerous small veins in altered sedimentary and volcanic rocks lying west of the peridotite and serpentinite.

Both hydraulic and drift methods were employed to work the Josephine Creek bench gravels. The upper bench of tightly cemented gravels has been mined by drifting along old channels in the bedrock. Most of the values are found in the gravel close to the serpentinite bedrock and only the lower 6 inches to 1 foot is considered ore. Other minerals associated with gold in the black sand are chromite, josephinite, platinum, and iridium.

Illinois River placers

Several placer operations on the Illinois River have been active from time to time at various points down stream from Josephine Creek. Early centers of activity during the 1860's and 1870's were nearby Sixmile Creek, Hoover Gulch, Rancherie and Briggs Creeks, and Oak Flat.

Upper Chetco placers

It is interesting that neither Diller (1914) nor Butler and Mitchell (1916) mention early placer-mining activity on the head of Little Chetco or Carter Creek where evidence of the early mining is still visible, especially in the vicinity of Emily Cabin and Ditch Creek along the Little Chetco. The Tertiary gravels of Gold Basin were tested in 1875 or 1876 by sinking shafts (now filled with water) and found to contain very little gold. Parks and Swartley (1916) reported some development, including a mile or more of ditch to collect snow melt, but that only a small amount of gravel was mined before work was suspended. Evidence of this early work is still visible.

Briggs Creek placers

Although there is a noticeable lack of lode mines between the Galice area and the Illinois-Chetco area as seen on figure 34, it should be pointed out that placer deposits along Briggs Creek and its tributaries (figure 35) have been significant producers. Unofficial estimates of the U.S. Bureau of Mines place the placer production of the Briggs Creek drainage at $100,000 since its discovery in 1868. Most of the placer-mining activity was concentrated in upper Briggs Creek valley at the Barr mine in sec. 7, T. 36 S., R. 8 W. and along lower Briggs Creek in the vicinity of Red Dog and Soldier Creeks. Mining was also done at the Elkhorn placer along the northwest side of Briggs Creek in sec. 24, T. 36 S., R. 9 W., and there is evidence of mining along the tributaries, Secret Creek and Onion Creek, as well as at various bars along main Briggs Creek. The bulk of all production in the area was during the late 1800's.

Principal Lode Mines

Eureka mine: The Eureka mine (map no. 6), situated at the head of Soldier Creek, a tributary of Briggs Creek, is reached by a steep, narrow jeep road which branches from the Illinois River road at the 14 mile (from Selma) post. The mine is known to have had a fair amount of production but no accurate record has been obtained. Its vein is situated in a shear zone in greenstone not far from a granodiorite stock and with closely associated serpentine. The mine was first operated in 1901 with a Huntington mill and later a 10-stamp. Some of the ore was quite rich, containing free gold and pyrite. Other areas in the vein have poor values. The workings aggregate about 1000 feet.

Peck mine: The Peck (Robert E.) mine (map no. 2) is situated near Babyfoot Creek on the head waters of the Chetco River in Curry County. Diller (1914) and Shenon (1933b) report fairly complete historical and geologic details of the Peck mine. Hydraulic mining was first done in the area about 1900. I. F. Peck located some claims in 1919 and the Robert E. Mining Co. was incorporated in 1922. In 1928 a small, exceedingly rich ore shoot produced ore and bullion worth $36,936.75 and in 1929, $42,203.35, or a total of $79,140.10. The ore was ground in a Huntington mill. Some tailings from the mill shipped in 1928 assayed $398.76 a ton. Total production to 1930 is estimated by Shenon (1933b, p. 52) to be more than $100,000. The mine was operated on a small scale by W. D. Bowser of Grants Pass during the period of 1939 to 1952. The additional production indicated by incomplete records was probably about $20,000. The main vein strikes N. 65° E. and dips steeply south. In the two lower levels the vein averages 30 inches thick. Values have been quite variable. According to Mr. Peck, a mill test of 300 tons from the No. 3 level yielded $22 to the ton of free gold, but $6 to the ton in gold was lost in the tailings (Shenon, 1933b, p. 54). Some confusion about the name has apparently existed. Present owners say that the main producer, situated on the Babyfoot side of the ridge, is called the Peck mine and a workings on the Miller Creek side is called the Robert E. The latter has had some production from ground sluicing and a 200-foot tunnel under the placer pit in north center section 23 near quarter corner of sec. 14, T. 38 S., R. 10 W. Mining water for both mines was brought around the hill in a ditch about a mile long from upper Miller Creek to a point on the ridge between the mines. This ditch was being built when Diller visited the area in 1911.

Lode Mines of the Illinois-Chetco Area

Becca and Morning Group — Illinois-Chetco Area, 5

Location: Josephine County, SW$\frac{1}{4}$ sec. 7, T. 38 S., R. 9 W., 3000 feet elevation.

Development: Three tunnels total a little more than 300 feet.

Geology: Northeast-trending gossan on serpentine-greenstone contact is leached, pyrite-bearing gouge zone which pans free gold at surface.

Production: Discovered 1915; prospected with small, homemade two-stamp mill. There was a small production, but the amount is not recorded.

References: Diller, 1914:64; Parks and Swartley, 1916:52; Department Bulletin 14-C (Josephine), 1942:123-124.

Calumet mine

Illinois-Chetco Area, 9

Location: Josephine County, secs. 5 and 8, T. 38 S., R. 9 W., between 1200 and 2400 feet elevation.

Development: About 2000 feet of tunnels into all sides of knoll between the forks of Rancherie Creek in $S\frac{1}{2}$ sec. 5; all are caved.

Geology: Country rock is serpentine and tuffaceous greenstone. The contact area is richly mineralized in places with pyrite, pyrrhotite, and some chalcopyrite and galena, and the knoll of tuffaceous greenstone contains numerous quartz veinlets, some of which carry chalcopyrite and galena. A prominent quartz ledge on the summit is said to assay $4 to $8 a ton (at $20 gold price) and Diller (1914) suggested a possible large, low-grade deposit.

Production: The bulk of exploration and development work was done before 1915. Production, if any, was not reported.

References: Diller, 1914:63-64; Parks and Swartley, 1916:49; Department Bulletin 14-C (Josephine), 1942:125.

Eureka mine

Illinois-Chetco Area, 6

Location: Josephine County, $NE\frac{1}{4}NW\frac{1}{4}$ sec. 22, T. 37 S., R. 9 W., at 2500 feet elevation.

Development: Underground workings, largely caved, aggregate about 1000 feet.

Geology: Ribbons, streaks, and bunches of vein quartz occur in a N. 50° west-striking, 75° northeast-dipping sheared zone of greenstone near its contact with serpentine and quartz diorite. Ore has spotty values, some quite rich, contains free gold and pyrite.

Production: Mine first operated in 1901 with Huntington mill, later used a 10-stamp mill. Production reportedly "considerable," but not definitely known.

References: Diller, 1914:62-63; Parks and Swartley, 1916:92; Department Bulletin 14-C (Josephine), 1942:129-130.

Frazier mine

Illinois-Chetco Area, 1

Location: Curry County, near center sec. 26, T. 38 S., R. 10 W., about 2800 feet elevation.

Development: One 80-foot tunnel and seven cuts expose the vein for 400 feet.

Geology: A 1- to 6-foot quartz vein lies along the north-trending contact of serpentine and greenstone. The main ore mineral is arsenopyrite with some pyrite and free gold.

Production: A $12,000 to $14,000 pocket was produced in 1935. To 1938 an additional $650 plus is reported.

Reference: Department mine file report, 1938.

Golden Eagle mine

Illinois-Chetco Area, 10

Location: Curry County, SW$\frac{1}{4}$NE$\frac{1}{4}$ sec. 24, T. 38 S., R. 10 W., at about 3100 feet elevation.

Development: One open cut 250 feet long, 50 feet wide, and 40 feet deep.

Geology: Crushed iron-stained vein, quartz and clay mixture, 6 inches to 2$\frac{1}{2}$ feet thick and containing some fine gold lying with the hill slope in landslide debris. Slide material is mixed sediments and volcanics of the Galice Formation, and underlying bedrock is serpentine.

Production: Located in 1935. Worked in a small way intermittently both as a hydraulic placer, concentrating gold in a sluice box, and open-pit hand operation recovering on an amalgam plate below a small scrubber. Total production is probably about $1000.

References: Department mine file report, 1963; Department Bulletin 14-C, vol. 1, 1940:59-60.

Golden Dream (Higgins) mine

Illinois-Chetco Area, 12

Location: Curry County, NE$\frac{1}{4}$ sec. 14, SE$\frac{1}{4}$ sec. 11, and W$\frac{1}{2}$ sec. 12, T. 38 S., R. 10 W., near the head of Slide Creek, from 2100 to 3350 feet elevation.

Development: L-shaped, ground-sluiced open cut with 150- and 200-foot legs and three or four short adits at about 3350 feet elevation. Other hydraulicked pits and a caved 100-foot tunnel which lie to the south in sections 11 and 14 were in the old Higgins group.

Geology: Sheared metasedimentary and metavolcanic rocks along serpentinite contacts contain numerous small quartz veinlets, some sulfide mineralization, and free gold in sheared talcose rock. Some diorite and gabbro dikes are found nearby.

Production: Higgins set up a three-stamp mill on Slide Creek about 2100 feet elevation. Rich eluvial material was reported sluiced at several sites in early 1900's, but production figures are not reported.

References: Diller, 1914:64-65; Parks and Swartley, 1916:120; Department Bulletin 14-C, vol. 1, 1940:59.

Gold Ridge (Pocket Knoll) mine

Illinois-Chetco Area, 7

Location: Josephine County, SW$\frac{1}{4}$ sec. 14, T. 38 S., R. 9 W., about 3900 feet elevation.

Development: Numerous shallow cuts, a 20-foot shaft, and short tunnel.

Geology: A 100-foot-wide zone of cherty slates with quartz veinlets trending N. 20° E. and dipping 50° SE. in pyrite-bearing greenstone, also with quartz veins. Serpentine is nearby.

Production: Was being actively prospected about 1900. Several small, rich pockets were reportedly recovered. Total production for the area is not known.

References: Diller, 1914:66-67; Department Bulletin 14-C (Josephine), 1942:131-132.

Hoover Gulch mine

Illinois-Chetco Area, 4

Location: Josephine County, sec. 10, T. 38 S., R. 9 W., about 2000 feet elevation.

Development: Early workings included a 40-foot shaft. More recent work is not reported.

Geology: Country rocks are greenstone intruded by small serpentine bodies and quartz veinlets.

Production: Formerly known as Williams and Adylott claims, probably worked first about 1900 and since then intermittently by subsequent claim holders. There has been a small production from highgrading. Total unknown.

References: Diller, 1914:66; Parks and Swartley, 1916:237.

Hustis mine

Illinois-Chetco Area, 13

Location: Curry County, near the center of sec. 14, T. 38 S., R. 10 W., from 2400 to 2600 feet elevation on the ridge between Slide and Miller Creeks.

Development: Workings include hydraulicked pits on both sides of the ridge and short caved adits.

Geology: Mineralization occurs in a north-striking shear zone injected by highly sheared talcose serpentinite lying between fractured graywacke to the west and a diorite dike and serpentinite on the east. A few narrow malachite-stained gossan zones indicate the presence of sulfides. Values were recovered as free gold in sluices.

Production: Formerly called Hustis and Anderson. Most activity about 1900. Production figures not reported.

Reference: Diller, 1914:65.

Mood mine

Illinois-Chetco Area, 8

Location: Josephine County, W½ sec. 34, T. 38 S., R. 9 W., near forks of Fiddler Gulch, about 1900 feet elevation.

Development: Nearly 2000 feet of underground workings.

Geology: Workings follow northeast shear zone parallel to the greenstone-serpentine contact containing gouge, irregular quartz veins, and silcified rock rich in pyrite.

Production: "Some thousands of dollars" were reportedly produced using an arrastra. Most of the work was done prior to 1927.

References: Diller, 1914:68; Department Bulletin 14-C (Josephine), 1942:136.

Peck (Robert E.) mine

Illinois-Chetco Area, 2

Location: Curry County, center sec. 23, T. 38 S., R. 10 W., about 2100 feet elevation.

Development: There are about 2000 feet of underground workings, including raises in four levels.

Geology: Ore occurs in northeast-striking steep south-dipping quartz fissure veins in greenstone, which average 30 inches thick where mined. Veins terminate at serpentine contact. Arsenopyrite, pyrite, and gold are the main ore minerals. Some of the ore mined was exceptionally rich.

Production: The mine was first located by I. F. Peck in 1919. A total of $79,140.10 was produced from a small, rich shoot in 1928 and 1929. Total production to 1952 is estimated at about $120,000. Some recent work has been done but production, if any, has not been reported.

References: Shenon, 1933b:51-55 (map); Department mine files.

Stumble claim

Illinois-Chetco Area, 11

Location: Curry County, NW$\frac{1}{4}$ sec. 2, T. 39 S., R. 10 W., at about 3200 feet elevation.

Development: Two short adits, one 40 feet with stope and the other caved, and several shallow surface cuts.

Geology: Thin, coarse-crystalline to vuggy and tapering multiple quartz veinlets in a zone $2\frac{1}{2}$ to $4\frac{1}{2}$ feet thick, striking N. 43° W. and dipping 60° SW., penetrate fine to medium, coarse-grained, thin-bedded indurated tuffaceous sediments. Ore minerals include gold, chalcopyrite, pyrite, and arsenopyrite. Gold assays from about half an ounce to nearly 8 ounces per ton have been obtained.

Production: A small amount of high grade has been mined, but no production records or history are available.

References: Wells and others, 1949 (map no. 109); Department mine file report, 1962.

Young mine

Illinois-Chetco Area, 3

Location: Curry County, near $\frac{1}{4}$ corner N. edge sec. 35, T. 38 S., R. 10 W., about 4200 feet elevation.

Development: Open cut in steep, rocky face is about 20 x 30 x 15 feet in size.

Geology: Country rocks are layered silicified sandstone and shale which strike north and dip 50° E. Several quartz veinlets occur in the metasediments. A 6-inch fissure, with breccia and crystalline quartz, which strikes west and dips 65° S. is exposed in the face of the cut. Values were reported to occur in a pocket. No details given.

Production: A $1000 pocket was reportedly recovered by R. D. Young in 1937.

Reference: Department Bulletin 14-C, v. 1, 1940:62.

GALICE AREA

Location

The Galice area is situated in the northwestern part of Josephine County in Tps. 33, 34, and 35 S., Rs. 8 and 9 W. (figure 42). It includes the Galice and Mount Reuben mining districts.

Lode-gold mines of the Galice area occur in a broad, northeast-trending zone about 5 miles wide and 15 miles long, extending from Mount Reuben southward across the Rogue River to the Howland mine near Cedar Mountain on the headwaters of Silver Creek. The area is steep and rugged. Elevations range from 600 feet on the Rogue River to 4000 or more feet on the mountain tops. Access to some parts of the area has been improved in recent years by the construction of numerous logging roads.

Geology and Mineralization

The principal host formation for mines in the Galice area is the Rogue Formation, which consists mainly of altered lavas, agglomerates, and tuffs ("greenstones") and amphibole gneiss, a more highly altered phase of the formation. The Rogue Formation lies stratigraphically between sedimentary rocks of the Dothan Formation to the west and the Galice Formation to the east. The Rogue Formation has been intruded by ultramafic rocks (peridotite and its alteration product, serpentinite), and it lies along the northern extension of the Illinois-Chetco gabbro-diorite complex (figure 37). This northern extension is largely granodiorite. The intrusive mass has undoubtedly been a contributing factor in the metamorphism of the rocks and the introduction of hydrothermal mineralizing solutions.

Wells and Walker (1953) describe geology of deposits in the area as follows:

> "The lode deposits occupy fractures in the gabbroic, dioritic, and dacitic rocks as well as in the volcanic rocks of the Rogue Formation and its metamorphosed derivatives. Fractures in the sedimentary rocks of the Galice and Dothan Formations are not mineralized. Although a few veins strike west, most of them strike north to north-northeast, and they dip steeply either to the east or to the west. Gold telluride is common at the Bunker Hill mine, but in most deposits the gold is free. Commonly it is associated with pyrite, pyrrhotite, and a little chalcopyrite. The gangue is quartz.
>
> "A later mineralization occurs at the Almeda mine. It is characterized by barite gangue, the presence of abundant chalcopyrite, a little sphalerite, and galena. This deposit occurs at the contact of the volcanic rocks of the Rogue Formation and the slate of the Galice Formation."

Youngberg (1947, p. 6) describes the hypogene minerals of the Mount Reuben district, which lies in the northern part of the Galice area, as follows:

> "The minerals in the veins of the district are similar, varying only in quantity. Those seen were pyrite, chalcopyrite, pyrrhotite, galena, free gold, molybdenite, quartz, barite, chlorite, sericite, and calcite. The principal minerals are quartz and pyrite with gold associated with the pyrite. Veins in the quartz diorite contain very small amounts of chalcopyrite and pyrrhotite. Minor amounts of molybdenite were noted in several veins. Chlorite, sericite, calcite, and minor amounts of barite are the common gangue minerals. Chalcopyrite is much more common in the veins in gabbro and greenstone, and pyrrhotite also occurs in larger amounts. Gold is almost completely absent in veins occurring in gabbro, and where it does occur it is found as small high-grade pockets. In greenstones gold is found associated with pyrite and chalcopyrite, and as free gold in quartz."

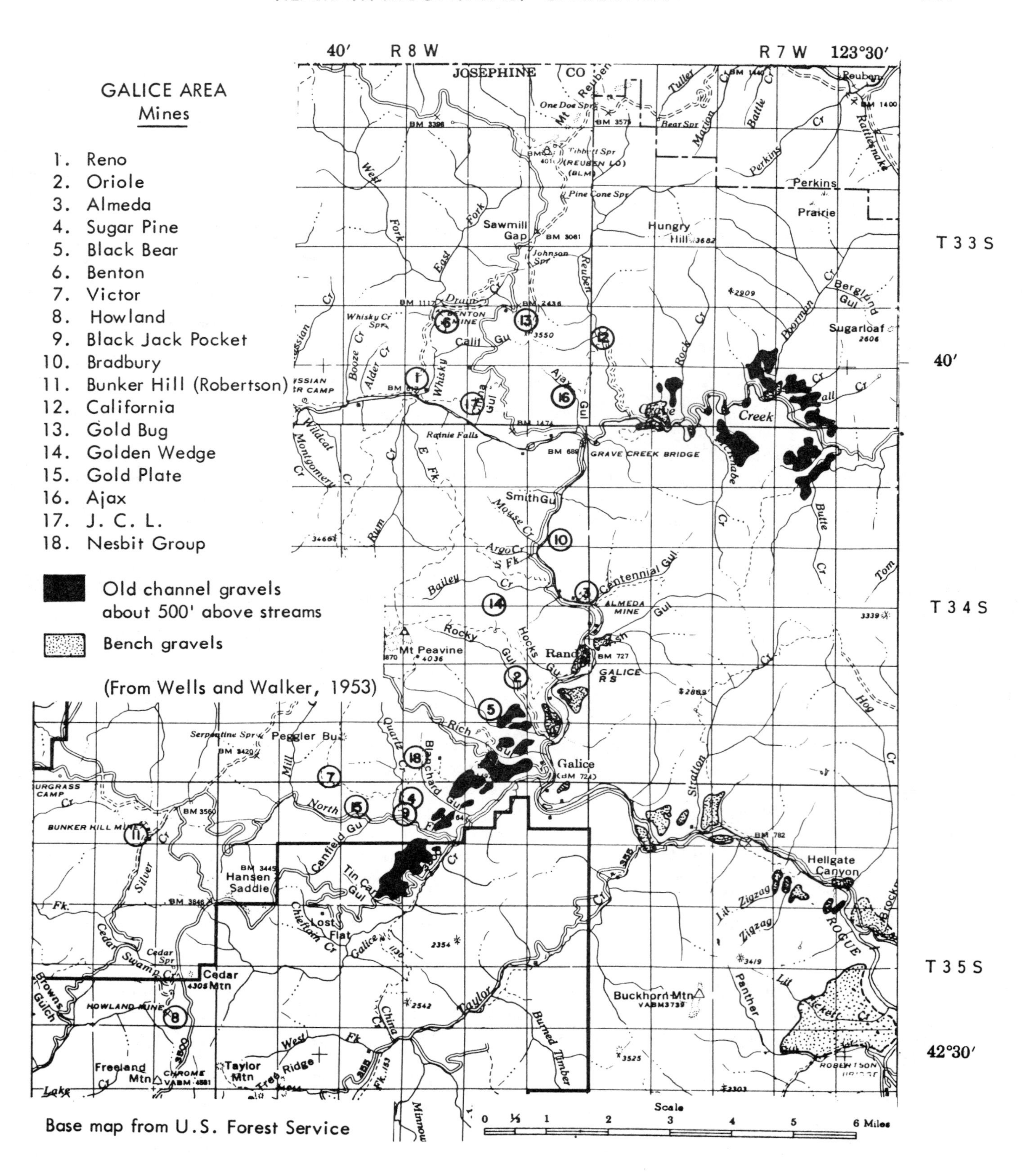

Figure 42. Index map of the Galice area.

Youngberg evaluates gold-bearing veins found in various rock types of the Mount Reuben area and indicates that the veins in quartz diorite, for example, the Benton mine, are most promising because of their greater size. Here the gold occurs intimately associated with pyrite and is not of the free-milling variety. Values range from 0.06 ounces to several ounces to the ton. The gold content of ore milled at the Benton mine has ranged from 0.22 to 0.35 ounce to the ton; however, this includes development ore which had sufficient value to pay for milling and was treated since it had to be trammed anyway. Generally higher grade ore shoots are found in veins penetrating Rogue Formation altered volcanic rock, consisting mainly of meta-andesite and associated schistose tuffs. The ore shoots occur along veins in prominent shear zones, usually at their junction with a fissure. Ores in the greenstone are largely free milling and have been successfully treated by amalgamation and were ground in an arrastra or a stamp mill. Copper stain resulting from oxidation of chalcopyrite has been found to be a good indicator for gold in these ore shoots.

Veins in gabbro are structurally somewhat like those in the quartz diorite and contain copper stains like those in the greenstone, but where sampled are persistently of lower grade and thus have been developed by only shallow surface cuts.

History and Production

Placer mining began on Galice Creek about 1854. Early work was done on the richest and most accessible deposits. Mining activity slackened during the 1860's, and by the 1880's small placers were being worked by Chinese. For the year 1883 the Galice Creek district had an estimated gold production of $8,000.

Quartz mining started in the Galice district about 1886. During the 1890's quartz mines of the Mount Reuben area were being developed. In 1898 the Gold Bug mine operated a five-stamp mill, and the Golden Wedge was grinding its ore in an arrastra. A 100-ton matting furnace was built at the Almeda mine in 1908 (figures 43 and 44), and in the same year 3000 feet of underground workings were driven at the mine. A 10-stamp mill erected at the Sugar Pine mine in 1908 was operated for a few months and later moved to the Oriole mine, which continued sporadic activity to 1942.

The Almeda smelter was operated for 30 days in 1912 and for about the same length of time the next year. Operations at the Almeda were suspended in 1916 and little work was done until 1940. Work was discontinued again in 1942 because of inability to secure priorities on mining equipment as a result of the onset of World War II. In March 1941, rich ore was encountered at the Bunker Hill property, and in just a few days the Robertson Brothers mined ore worth $20,480. The Benton mine on a tributary of Whiskey Creek was the largest underground mining operation in southern Oregon that was closed by the war emergency of 1942 (W.P.B., L-208). Since that time gold-mining activity in the area has been restricted to a few seasonal placers.

Total gold production for the area between 1854 and 1912 is estimated at $3,000,000, and the U.S. Bureau of Mines records indicate that 7,890.74 ounces of placer gold and 21,441.29 ounces of lode gold were produced from 1912 to 1948. Later production has been small.

Libbey (1967) compiled a comprehensive report on the Almeda mine, including history of mining activity, production statistics, mine maps, assay data, and geologic summary.

Placer Mines

Old channel gravels

Old channel gravel deposits lying on the hillside just west of Galice range from a quarter to more than half a mile wide and extend nearly 4 miles in a southwesterly direction. These patches of stream gravel lie on benches of bedrock about 500 feet above the present stream beds and have been dissected by tributaries of Galice Creek and Rich Gulch and Rocky Gulch. The deposits have been mined by hydraulic methods in several places, including the Ankeny, Courtney, Carnegie, California-Oregon, Last Chance, and Old Channel mines. The Old Channel mine has the more extensive work and consists of a

Figure 43. View of Almeda mine near Galice about 1908, showing smelter under construction and low-water bridge across Rogue River.

Figure 44. Almeda mine about 1910 showing 100-ton matte furnace building. (Photograph courtesy of Grants Pass Courier.)

pit about a third of a mile across and in places more than 100 feet deep. Mining here began about 1860 when the first high ditch was built. Available reports do not list production of old channel placer mines, but it is estimated that they produced more than a million dollars. Much of the placer gold taken from Galice Creek placers is probably a reconcentration eroded from the old channel deposit. Nuggets derived from the old channel gravels are characteristically well smoothed and flattened from having been transported a long distance by streams. They are also purer (better than 900 fine) than most gold from present stream deposits. The remaining patches of old channel gravel mapped in the vicinity of Galice by Wells and Walker (1953) appear to aggregate about 1 square mile. Thickness of the gravels ranges from 15 to 210 feet (Diller, 1914, p. 99). Average thickness is probably less than 100 feet. The over-all average value recovered per cubic yard mined was between 10 and 15 cents (N. L. Lewis, oral communication, 1966). The richer portions of the old channel have been mined, but some good ground remains.

Several patches of old channel gravel are also mapped by Wells and Walker (1953) along the sides of lower Grave Creek canyon. These appear to aggregate about the same area as is found in the vicinity of Galice.

High bench gravels

Bench gravels ranging from 100 to 400 feet above the present streams are mapped as "high bench gravels" by Wells and Walker (1953). These gravels are found at various places on both sides of the Rogue River in the vicinity of Galice, Hellgate Canyon, and Picket Creek. The deposits are capped by a layer of brick-red soil. They have been mined to a limited extent as at the Dean & Dean and Rocky Gulch placers just downstream from Galice and at the Hellgate placers in Hellgate Canyon.

Principal Lode Mines

Benton mine: Libbey (1963) describes the Benton mine (figure 45) as follows:

"The mine, owned by the Lewis Investment Co., Portland, is on Drain Creek about 21 miles southwest of Glendale in secs. 22, 23, 26, and 27, T. 33 S., R. 8 W., Josephine County. Eight patented and 16 unpatented claims are included in the Benton Group. Joe Ramsey made the discovery in 1893. Mr. J. C. Lewis acquired the property in 1894 and developed it until 1905, completing approximately 5,000 feet of development work, at which time the mine was shut down. When the price of gold was increased in 1934, the mine was reopened and development work was resumed. A cyanide plant was installed and production maintained until April 15, 1942, when government regulations forced the closing down of mining and milling operations. Between 1935 and 1942, including time spent on exploration and construction, ore mined and milled totaled 64,282 tons averaging $8.55 for a gross value of $549,414.00. All development rock high enough in value to pay milling cost was sent to the mill rather than to the waste dump. About 10,000 lineal feet of work was done in the Benton mine proper, and about 1,150 feet on adjacent claims.

"Ore bodies were formed in quartz veins by replacement in a quartz diorite or granodiorite stock which is in contact with metavolcanics and greenstone on the east. Eight veins have been found on the property. The main Benton vein has been explored and mined through the Kansas adit for an over-all strike length of 2,000 feet trending N. 20° to 40° E., and for 600 feet in depth. The main ore shoots were formed within a network of intersecting veins related to premineral faulting, and their emplacement was governed by structural control. The ore bodies have a pronounced rake (inclination in the plane of the vein) to the south. Minor postmineral faulting has been encountered but nothing that presented a serious problem.

"On the bottom level (1,020) development revealed ore of better grade than the average value of ore mined in upper levels. A drift on the Louisiana No. 1 vein, 200 feet long, with a strike N. 80° E. and dip of 55° N.W., to its junction with the Benton vein showed ore which averaged $25 a ton for widths of from $2\frac{1}{2}$ to 3 feet, with the face still in ore when work stopped. A winze on the 1,020 level sunk on the Benton vein from a point

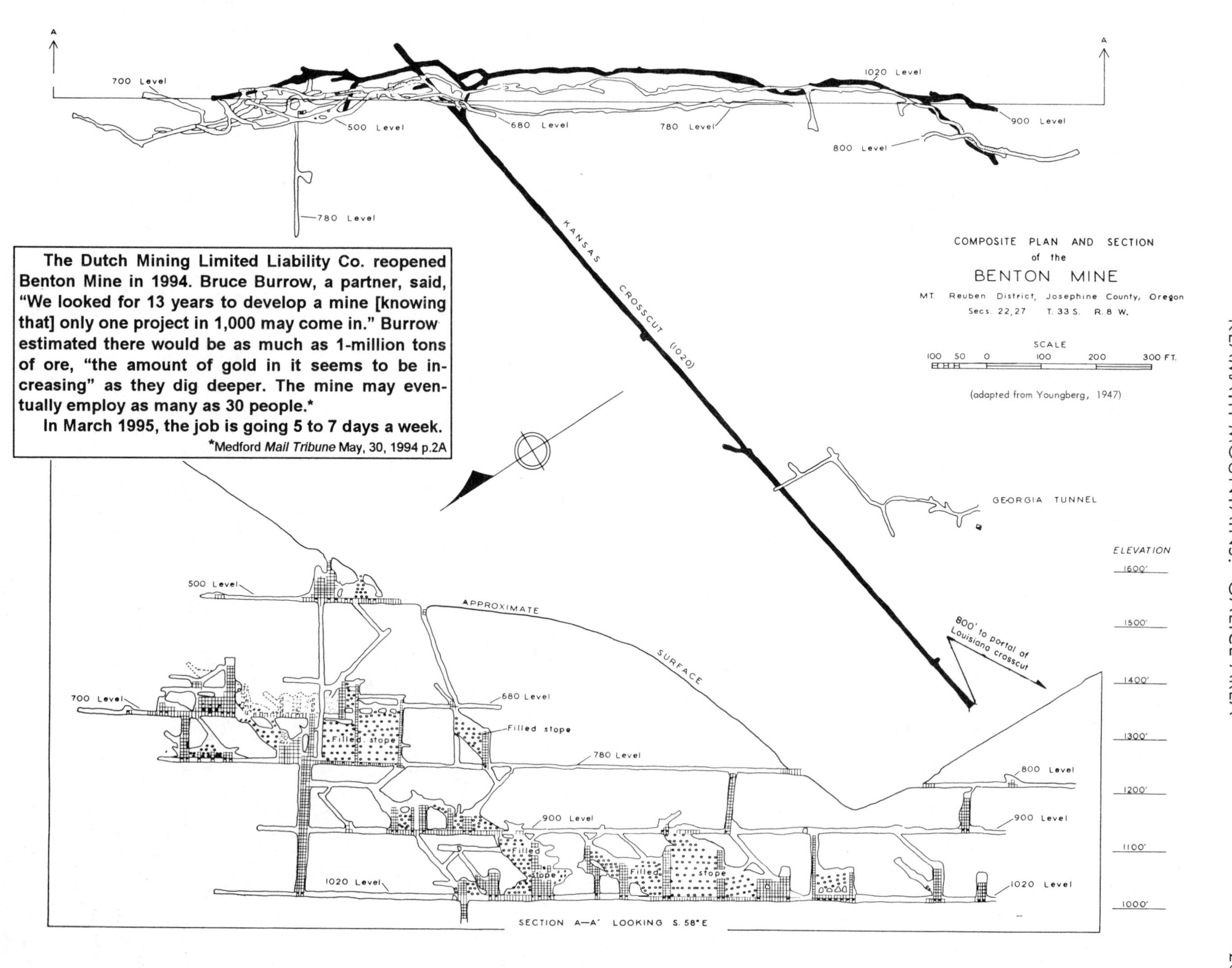

The Dutch Mining Limited Liability Co. reopened Benton Mine in 1994. Bruce Burrow, a partner, said, "We looked for 13 years to develop a mine [knowing that] only one project in 1,000 may come in." Burrow estimated there would be as much as 1-million tons of ore, "the amount of gold in it seems to be increasing" as they dig deeper. The mine may eventually employ as many as 30 people.*

In March 1995, the job is going 5 to 7 days a week.

*Medford *Mail Tribune* May, 30, 1994 p.2A

Figure 45. Map of workings at the Benton mine, Galice area.

50 feet south of the Kansas crosscut to a depth of 64 feet was channel sampled at about 5-foot intervals in both compartments of the winze. The north compartment samples averaged about $40 a ton for $4\frac{1}{2}$ feet average; the samples from the north compartment averaged about $18 a ton for approximately 5 feet average width (Elton A. Youngberg, written communication, 1963).

"The cyanide plant of 40 tons capacity was completed in 1937 and enlarged to 60 tons capacity in 1940. It incorporated a counter-current system using Dorr thickeners, Dorr agitators, an Oliver continuous filter, together with Merrill-Crowe precipitation equipment. Reportedly mill recovery was about 85 percent, which could be increased to 90 percent if changes indicated by the operating experience were made. An adequate water supply was obtained from Drain Creek. Diesel power was used.

"It may be noted that in 1941 the Benton mine had the largest individual payroll in the county."

Bunker Hill mine: The Bunker Hill, or Robertson, mine (map no. 11) has been one of the more important producers in the Galice area. It was located by John Robertson and sons in 1914, developed gradually until rich ore was struck in 1925. During the peak of operation in 1930 the mine employed 30 men. Ore was concentrated by a 10-stamp mill equipped with amalgamation plates and tables. Ore mined in 1925 averaged 19.50 ounces to the ton; in 1929 ore averaged 1.97 ounces and in 1930, 2.04 ounces to the ton. Total production from the mine has been about $160,000. The veins are narrow and often pinch to a fraction of an inch. In Number 1 vein, the most extensively mined, average thickness over the 140 feet of strike length is estimated by Shenon (1933b, p. 46) to be 1 foot. Other veins exposed in the workings are parallel, striking northwest and dipping steeply southwest. Gold values of the other veins are low. William Robertson reported an average mill recovery from the 10-foot vein of $6.00 per ton.

Ore minerals reported by Shenon (1933b) are gold, pyrite, and petzite. Predominant gangue minerals in the veins are quartz, epidote, and calcite.

The more recent work done at the mine in 1945 consisted of putting in a crosscut 200 feet below the main workings. Total length of the new crosscut is about 1300 feet. A few small veins parallel to those mined in the upper workings were encountered, but no further development or mining has resulted to date (1965).

Almeda mine: The Almeda mine (map no. 3) is located in the altered contact zone of the Rogue Formation and Galice slates known as the Big Yank lode and has a distinctly different type of mineralization (see Silver Peak area). Diller (1914) indicated that this mineralized zone could be traced intermittently for more than 20 miles in a N. 30° E. direction from Briggs Creek valley to Cow Creek at Reuben siding. By the presence of similar mineralization and alteration this zone can be traced about 16 miles farther in a northeasterly direction to include the Silver Peak mine and others southwest of Canyonville in Douglas County. The ore zone at the Almeda is associated with a highly silicified porphyritic dacite dike impregnated with variable amounts of pyrite. Shenon (1933a, p. 27-28) states: "The Big Yank lode, for the most part, consists of intensely silicified rock with variable amounts of pyrite, but in places masses of silicified rock have been partly or wholly replaced by barite and sulfides, which constitute the richer ore shoots. The mineralized zone constituting the Big Yank lode varies in width from place to place but the Almeda mine is about 200 feet wide. Two types of ore have been previously described: 'siliceous gold-silver ore' and 'copper ore with barite'. . . ." The latter type of ore has been partly or wholly replaced by barite and sulfides, some of which is nearly massive chalcopyrite.

Lode Mines of the Galice Area

Ajax mine

Galice Area, 16

Location: Josephine County, N$\frac{1}{2}$ sec. 36, T. 33 S., R. 8 W.; 2500 feet elevation.

Development: Total about 3000 feet.

Geology: Country rock is greenstone with northeast-striking, 80° NW-dipping shear zone 140 feet wide, with small lenticular bodies of quartz and some pyrite.

Production: Early history unknown; about $12,000 total production from old steam-driven stamp mill concentrates and hand-sorted high-grade ore shipments.

Reference: Youngberg, 1947:11.

Almeda mine

Galice Area, 3

Location: Josephine County, SE$\frac{1}{4}$ sec. 13, T. 34 S., R. 8 W., about 750 feet elevation.

Development: Extensive, six levels above the river and shaft to 450 feet below river, with 100-foot interval levels.

Geology: Altered dacite porphyry dike in Rogue volcanics (greenstone) adjacent to Galice slates. Ore occurs in a zone of intense silicification, barite, and disseminated to massive sulfides.

Production: Small smelter, gross value - gold, silver, copper, and lead between 1911 and 1916 was $108,000.

References: Diller, 1914:72-81; Parks and Swartley, 1916:8; Shenon, 1933a:24-35; Winchell, 1914:207-214; Department Bulletin 14-C (Josephine):17-24; Libbey, 1967.

Benton mine

Galice Area, 6

Location: Josephine County, sec. 22, 23, and 27, T. 33 S., R. 8 W., at 1000 feet elevation.

Development: Underground workings total 10,000 feet.

Geology: Ore occurs in a system of steeply dipping quartz fissure veins in normal tension faults and fractures in a quartz diorite stock. The principal ore mineral is gold-bearing pyrite. At vein intersections replacement-type ore bodies are from 20 to 30 feet in width.

Production: The mine was discovered in 1893. During the major period of production, 1934-1942, the cyanide plant processed 64,282 tons of average $8.55 ore having a gross value of $549,414.00.

References: Winchell, 1914:193-195; Parks and Swartley, 1916:28-29; Department Bulletin 14-C (Josephine):26-28; Youngberg, 1947; Libbey, 1963.

Black Bear mine

Galice Area, 5

Location: Josephine County, SW$\frac{1}{4}$ sec. 26, T. 34 S., R. 8 W., about 1900 feet elevation.

Development: Eight tunnels aggregating more than 2200 feet.

Geology: Mineralization occurs in a $2\frac{1}{2}$-foot-wide nearly vertical north-trending fault zone with gouge, quartz lenses, and rich pyrite lenses, some chalcopyrite. Country rock is amphibolite bordered by dunite to the southwest. Oxidized ore selectively mined.

Production: Early history is spotty. Total production not known. Small ball mill powered by water during wet season used in recent years.

References: Winchell, 1914:204-205; Parks and Swartley, 1916:33; Department Bulletin 14-C (Josephine), 1942:29-30.

Black Jack Pocket mine

Galice Area, 9

Location: Josephine County, secs. 3 and 4, T. 35 S., R. 8 W., about 1500 feet elevation.

Development: Small.

Geology: Rogue Formation amphibole gneiss (see Sugar Pine mine).

Production: In early 1900's, between $6,000 and $7,000 pocket recovered by hand mortaring.

References: Diller, 1914:57; Parks and Swartley, 1916:34; Department Bulletin 14-C (Josephine),1942:30.

Bradbury mine

Galice Area, 10

Location: Josephine County, SW$\frac{1}{4}$ sec. 12, T. 34 S., R. 8 W., about 800 feet elevation.

Development: Three adits 150, 420, and 525 feet above river probably aggregate more than 500 feet.

Geology: Country rock is Rogue volcanics "schistose greenstone." Fissure vein strikes N. 8° to 17° E., contains white quartz as much as 50 inches wide, with a little pyrite and rare free gold.

Production: Last operated in 1936. It was equipped with 5- to 9-ton ball mill, small table, and two flotation cells. Production not reported. One patented claim.

References: Winchell, 1914:198; Parks and Swartley, 1916:41; Department Bulletin 14-C (Josephine), 1942:31.

Bunker Hill (Robertson) mine — Galice Area, 11

Location: Josephine County, SE¼ sec. 2, T. 35 S., R. 9 W., elevation 4500 feet.

Development: About 2200 feet.

Geology: Group of small northwest-striking, steeply southwest-dipping quartz veins formed in fractures in greenstone near a tongue of quartz diorite. Ore minerals are gold, pyrite, and petzite.

Production: Located in 1914 and rich ore discovered 1925. Total production was about $150,000 from 1924 to 1942. Equipment included a 10-stamp mill.

References: Shenon, 1933c:2-45; Department Bulletin 14-C (Josephine), 1942:54-57.

California mine — Galice Area, 12

Location: Josephine County, secs. 25 and 30, T. 33 S., R. 8 W., between 1200 and 3150 feet elevation.

Development: Total development, including shaft, raises, crosscuts, and drifts in the old and later workings, is about 8500 feet. The lower crosscut is 7364 feet long, running west from Reuben Creek at 1179 feet elevation. See maps plate 5, figures 3 and 4, Department Bulletin 34.

Geology: Country rock is greenstone with few shears and small, north-trending, nearly vertical quartz veins. Small ore shoots were found in the old upper workings. Gold is largely free, with some associated pyrite. No ore was found in the lower tunnel.

Production: Discovered in 1890. Total production was estimated to have been from $3000 to $10,000.

References: Department Bulletin 14-C (Josephine), 1942:33; Youngberg, 1947:29.

Gold Bug mine — Galice Area, 13

Location: Josephine County, NE¼ sec. 26, T. 33 S., R. 8 W., at about 2500 feet elevation.

Development: Two adits, two shafts 150 and 300 feet deep; 125-foot inclined winze, four levels below No. 2 adit, and about 600 feet of stoped area on dip of ore shoot.

Geology: Vein is a northwest-striking, northeast-dipping mineralized shear zone in greenstone containing quartz, calcite, chlorite, pyrite, chalcopyrite, and free gold. Ore shoot formed at junction with north-trending shear zone. Stope widths from 2 feet to as much as 20 feet are reported; gold content is about 0.60 ounces per ton.

Production: Production has reportedly been about $750,000. A steam-powered five-stamp mill was used. Dates of operation not reported, but it was probably during the late 1800's and early 1900's.

References: Winchell, 1914:195-197; Parks and Swartley, 1916:102; Diller, 1914:52; Youngberg, 1947:17-19.

Gold Plate mine

Galice Area, 15

Location: Josephine County, NW$\frac{1}{4}$ sec. 4, T. 35 S., R. 8 W., at 1500 feet elevation.

Development: Nearly 500 feet in two tunnels.

Geology: Country rocks are highly altered basic volcanics of Rogue Formation, in part amphibole gneiss. Workings cut a flat-lying portion of quartz fissure vein about 2 feet thick including chloritic gouge. Ore minerals include pyrite and gold.

Production: Located about 1900 and was equipped with Straub circular eight-stamp mill. Production is not reported, and probably small.

References: Diller, 1914:59; Department mine file report, 1936; Department Bulletin 14-C (Josephine), 1942:40-41.

Golden Wedge mine

Galice Area, 14

Location: Josephine County, S. edge sec. 14, T. 34 S., R. 8 W., at 2000 feet elevation.

Development: About 1200 feet underground workings.

Geology: Quartz veins and lenses form an ore body in greenstone which strikes N. 20° E., and dips 50° to 60° E., contains pyrite and graphitic material, and assays $10 to $20 gold per ton.

Production: Discovered in 1893. Total production may have been $50,000. Mine was equipped with 10-stamp mill.

References: Diller, 1914:51; Winchell, 1914:199-200; Parks and Swartley, 1916:108.

Howland mine

Galice Area, 8

Location: Josephine County, SW$\frac{1}{4}$ sec. 24, T. 35 S., R. 9 W., at 3600 feet elevation.

Development: Several hundred feet of tunnel.

Geology: Country rocks are mapped as amphibolite of gabbroic habit near its contact with serpentine. Values occur in quartz veins, with abundant iron oxide near the surface.

Production: The mine was worked continuously between 1929 and 1937, mainly development with a little production from a small, water-powered ball mill.

Reference: Grants Pass Courier, January 27, 1937; Wells and Walker, 1953.

J. C. L. mine

Galice Area, 17

Location: Josephine County, W$\frac{1}{2}$ sec. 35, T. 33 S., R. 8 W., about 1000 feet elevation.

Development: Total workings are about 3000 feet in six adits on eight patented claims.

Geology: Country rocks are greenstone and gabbro. A northeast-striking, west-dipping ore shoot 20 to 50 feet long and 18 inches thick was stoped from surface to No. 5 level. Gold occurred free in quartz lenses in shear zone with small sulfide content. An east-dipping 3- to 5-foot banded quartz vein was exposed in China Gulch adit.

Production: Purchased by John C. Lewis about 1900. About $100,000 gold was recovered by stamp milling and amalgamation.

Reference: Youngberg, 1947:13; and workings map.

Nesbit Group

Galice Area, 18

Location: Josephine County, NE$\frac{1}{4}$SW$\frac{1}{4}$ sec. 34, T. 34 S., R. 8 W., about 2000 feet elevation.

Development: Was developed by three short adits.

Geology: Country rock is described as rotten greenstone a short distance east of a narrow serpentine body. Average values of free gold in yellowish-red mantle were said to be $6.50 per ton (1914).

Production: The property was worked in a small way in early 1900's. No production figures are reported.

References: Diller, 1914:57; Winchell, 1914:205-206; Parks and Swartley, 1916:161; Department Bulletin 14-C (Josephine), 1942:47.

Oriole mine

Galice Area, 2

Location: Josephine County, NE$\frac{1}{4}$ sec. 26, T. 34 S., R. 8 W., about 1300 feet elevation.

Development: Approximately 3300 feet of underground workings on four levels were mapped in 1911. Some later development makes the total more than 3500 feet.

Geology: Ore lies in a 3- to 10-foot zone in the sheared greenstone footwall of a north-trending, 75° east-dipping fault contact with dacite porphyry. The ore carries a little pyrite and chalcopyrite, and occurs as quartz lenses in the shear.

Production: Discovery date is not reported, but was probably during the 1890's. In 1908 a 10-stamp mill from the Sugar Pine mine was moved in. Sporadic operations continued to 1942. Estimated production from smelter receipts to 1937 was about $65,000.

References: Diller, 1914:48-50; Winchell, 1914:201-202; Parks and Swartley, 1916:175; Department Bulletin 14-C (Josephine), 1942:50-51; and Grants Pass Courier, January 27, 1937.

Reno mine

Galice Area, 1

Location: Josephine County, NW$\frac{1}{4}$ sec. 34, T. 33 S., R. 8 W., 1250 feet elevation.

Development: About 850 feet in three adits.

Geology: Workings explore a main northwest-striking 10° to 20° east-dipping quartz vein which penetrates shears in hornblende-magnetite-rich gabbro near serpentine. Vein is mostly from 3 to 10 inches thick, with local swells to several feet. Values are erratic to high grade. Ore minerals are pyrite and gold.

Production: Location date probably early 1900's. There have been numerous owners. Was active in 1913 (Diller, 1914). Estimated production about $5,000. A 15-ton mill was installed in 1930's. Operated in small way by Quinton Stone to 1964.

References: Youngberg, 1947; Diller, 1914:52; Parks and Swartley, 1916:90; Department Bulletin 14-C (Josephine), 1942:53.

Sugar Pine mine

Galice Area, 4

Location: Josephine County, NW$\frac{1}{4}$ sec. 3, T. 35 S., R. 8 W., about 1750 feet elevation.

Development: About 3000 feet of workings in two adits.

Geology: Country rocks are mapped by Wells, Walker, and other (1953) as amphibole gneiss recrystallized from the Rogue volcanics. Quartz veinlets occur in a zone 1 to 5 feet wide which strikes about north and dips steeply west. Ore minerals are pyrite, chalcopyrite, and galena. Ore from 6 to 36 inches of good grade was found in the shoots.

Production: One of earliest mines discovered in Galice district and was opened in 1860. Ore from a rich shoot reportedly yielded more than $25,000 when milled in an arrastra. Total production not reported.

References: Winchell, 1914:206-207; Diller, 1914:58-59; Parks and Swartley, 1916:215-216; Department Bulletin 14-C (Josephine), 1942:60.

Victor mine

Galice Area, 7

Location: Josephine County, S. edge SE$\frac{1}{4}$ sec. 32, T. 34 S., R. 8 W., about 2750 feet elevation.

Development: Not reported.

Geology: Country rocks are mapped as amphibole gneiss. Workings reported to be on a small, rich vein.

Production: Diller reported he did not visit mine, but received word in 1911 that operators recovered about $2500 in a month with hand mortar. In 1912, five men working the mine were averaging $4 per man day.

References: Diller, 1914:59; Department Bulletin 14-C (Josephine), 1942:63.

SILVER PEAK AREA

Location

The Silver Peak area lies in Douglas County southwest of Canyonville in T. 31 S., Rs. 5 and 6 W. (figure 46). It is generally steep and mountainous, with narrow ridges and deep canyons. Elevations range from about 800 feet at Canyonville to 3973 feet at Silver Butte summit 6 miles to the southwest. The area is drained by tributaries of Cow Creek and the South Umpqua River. It is covered by the Canyonville 15-minute quadrangle map.

Geology and Mineralization

Gold-silver mineralization is distributed along a northeast-trending zone of rock alteration and associated sulfide mineralization that, because of position and similarity of mineral content, is considered to be a northern extension of the Big Yank lode at the Almeda mine in the Galice area to the southwest. Principal mines in the area include the Silver Peak, Golden Gate, Gold Bluff, Huckleberry and Levens Ledge.

The mineralized area is situated in the contact zone between the indurated Upper Jurassic marine sediments of the Dothan Formation to the west and the altered volcanic rocks of the Rogue Formation to the east. The altered volcanic rocks are mapped as "greenstone" by Diller (1924). The contact strikes northeast and dips steeply southeast in this area. Alteration in the mineralized zone according to Shenon (1933a) consists of thin-bedded black chlorite schist and highly altered, fine-grained argillite. Near the ore bodies the schist is bleached to light gray, almost white, and contains abundant sericite. Where mineralized, the sericite schist contains considerable quartz, barite, and disseminated-to-massive sulfides. The mineralization (according to Shenon) occurs in altered Dothan Formation. Greenstones in the vicinity contain abundant epidote, fine-grained quartz, chlorite, zoisite, saussurite, and other alteration products.

Dole and Baldwin (1947) traced the Big Yank lode between the Almeda and Silver Peak mines. They concluded that the mineralized belt is made up of independent shear zones lying en echelon. From south to north the zones are situated progressively westward across the section from the Galice Formation-Rogue Formation ("greenstone") contact at the Almeda mine into the eastern portion of the Dothan Formation at the Silver Peak mine. Dole and Baldwin suggest that since only a few of the numerous shear zones have been mineralized, possible location of intrusives at depth may determine the location and extent of mineralization. Barite, one of the characteristic gangue minerals of the Big Yank lode, appears to be but one phase of regional mineralization and seems to be associated with the more intensely mineralized portions. All who have studied these related mineralized zones have the opinion that further investigation is needed.

White and Wolfe (1950) pointed out interesting occurrences of sulfide mineralization in greenstones along Canyon Creek and its West Fork, as well as other prospects along the mineralized zone extending southwest from the Silver Peak mine.

History and Production

According to Shenon (1933a), ore was discovered at the Silver Peak mine in 1910 by Robert Thomason, on land later owned by Silver Peak Copper Co. Two years after Thomason's discovery, J.E. Reeves purchased adjacent patented timber land which included a large portion of the ore subsequently developed. In 1920 the Oregon Exploration Co. located claims over part of the patented timber land and from 1922 to 1929 the property was in litigation. During this period 3256 tons of ore were shipped from the property.

The Oregon Exploration Co. was reorganized as the Umpqua Consolidated Mining Co. in 1929 and shipped one car (38 tons) in 1930. Gross value of ore shipped, excluding zinc, is estimated at $73,000.

Some time during the early 1890's a Swedish citizen is reported to have discovered the Gold Bluff mine (no. 4). From it was produced $7,000 in gold dust, but since he was not a citizen of the United States he was told he could not locate a claim. Others located claims at the Gold Bluff and produced $20,000 to $40,000. A Mr. Jennings purchased the mine and produced $28,000, then he sold it to a promotion company that installed a 50-ton mill. The property a few years later was lost for non-payment of taxes and was sold to H. Q. Brown of Nickel Mountain, who produced about $3,000. All of this reportedly took place during the 1890's (Department mine file report, no date).

Another mine in the area, the Levens Ledge (no. 2), is recorded as having produced $75,000 or $80,000 in gold in the early days. No dates of operation or source of information is given (Department Bulletin 14-C, v. 1, 1940, p. 106).

Judging from available reports, a total gold production of about $216,000 may be credited to the Silver Peak area prior to 1930.

Placer Mines

Some evidence of placer mining can be found on Jordan, Mitchell, Russell, and West Fork Canyon Creeks and on Middle Creek, which drains the west side of Silver Butte, but no records of this activity have been published and no major producing mines were developed. The most extensive placer-mining operations were in bench gravels along Cow Creek southwest of the area. Of these, the Victory placer in sec. 33, T. 32 S., R. 7 W. (about 6 miles west of Glendale) was probably the largest producer, but the amount of its output has not been reported.

Diller and Kay (1924) describe several placer operations on small patches of Quaternary bench gravels at about 1200 feet elevation (500 feet above main streams) in the area between Riddle and Canyonville. They suggest that at least a part of the gold is derived from the decomposition of the Cretaceous sediments on which the gravel rests.

Principal Lode Mines

Silver Peak mine: The Silver Peak mine (map no. 1), which is situated at about 3280 feet elevation, is mainly a copper-zinc mine, although some of the values are in silver and gold, as shown in the following production table from Shenon (1933a, p. 21).

Average metal content of ore from Silver Peak and Umpqua Consolidated mines.

Year	Ore produced tons	Gold oz./ton	Silver oz./ton	Copper Percent
Silver Peak				
1926	389	0.10	7.3	6.0
1928	937	0.044	2.7	6.7
1929	1666	0.07	3.6	5.6
1930	264	0.057	3.0	4.4
Umpqua Consolidated				
1930	38	0.24	2.2	3.9

Assays of four carefully cut samples taken at selected places in the mine workings by Shenon show that the zinc content of the ore is about equal to the copper content. Shenon describes the sequence of mineralization as follows:

"Quartz was the first gangue mineral to be deposited. It is everywhere fine-grained but tends to be coarser in the fractures along which it was introduced. Barite was introduced next, then fracturing occurred, and pyrite was deposited. After a second fracturing sphalerite, tennantite, chalcopyrite, bornite, galena, and chalcocite were deposited as an overlapping series and probably in the order named, although the relation of galena and chalcocite was not well established."

Galena, tennantite, and chalcocite occur only in relatively small amounts. Gangue minerals, in addition to quartz and barite, are sericite and minor epidote. Ore minerals occur both as massive tabular-shaped bodies from 10 inches to more than 20 feet wide and disseminated in the highly foliated schist. The massive ore is distinctly banded parallel to the direction of foliation. The combined workings of the Umpqua Consolidated Mining Co. and the Silver Peak Copper Co. have exposed a mineralized zone more than 100 feet wide, 450 feet long, and 270 feet deep. The zone strikes northeast and dips from 30° to 80° SE. Later work done in 1952 consisted of driving 70 feet of tunnel S. 15° W. from the west end of the old Umpqua Consolidated drift. The new work exposed gray schist with disseminated sulfides over the entire distance. Strike and dip of the schistosity parallels the ore zone.

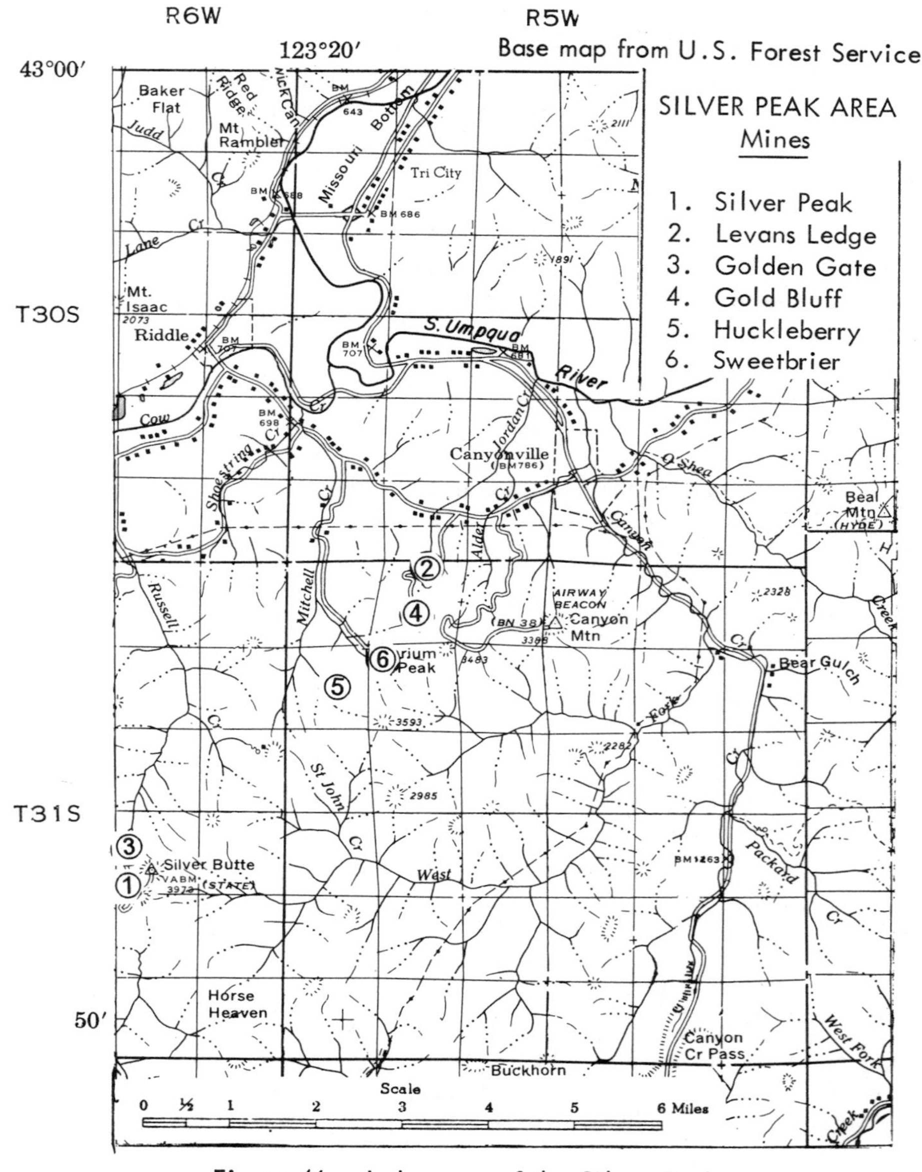

Figure 46. Index map of the Silver Peak area.

Lode Mines of the Silver Peak Area

Gold Bluff mine

Silver Peak Area, 4

Location: Douglas County, S½ sec. 5, T. 31 S., R. 5 W., at 2400 feet elevation.

Development: Several tunnels have been driven. One is 100 feet long; others not reported.

Geology: A 15-foot, iron-stained, bleached schist zone in foliated greenstone alongside of serpentine with disseminated sulfides strikes northeast and dips steeply southeast.

Production: Total production about $70,000 in 1890's; milling methods not reported. Property contains 40 acres patented land.

References: Department mine file report, undated; White and Wolfe, 1950:73.

Golden Gate mine

Silver Peak Area, 3

Location: Douglas County, secs. 22 and 23, T. 31 S., R. 6 W., at about 3000 feet elevation.

Development: About 600 feet under ground, and few surface cuts.

Geology: Mineralization similar to Silver Peak mine. Northeast-striking, altered shear zone with disseminated pyrite cubes and few small stringers of chalcopyrite.

Production: Located 1919 by N. A. Bradfield. Two cars of ore were shipped. Gross smelter returns were $1,000, mostly gold for one car of 36 tons. The second car shipped averaged $1.76 a ton.

Reference: Shenon, 1933a:23-24.

Huckleberry mine

Silver Peak Area, 5

Location: Douglas County, center sec. 7, T. 31 S., R. 5 W., at about 1850 feet elevation.

Development: Two tunnels with several hundred feet of workings.

Geology: North extension of "Big Yank Lode" described as disseminated pyrite in chloritic schist and greenstone. Zone strikes N. 40° E. and dips 75° SE.

Production: Produced $2000 from 1912 to 1915, and about $16,000 between 1931 and 1936, using a two-stamp mill equipped with amalgam plates and a table.

References: White and Wolfe, 1950:72-74; Department Bulletin 14-C, v. 1, 1940:105.

Levens Ledge mine

Silver Peak Area, 2

Location: Douglas County, NE$\frac{1}{4}$ sec. 5, T. 31 S., R. 5 W., about 1700 feet elevation.

Development: On 169 acres patented land there are seven tunnels totaling about 2400 feet, as well as raises, shafts, and stoped areas.

Geology: Country rocks are greenstone, serpentine, and a later diorite dike. The vein consists of overlapping quartz lenses in a N. 70°-80° E., 2- to 8-foot shear zone which dips 60° to 70° SE. Ore minerals are pyrite, chalcopyrite, and gold.

Production: Dates of production are not reported, but there are estimates of between $70 and $80 thousand using a stamp mill with amalgamation plates and table.

References: Department mine file report, 1945; Department Bulletin 14-C, v. 1, 1940:106.

Silver Peak mine

Silver Peak Area, 1

Location: Douglas County, secs. 23 and 26, T. 31 S., R. 6 W., about 3300 feet elevation.

Development: There are approximately 2500 feet of workings in three adits.

Geology: Massive to disseminated sulfides including pyrite, sphalerite, chalcopyrite, and less common bornite, galena, tennantite, chalcocite, and covelite occur in a gangue of quartz and barite in a northeast-striking altered zone of sericite-chlorite schist along the contact of greenstones with the Dothan Formation. The ore occurs in lenses and bands which reach more than 100 feet in width.

Production: The deposit was discovered in 1910. Production of copper, silver, and gold to 1930 is estimated to be $73,000.

References: Shenon, 1933a:15-23; Department Bulletin 14-C, v. 1, 1940:110-116.

Sweetbriar mine

Silver Peak Area, 6

Location: NW$\frac{1}{4}$NW$\frac{1}{4}$ sec. 8, T. 31 S., R. 5 W., at about 1800 feet elevation on east fork of Mitchell Creek.

Development: An open cut and a 35-foot drift.

Geology: Workings are on a pale green to white schist with disseminated pyrite and chalcopyrite and small amount of malachite.

Production: None reported.

Reference: White and Wolfe (1950:73).

GREENBACK - TRI-COUNTY AREA

Location

The Greenback-Tri-County area extends from Grants Pass northeastward to include a group of mines which lie in northeastern Josephine County and along the adjacent margins of Douglas and Jackson Counties where the three counties join (figure 47). The area incorporates parts of the Grants Pass, Greenback, Riddle, and Gold Hill mining districts. Several lode-gold mines in the area have been significant producers, the most important of which is the Greenback mine.

The topography of the area is mountainous and in places steep, but not especially rugged. Elevations range from about 1000 feet on Grave Creek near the mouth of Wolf Creek to 5264 feet on King Mountain. The area is drained by tributaries of the Rogue and South Umpqua Rivers. The main creeks are Cow Creek, Grave Creek, Evans Creek, and Jumpoff Joe. Smaller streams are Wolf Creek, Starvout, Quines, Coyote, and Louse Creeks. All have been placered to some extent.

Geology and Mineralization

Gold deposits in the Greenback - Tri-County area lie between the Grants Pass diorite batholith and the elongate Evans Creek-Cow Creek diorite body which extends southwest from the vicinity of Tiller across the southern boundary of Douglas County into Jackson County (figure 37). In general, the deposits are situated along or adjacent to a prominent northeast-trending zone of structural weakness which is marked by serpentinite intrusions.

The host rocks are altered sediments and volcanics of the Galice Formation as well as slightly higher grade metamorphic rocks of the Applegate Group. In a few cases, such as at the Oro Grande and Little Arctic, the gold is deposited entirely in serpentinite. Quartz veins carrying gold mineralization have variable strikes, ranging from northeast, parallel to the major northeasterly regional structural trend, to east. The Greenback and a few other veins strike in a northwesterly direction at right angles to the regional structure. These northwest-striking veins have apparently penetrated transverse fractures in greenstone which opened up at right angles to major northeast-trending faults.

The usual ore mineral assemblage includes native gold, pyrite, chalcopyrite, and arsenopyrite, and occasional galena and sphalerite. Quartz, calcite, chlorite, talc, and epidote are the main gangue minerals. Some of the richer deposits contain arsenopyrite as the principal ore mineral.

A few of the deposits lying along the northern and eastern "halo" of the Grants Pass batholith contain some copper and significant zinc; in this respect they are of higher temperature origin and differ from the other deposits in the area. The Oak mine and the Lucky Queen are examples of zinc mineralization, whereas the Copper Queen is a pyrrhotite-chalcopyrite type. More complete evidence is needed to establish the presence of temperature mineral zoning due to the proximity of the Grants Pass batholith, but this is a reasonable explanation for the occurrence of the higher temperature mineralization.

History and Production

Placer mines

Mining in the Greenback area began with the working of rich placer deposits along Grave Creek and its tributaries. The earliest production record was for 1883, when Grave Creek placers produced $20,000 in gold. In 1895 several small placer mines were active on Grave Creek and its tributaries, Coyote and Wolf Creeks. Winchell (1914, p. 184) estimated that the Columbia placer (figure 48) on Tom East Creek below the Greenback mine produced more than $400,000. The Columbia placer continued to

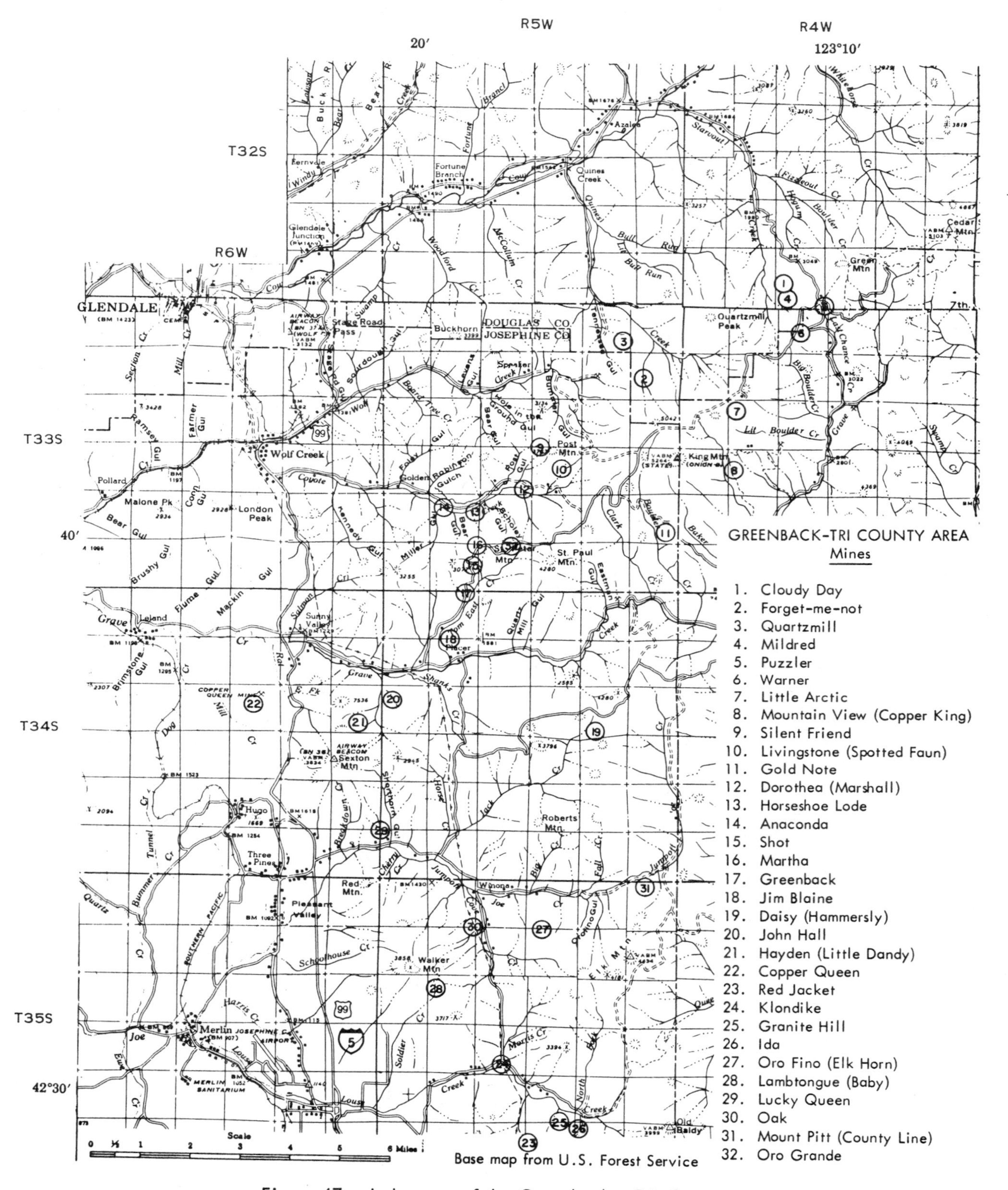

Figure 47. Index map of the Greenback - Tri-County area.

be a substantial producer through 1941.

Considerable placer mining was also done on Jumpoff Joe Creek. The Sexton placer on Bummer Gulch near the head of Jumpoff Joe and the Cook and Howland placer mine just below Brass Nail Gulch, also on upper Jumpoff Joe Creek, were active during the 1930's. The northern California Dredging Co. set up a dragline "doodlebug" dredge with a rated capacity of 1500 cubic yards per day on the Jumpoff Joe placer in 1941. The period of operation is not reported, but it is believed to have been short lived. The Swastika placer on Jumpoff Joe Creek near the mouth of Jack Creek was operated for several years before 1910, and placering has been done on Jack Creek and nearby Horse Creek. The main placers on Louse Creek have been the Forest Queen, Granite Hill, and Red Jacket. The Red Jacket is reported to have yielded $65,000 production from about five acres. The Forest Queen placer was last worked during the winter of 1959-60. Although good values were present, the operation was discontinued, owing to difficulty in breaking up the tightly cemented gravel.

The largest dredging operation in the history of Josephine County was carried on from September 1935 to November 1938 by the Rogue River Gold Co., which operated a dredge on the south side of Grave Creek just upstream from Leland. An estimated 115 acres was worked. Average depth of the gravels is not reported, but the reason for discontinuing operations in 1939 was that the gravels extended so deep it was impossible to clean bedrock. Capacity of the bucketline electric dredge was 5000 cubic yards in 24 hours. Rogue River Gold Co. dredge was listed as the largest producer in the county during the period of operation. The dredge was a special design of the company and was equipped with sixty-five $7\frac{1}{2}$-cubic-foot buckets. The company had perfected a system of water clarification, using a series of large settling ponds to meet the demands of recreational interests on the lower Rogue River.

Carlson and Sandberg operated a dragline dredge on lower Coyote Creek for a short time in the winter of 1937-1937.

Lode mines

In 1898 ore from the Greenback mine was being treated in an arrastra at the settlement called Placer on Grave Creek. The mine was owned by the Victor Junior Gold Mining Co. in 1900, and by 1905 an electric-powered 40-stamp mill and 100-ton cyanide plant were installed by the Greenback Gold Mining & Milling Co. (figure 49).

The major lode-mine producers in the area have been the Greenback, $3\frac{1}{2}$ million; Warner, $200,-000+; Daisy, $200,000; John Hall, $90,000; Granite Hill, $75,000; Dorthea, $50,000; and Silent Friend, $41,000+.

A few pocket-type enrichments such as the Hardscrabble Pocket (Gilbert Eri mine), said to have produced $80,000, and the nearby Sluter Pocket, $16,000, were situated near the head of Wolf Creek.

Principal Lode Mines

Greenback mine: Libbey (1963) describes the Greenback mine (figure 50) (map no. 17) as follows:

"The location is on Tom East Creek about 1.5 miles north of the old settlement of Placer and about 5 miles east of U.S. Highway 99 at the Grave Creek bridge. The property includes 243 acres of patented ground and 76 acres held by location. Legal description is secs. 32 and 33, T. 33 S., R. 5 W., and sec. 4, T. 34 S., R. 5 W., Josephine County.

"Parks and Swartley (1916, p. 112-114) reported that the property was owned and operated by a New York group. In 1924 it was acquired by L. E. Clump, who held the mine until 1954. During part of that time the mine was operated by the following lessees: Finley and McNeil of San Francisco in 1937; P. B. Wickham in 1939; and in 1941 Anderson and Wimer, who discontinued work in 1942. The mine was purchased from Clump in 1954 by Wesley Pieren, Grants Pass, the present owner, who is carrying on some exploration.

"The early history began with a rich surface discovery in 1897. The ore was first worked in an arrastra and later, after mine development work, a 40-stamp mill was installed, together with concentration tables and cyanide tanks. Capacity was rated at 100 tons per day. Electric

Figure 48. Columbia placer on Tom East Creek below the Greenback mine (photograph courtesy of Grants Pass Courier).

Figure 49. View at the Greenback mine about 1906 showing 40-stamp mill (photograph courtesy of Grants Pass Courier).

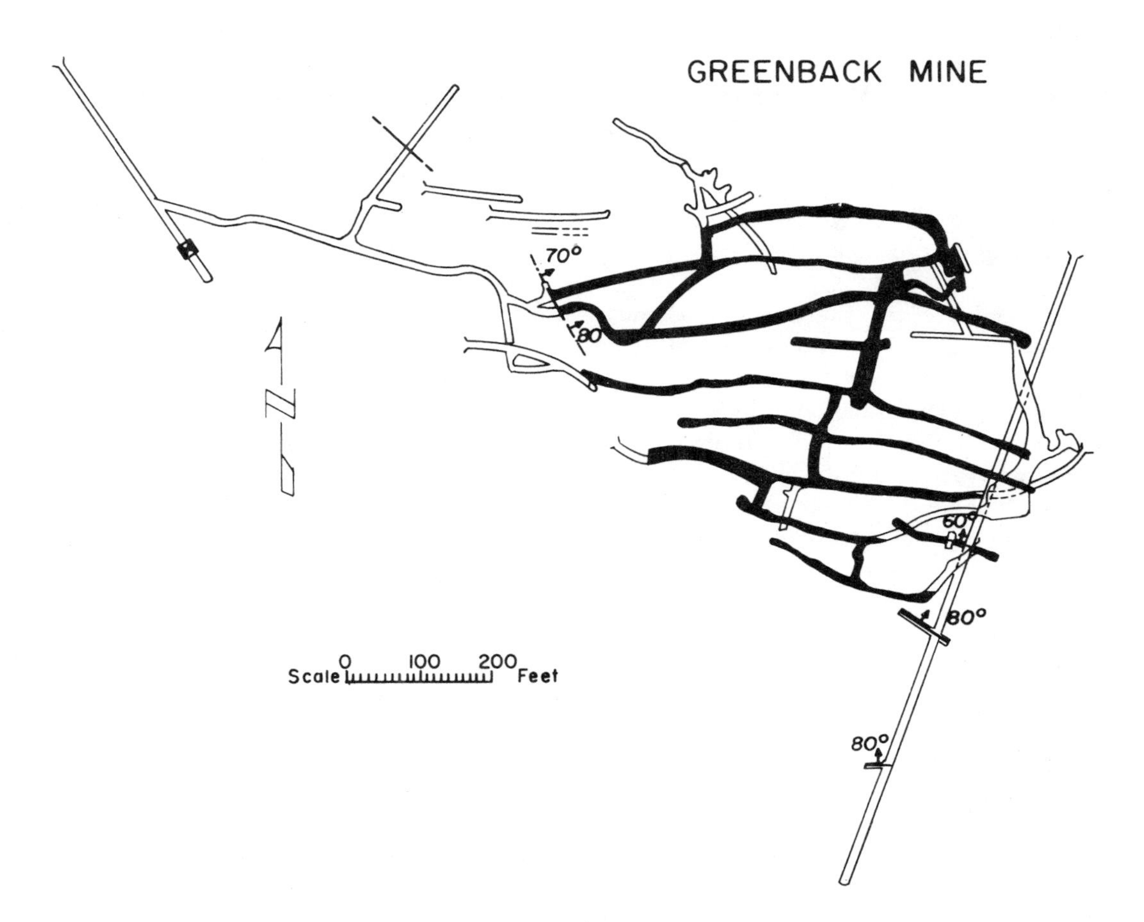

Figure 50. Greenback mine, workings and ore body; workings on ore in solid black (from Winchell, 1914).

power was brought in from the Savage Rapids Dam on the Rogue River.

"Total underground development work aggregated about 7000 lineal feet on 12 levels to a depth of 1000 feet on the dip of the vein, which strikes about east and dips about 45° N. to the ninth level and 55° to 60° below the ninth. The country rock is greenstone and the quartz vein was productive for about 600 feet in length along the strike. Thickness averaged about 3 feet. Value was reported to average somewhat more than $8 a ton (gold at $20 an ounce). The vein was cut off by a fault on the west and by serpentine on the east. Commercial values were principally gold partly recovered by amalgamation. Concentrates made up of chalcopyrite, pyrite, and some arsenopyrite averaged about $75 a ton (gold at $20 an ounce).

"According to Mr. Wickham, production amounted to 3½ million dollars. Mr. Pieren reports that about $100,000 was produced during L. E. Clump's ownership and that the average mill ore was $13.70 a ton.

"One of the largest placer deposits in the state was formed on Tom East Creek below the outcrop of the Greenback mine. It operated as Columbia Placers for many years."

Warner mine: The Warner mine (map no. 6) is a small, high-grade property situated at about 3750 feet elevation. It lies near the head of Last Chance Creek just inside Jackson County and may be reached by way of the Starvout-Grave Creek road about 8 miles southeast of Azalea on Interstate 5 (U.S. Highway 99). The mine was owned by H. B. Warner of Azalea in 1938. It has been worked intermittently for a period of about 40 years by various owners and lessees. Total workings, when visited in June 1963, was about 570 feet, including shafts and raises.

High-grade ore has been mined from a narrow, northwest-trending, nearly vertical contact zone between serpentine and metagabbro. The ore consists of fractured metagabbro that has been penetrated by quartz and calcite veinlets and contains disseminated to nearly massive streaks of arsenopyrite, and some free gold in the quartz. Minor amounts of pyrite and chalcopyrite have also been detected. Values are variable. Samples assayed by the Department range from a few dollars to more than $4000 per ton in gold. The gold-silver ratio is generally greater than 5 to 1. The main gold values appear to be in the arsenopyrite. Attempts at milling the ore have been generally unsuccessful due to poor recovery. Total production is not accurately known buf estimated at about $200,000, which was shipped both as milled concentrates and hand-sorted lump ore.

Daisy (Hammersly) mine: The Daisy mine (map no. 19) is located at the head of Jumpoff Joe Creek at about 3800 feet elevation. It is about 10 miles east of Interstate 5 from the Hugo interchange. The mine was discovered in 1890 and, according to estimate of the owners, has produced $250,000 to $300,-000 in gold. Workings were developed by an inclined shaft (58° N.) which opens 80-foot, 110-foot, and 175-foot levels. The 80-foot level drains the shaft to that level. Ore has been stoped around the shaft from near the surface to the lower level. A proposed lower drainage tunnel about 1000 feet east of the shaft was started about 1939 and 540 feet driven but never completed. Retimbering of the lower caved 540-foot drainage tunnel was in progress when the mine was visited by the Department in 1941. The country rocks are greenstone, altered diorite, and serpentine. There are reported to be five veins, three of which are parallel and two cut the others at right angles. The workings trend N. 50° W. and dip about 55° NE. The ore is reported to be "base" but other details are lacking in reports.

Granite Hill mine: The Granite Hill property (map no. 25) includes 400 acres of patented land on Louse Creek about 9 miles by road northeast of Grants Pass. Parks and Swartley (1916) reported that the Granite Hill mine was purchased by the American Goldfields Co. in 1901, and that between 1902 and 1907 production amounted to $75,000. The mine was closed early in 1908 and later it belonged to the Oregon Gold Mines Co. It was equipped with a 20-stamp mill having amalgamation plates and Frue vanners (see figure 10, pages 34, 35). Ore was mined from a vertical shaft, said to be 430 feet deep, which was later filled to 165 feet with water. The large flow of water encountered in the lower workings is said to have flooded the mine and resulted in its closure in 1908. There are reported to be about 5000 feet of workings on the second level and 7000 feet on the other levels. The vein, which strikes west and dips 70° S. and is said to attain a width of 12 feet on the third level and 14 feet on the fourth level, has an average width of about 5 feet. The country rock is a quartz diorite. The vein is quartz carrying chalcopyrite, galena, pyrite, and gold. Sulfide concentrates from the mill assayed $75 to $100 a ton and were shipped to the smelter at Selby, Cal. The average value of ore milled in 1907 was about $5 a ton ($\frac{1}{4}$ ounce to the ton). Former mine superintendent, C. W. Morphy, is quoted by Parks and Swartley (1916) as reporting three rich ore shoots, each having a length along the vein of about 150 feet and a pitch to the west of south. Oxidized ores found above the 200-foot level were reported to be more valuable than the deeper sulfide ores. Some of the production from the Granite Hill mill came from the nearby Red Jacket and Ida Group.

Lode Mines of the Greenback - Tri-County Area

Anaconda mine

Greenback - Tri-County Area, 14

Location: Josephine County, NE$\frac{1}{4}$ sec. 29, T. 33 S., R. 5 W.

Development: Three or four levels, mostly caved.

Geology: Gold with arsenopyrite occurs in a small, high-grade, rather flat-lying fissure vein in sheared Galice slate near "diorite" greenstone contact.

Production: Worked in small way since 1890's. A 25-ton Huntington mill was installed in 1941. No record of total production.

Reference: Department Bulletin 14-C (Josephine), 1942:96-97.

Cloudy Day mine

Greenback - Tri-County Area, 1

Location: Douglas County, N$\frac{1}{2}$ sec. 33, T. 32 S., R. 4 W., elevation 3000 feet.

Development: Three tunnels, each a little less than 100 feet long.

Geology: Fractured quartz vein in 15-foot, northeast-striking, 45°-southeast-dipping black, slaty zone in "greenstone" carries pyrite, chalcopyrite, arsenopyrite, and free gold. Gold to silver ratio, 5:1.

Production: Claims located in 1928 and worked until 1936. Production not recorded. One 1500-pound stamp used with amalgamation plate and table run by 8-hp. steam engine.

Reference: Department mine file report, 1939.

Copper Queen mine

Greenback - Tri-County Area, 22

Location: Josephine County, sec. 15, T. 34 S., R. 6 W., 1950 feet elevation.

Development: Seven tunnels and many test pits. Total underground development is about 1000 feet.

Geology: Irregular masses of gossan-capped pyrrhotite with some chalcopyrite and pyrite occurring in a 30-foot, northeast-striking, southeast-dipping zone in or near the contact of Galice sediments and volcanics (greenstone).

Production: Two carloads of sorted copper ore were shipped in 1916. Gold values averaged about 0.44 oz./ton, and silver 2.7 oz./ton. Copper percentages were low.

References: Parks and Swartley, 1916:73; Department Bulletin 14-C (Josephine), 1942:99; Department mine file report, 1939.

Daisy (Hammersly) mine

Greenback - Tri-County Area, 19

Location: Josephine County, S½ sec. 14, T. 34 S., R. 5 W., between 3600 and 3800 feet elevation.

Development: About 1700 feet including four levels, 175-foot inclined shaft, and unfinished lower drainage tunnel. Area around shaft is stoped to surface from 175-foot level.

Geology: The main vein trends west and dips north about 50°. The ore is base. Country rocks are greenstone, diorite, and serpentine.

Production: Mine was discovered in 1890; produced an estimated $250,000. Used five-stamp mill, amalgamation plate, and table.

References: Winchell, 1914:228; Parks and Swartley, 1916:84; Department Bulletin 14-C (Josephine), 1942:99-100.

Dorothea (Marshall) mine

Greenback - Tri-County Area, 12

Location: Josephine County, SW¼ sec. 22, T. 33 S., R. 5 W., at 2400 feet elevation.

Development: Total including main level, shafts, raises, and sublevels amounts to about 1500 feet.

Geology: An 8- to 20-foot west-striking and steep north-dipping quartz fissure vein has greenstone footwall and serpentine hanging wall. Values reportedly range from $5 to $12 per ton free-milling gold.

Production: History - not complete. Produced about $50,000 using steam-powered five-stamp mill.

References: Winchell, 1914:186; Parks and Swartley, 1916:87; Department Bulletin 14-C (Josephine), 1942:100-101.

Forget-Me-Not mine

Greenback - Tri-County Area, 2

Location: Douglas County, sec. 12, T. 33 S., R. 5 W.

Development: Several short tunnels, one shallow shaft, and trenches have explored a 300 x 3000-foot area.

Geology: Pyrite and minor chalcopyrite occur in an east-striking, south-dipping, silicified zone of metavolcanic rock that is about 300 feet wide and bounded on the east by serpentine. A minor amount of free gold is reported.

Production: Dates are not reported, but $1300 was reportedly milled from 400 pounds of selected high-grade ore. Preliminary sampling to prove a large, low-grade deposit has indicated that not all of the altered zone carries values.

Reference: Department mine file report, 1945.

Gold Note mine

Greenback - Tri-County Area, 11

Location: Jackson County, west edge sec. 30, T. 33 S., R. 4 W., about 3200 feet elevation.

Development: Nine tunnels, one raise, and numerous cuts.

Geology: Host rocks are west-striking, south-dipping, altered slates, presumably Galice Formation underlain by greenstone on the north. Fine gold occurs with limonite in fractured, weathered slate. Some chalcopyrite with pyrrhotite is found in the lower workings.

Production: Some copper was produced using a small matte smelter during World War I. Has been operated mainly as a gold mine. Total production is not reported. The mine is under exploration by a major company at the time of this printing.

References: Parks and Swartley, 1916:109; Department Bulletin 14-C (Jackson), 1943:71-72.

Granite Hill mine

Greenback - Tri-County Area, 25

Location: Josephine County, SW$\frac{1}{4}$ sec. 26, T. 35 S., R. 5 W., about 2100 feet elevation on 460 acres patented land.

Development: 430-foot vertical shaft, 5000 feet on second level, and about 7000 feet on other levels.

Geology: A west-striking quartz vein in quartz diorite dips 70° S., averages 5 feet in width, and is as much as 14 feet wide. Ore minerals are pyrite, chalcopyrite, galena, and gold. Average mill ore assayed $\frac{1}{4}$ oz. per ton.

Production: Between 1902 and 1907 produced $75,000 with a 20-stamp mill. Flooding of shaft caused closure in 1908. (See photographs, figure 10.)

References: Diller and Kay, 1909; Diller, 1914:42; Winchell, 1914:224 and 226; Parks and Swartley, 1916:171-172; Department Bulletin 14-C (Josephine), 1942:75.

Greenback mine

Greenback - Tri-County Area, 17

Location: Josephine County, SE$\frac{1}{4}$ sec. 32, SW$\frac{1}{4}$ sec. 33, T. 33 S., R. 5 W., and extending into secs. 4 and 5, T. 34 S., R. 5 W., about 2000 feet elevation.

Development: Workings aggregate about 7000 feet on 12 levels mined to 1000 feet in depth on dip of vein.

Geology: The main Greenback vein averages about 20 inches wide, strikes from west to a little north of west, and dips about 50° N. The country rock is called greenstone. The vein terminates in a northwest-trending fault to the west and to the east against a serpentine body. Its length is more than 500 feet. Principal vein filling is quartz, calcite, and pyrite. Gold was 75 percent free milling. Some very high-grade areas were found in the vein. The Irish Girl, a smaller vein paralleling the Greenback vein and about 80 feet to the south, has had little development due to its lower grade.

Production: Discovered in 1897. Reported total production is $3\frac{1}{2}$ million dollars. Most of the production was between 1898 and 1912. Ore was ground in an electrically operated, 100-ton capacity 40-stamp mill installed in 1905. Average grade of mill ore was about $9 per ton ($20 value gold). An unrecorded quantity of ore also came from the other nearby mines. (See photograph, figure 49.)

References: Diller, 1914:31-34; Winchell, 1914:186-188; Parks and Swartley, 1916:112-114; Department Bulletin 14-C (Josephine), 1942:104-106; Libbey, 1963.

Hayden (Little Dandy) mine

Greenback - Tri-County Area, 21

Location: Josephine County, east edge sec. 13, T. 34 S., R. 6 W., elevation about 1600 feet.

Development: Two tunnels on property, lower 375 feet on vein, upper caved.

Geology: Small, rich quartz vein running slightly north of east in greenstone near serpentine contact.

Production: Discovered 1897 by John Hayden and John Hall. To 1905, Hayden produced about $10,000, using an overshot waterwheel-driven arrastra.

References: Department Bulletin 14-C (Josephine), 1942:108; Grants Pass Courier, sec. 2, p. 6, Jan. 27, 1937.

Horseshoe Lode mine

Greenback - Tri-County Area, 13

Location: Josephine County, NW$\frac{1}{4}$ sec. 28, T. 33 S., R. 5 W., at 1900 feet elevation.

Development: In 1940 there were two tunnels, 60 and 94 feet long, and three open cuts.

Geology: Country rocks are serpentine and greenstone. Two veins, one striking east and dipping north, and another striking northwest and dipping southwest, intersect to form ore shoot. Minerals are quartz, rhodochrosite, pyrite, chalcopyrite, gold, silver, and a reported trace of tellurides.

Production: About $5,000 had been produced by 1940. Dates of operation were not given.

Reference: Department Bulletin 14-C (Josephine), 1942:108-109.

Ida mine

Greenback - Tri-County Area, 26

Location: Josephine County, SE$\frac{1}{4}$ sec. 26, T. 35 S., R. 5 W., about 2200 feet elevation.

Development: About 2000 feet of underground work and numerous surface cuts on seven claims, one patented.

Geology: Country rocks include greenstone, serpentine, diorite, and porphyritic dike rocks. Quartz fissure veins occur in east-west shear zones. Mineralization also occurs in a broad, north-trending serpentine shear zone with no quartz gangue. Eight different veins are mentioned in the reports. Ore minerals are pyrite, small amounts of chalcopyrite, galena, and free gold. In places, later low-temperature minerals such as arsenic, antimony, and mercury are reported.

Production: Original location was about 1890. Was equipped with a 50-ton flotation mill by 1931, after a history of intermittent activity and several different operators. Very little mining has been done since 1932. Total production is not reported.

References: Parks and Swartley, 1916:171-172; Department Bulletin 14-C (Josephine), 1942: 79-81.

Jim Blaine mine — Greenback - Tri-County Area, 18

Location: Josephine County, secs. 4 and 5, T. 34 S., R. 5 W., half a mile southwest of Greenback.

Development: Several adits, total not reported.

Geology: The country rock is greenstone; 3-foot quartz vein strikes N. 58° W., dips 65° NE.

Production: Was equipped with small, water-powered stamp mill when examined by Parks and Swartley. Operated during early 1900's. Production not reported.

References: Winchell, 1914:188; Parks and Swartley, 1916:134-135.

John Hall group — Greenback - Tri-County Area, 20

Location: Josephine County, NE$\frac{1}{4}$ sec. 18, T. 34 S., R. 5 W., about 1800 feet elevation.

Development: Six tunnels totaling 1400 feet and much surface work.

Geology: Several veins, the majority striking westerly. Country rocks are greenstone, serpentine, slates, and some intrusives.

Production: Operation dates not given. Production of $90,000 reported from high-grade enrichments.

Reference: Department Bulletin 14-C (Josephine), 1942:109.

Klondike mine — Greenback - Tri-County Area, 24

Location: Josephine County, W$\frac{1}{2}$ sec. 22, T. 35 S., R. 5 W., about 2100 feet elevation.

Development: About 660 feet in a number of short adits (1940).

Geology: Country rocks are mapped as greenstone. A quartz vein which strikes northwest and dips 75° NE. is reported to average $21.00 to the ton. One of the upper tunnels exposes a 6- to 10-inch vein.

Production: Located in 1905 by a Mr. Jordan, who mined a small pocket. Later owners milled a small tonnage of $20+ ore in first a two-stamp mill and later a small ball mill. Total production is small, but not recorded.

References: Department Bulletin 14-C (Josephine), 1942:84; Diller and Kay, 1924.

Lambtongue (Baby) mine — Greenback - Tri-County Area, 28

Location: Josephine County, NE$\frac{1}{4}$ sec. 17, T. 35 S., R. 5 W., elevation 2200 feet.

Development: About 1500 feet in two adits, plus a connecting raise and stopes.

Geology: A narrow fissure vein in metagabbro strikes northwest and dips steeply northeast; contains quartz, pyrite, chalcopyrite, and gold.

Production: Located in 1897, produced $20,000 prior to 1916; during 1937 and 1938 produced about $6000.

References: Parks and Swartley, 1916:18; Diller, 1914:34-35; Winchell, 1914:225; Department Bulletin 14-C (Josephine), 1942:84-85.

Little Arctic mine — Greenback - Tri-County Area, 7

Location: Jackson County, SW$\frac{1}{4}$ sec. 8, T. 33 S., R. 4 W., at 4700 feet elevation.

Development: About 500 feet of workings from the main adit plus a 60-foot shaft to surface, a 30-foot shaft below the main adit, and a stoped area.

Geology: The workings lie entirely in serpentine. Streaks and coatings of gold as much as a half inch thick occur in a west-striking zone of dark-green sheared serpentine. Calcite is sometimes present and some auriferous arsenopyrite is found below the zone of oxidation.

Production: Mine was located in 1936. To 1965 estimated production has been about $10,000 from intermittent periods of operation.

Reference: Department mine file report, 1962.

Livingstone mine (Spotted Fawn) — Greenback - Tri-County Area, 10

Location: Josephine County, E$\frac{1}{2}$ sec. 22, T. 33 S., R. 5 W., at about 3000 feet elevation.

Development: Four short tunnels in addition to main workings; total not reported.

Geology: Country rocks are greenstone and serpentine. Veins formed in sheared contacts strike N. 10° E. and dip 50° E. Gangue minerals are quartz, calcite, chlorite, and serpentine. Ore minerals are pyrite, arsenopyrite(?), chalcopyrite, and gold.

Production: Discovered in 1901. Total production to 1937 about $20,000 from ore shoot 100 feet long and 10 feet thick. Mine has been highgraded. Was equipped with five-ton Chilean mill.

Reference: Department Bulletin 14-C (Josephine), 1942:110-111.

Lucky Queen mine — Greenback - Tri-County Area, 29

Location: Josephine County, secs. 30 and 31, T. 34 S., R. 5 W., about 1600 feet elevation.

Development: About 1500 feet in six tunnels and two raises.

Geology: Ore is in quartz veins in argillaceous quartzite. Broad mineralized shear zones are also described associated with diorite, talc, and sheared country rock. A 6- to 30-inch northeast-striking, southeast-dipping vein is said to average about half an ounce per ton gold.

Production: First located 1879. A 10-stamp mill built in 1886 and since removed. Production unreported.

References: Winchell, 1914:226; Parks and Swartley, 1916:145; and Department Bulletin 14-C (Josephine), 1942:86-87.

Martha mine — Greenback - Tri-County Area, 16

Location: Josephine County, SW$\frac{1}{4}$ sec. 28, T. 33 S., R. 5 W., between 2400 and 3000 feet elevation.

Development: It is opened by four adits having a total length of about 3000 feet.

Geology: Country rock is greenstone. Ore is similar to Greenback. Vein strikes N. 70° W., dips 55° to 60° N. Average width about 2 feet.

Production: First operated about 1900. In 1904 purchased by Greenback Company. In 1906 connected to Greenback by aerial tram. After closure of Greenback, a five-stamp mill was erected at the mine. Production not reported. Four claims are patented.

References: Winchell, 1914:185-186; Parks and Swartley, 1916:149; and Department Bulletin 14-C (Josephine), 1942:112.

Mildred mine — Greenback - Tri-County Area, 4

Location: One patented claim NW$\frac{1}{4}$ sec. 33, T. 32 S., R. 4 W., about 2500 feet elevation.

Development: Five caved tunnels total about 200 feet.

Geology: Values are in quartz stringers in andesite porphyry (greenstone).

Production: Worked in small way for about 20 years prior to 1936. A mill was built in 1936 and $630 pocket was mined in 1937.

Reference: Department Bulletin 14-C, v. 1, 1940:106-107.

Mountain View (Copper King) mine — Greenback - Tri-County Area, 8

Location: Jackson County, secs. 17 and 20, T. 33 S., R. 4 W., about 4300 feet elevation.

Development: Old workings include 900-foot adit with 125-foot-deep winze at face, and a 68-foot adit to the "copper ledge."

Geology: Chalcopyrite and gold occur in a 12-foot-wide fissure vein in a shear zone in greenstone and serpentine. Owner reported 4 to 15 percent copper and $2 to $4 gold per ton.

Production: Located 1913. Equipped with 16-ton ball mill and small table. No production reported.

References: Parks and Swartley, 1916:226; Department Bulletin 14-C (Jackson), 1943:96-97.

Mount Pitt (County Line) mine

Greenback - Tri-County Area, 31

Location: Josephine-Jackson County line, southeast corner sec. 36, T. 34 S., R. 5 W. and southwest corner sec. 31, T. 34 S., R. 4 W., at about 3050 feet elevation.

Development: About 800 feet of underground work in 1916.

Geology: The ore consists of pyrite in quartz and calcite in sheared argillite and greenstone associated with serpentine. The "vein" (shear zone) strikes about N. 10° W. and is about 3 feet wide.

Production: Diller (1914) reports that the claim was located by H. G. Rice of Grants Pass (no date) and owned by A. C. Cooper. Parks and Swartley (1916) report the owner as G. E. Howland. Ore was ground in a steam-powered five-stamp mill and concentrated by gravity, amalgamation, and cyanidation. Production and dates were not reported.

References: Diller, 1914:36; Winchell, 1914:227-228; Parks and Swartley, 1916:158; Diller and Kay, 1924:7 and geologic map; Department Bulletin 14-C (Josephine), 1942:88.

Oak mine

Greenback - Tri-County Area, 30

Location: Josephine County, SW$\frac{1}{4}$ sec. 4, T. 35 S., R. 5 W., about 1600 feet elevation.

Development: About 800 feet in two adits with a 50-foot winze and new raise (1966).

Geology: Mineralization occurs in a 3- to 12-foot-wide, N. 25° to 30° W.-striking, 70° to 75° W.-dipping shear zone in greenstone and consists of replacement by sphalerite, chalcopyrite, pyrite, pyrrhotite, small amounts of galena, and quartz. Gold and silver values are also present.

Production: Early work was for gold only. A few hundred tons of gossan ore yielding about $3.00 gold per ton were milled. Depth of free-milling gold is to 50 feet. Recent exploration and development work by the Oak Mining Co. (1965-1967) has re-opened and extended the workings and put in a raise from the main adit and run a new lower crosscut to the vein.

References: Winchell, 1914:218-220; Parks and Swartley, 1916:165; Department Bulletin 14-C (Josephine), 1942:88-89.

Oro Fino (Elk Horn) mine

Greenback - Tri-County Area, 27

Location: Josephine County, southeast corner sec. 3, southwest corner sec. 2, and northeast corner sec. 10, T. 35 S., R. 5 W., about 2800 feet elevation.

Development: About 1400 feet of workings, largely in four levels connected to the main adit, plus several shafts (map in Department files).

Geology: Workings are on a west-striking, branching quartz-vein system which dips steeply south and occurs in greenstone. Fractures in the quartz are filled with calcite and pyrite; vein widths are from a few inches to more than 3 feet.

Production: The first work of importance at the mine was done in 1898. Total production is unknown, but 14 carloads of high-grade ore were reportedly shipped to the smelter before 1914, and some lower grade was concentrated in a small mill prior to 1929. In 1964 L. E. Frizzell of Grants Pass reopened the mine and did limited exploration.

References: Diller, 1914:36-37; Winchell, 1914:227; Parks and Swartley, 1916:176; Department Bulletin 14-C (Josephine), 1942:89-90; Department mine file reports and maps, 1945.

Oro Grande mine

Greenback - Tri-County Area, 32

Location: Josephine County, southeast corner sec. 28, T. 33 S., R. 5 W., at 3150 feet elevation in saddle.

Development: Shallow surface workings only.

Geology: Country rock is a highly sheared serpentine. Gold occurs as thin coatings smeared out on shears of talcose serpentine in a northeast-striking vertical zone about 4 feet wide. Some fine-grained, gold-bearing arsenopyrite was also found.

Production: The claim was located in February 1959. Early production included numerous high-grade specimen pieces. Total production to 1966 is estimated to be about $1000.

References: Department mine file report, 1959: C. Wiles, oral communication, 1966.

Puzzler mine

Greenback - Tri-County Area, 5

Location: Douglas County, sec. 34, T. 32 S., R. 4 W., elevation 4000 feet at head of Last Chance Creek.

Development: Twelve-foot shaft, open cut, and 44-foot tunnel in 1938.

Geology: Geologic map by Diller and Kay (1924) shows argillite, greenstone, and serpentine. Nearby mineralization is located in sheared serpentine contact zones.

Production: Ore has been milled in an arrastra driven by a 16-foot, overshot water wheel. No production figures are reported.

Reference: Department Bulletin 14-C, v. 1, 1940:110.

Quartzmill mine — Greenback - Tri-County Area, 3

Location: Douglas County, W½NW¼ sec. 1, T. 33 S., R. 5 W., on Quines Creek about 2150 feet elevation.

Development: A 250-foot drift heads S. 55° W.

Geology: Country rock is greenstone. Values in small quartz veinlets, and occur as free gold and with pyrite.

Production: Property was equipped with a water-powdered, two-stamp mill in 1939. Was first located in the 1860's. Production not reported.

Reference: Department Bulletin 14-C, v. 1, 1940:110.

Red Jacket mine — Greenback - Tri-County Area, 23

Location: Josephine County, NE¼ sec. 34, T. 35 S., R. 5 W., at 2240 feet elevation.

Development: Not described.

Geology: Quartz ore in greenstone carries chalcopyrite, galena, and pyrite. Vein strikes northeast and dips 45° NW., is about 18 inches wide and high grade.

Production: Was mined by Oregon Gold Mines Co. about 1905, together with Granite Hill mine. Placer production below lode produced about $65,000.

References: Winchell, 1914:224-225; Parks and Swartley, 1916:172; Department Bulletin 14-C (Josephine), 1942:91.

Shot mine — Greenback - Tri-County Area, 15

Location: Josephine County, NW¼ sec. 33, T. 33 S., R. 5 W., at 2800 feet elevation.

Development: About 750 feet of workings, including raises, in three tunnels.

Geology: Three west-striking, steep, north-dipping, roughly parallel quartz veins in 120 feet distance in greenstone range from few inches to 2 feet width. Reported average assay is $40 per ton. Ore minerals are pyrite, galena, and gold.

Production: Operated from about 1890 to 1940. Some ore originally treated in an arrastra. Total production not reported.

Reference: Department Bulletin 14-C (Josephine), 1942:115.

Silent Friend mine — Greenback - Tri-County Area, 9

Location: Josephine County, S½ sec. 15, T. 33 S., R. 5 W., about 3200 feet elevation.

Development: About 500 feet in two adits.

Geology: Ore occurs in veinlets which trend various directions - mainly southwest and west - in greenstone, and is composed of quartz, calcite, chlorite, pyrite, arsenopyrite, and locally chalcopyrite and gold.

Production: First worked in late 1800's; 120 acres are patented. Total production from various operators, including a reported $30,000 pocket, has been about $41,000 to 1940.

References: Winchell, 1914:185; Parks and Swartley, 1916:202; Diller and Kay, 1908:143; Department Bulletin 14-C (Josephine), 1942:116.

Warner mine — Greenback - Tri-County Area, 6

Location: Jackson County, NE$\frac{1}{4}$ sec. 4, T. 33 S., R. 4 W., about 3700 feet elevation.

Development: About 600 feet, including shafts, drifts, and raises.

Geology: Mineralization occurs along a northwest-trending, nearly vertical contact between metagabbro and serpentine. Ore is fractured metagabbro with quartz and calcite veinlets, with disseminated to massive auriferous arsenopyrite and minor pyrite and chalcopyrite. Occasional free gold is seen in quartz and the oxidized zone. Assays are often quite high.

Production: Owned by H. B. Warner in 1938. 160 acres patented. Worked off and on by various operators to present. Estimated total production is about $200,000.

Reference: Department mine file report, 1963.

GOLD HILL-APPLEGATE-WALDO AREA

Location

The Gold Hill-Applegate-Waldo area is a broad region covering more than 900 square miles in western Jackson and southeastern Josephine Counties (figure 51). The numerous gold mines and prospects extend from the Rogue River drainage north of Gold Hill southward across the Applegate River system and westward into the upper Illinois River drainage. The area includes parts of the Gold Hill, Upper and Lower Applegate, Grants Pass, and Waldo mining districts.

The diverse topography includes upland alluviated valleys, high ridges, and peaks. Elevations range from about 1000 feet on the Rogue River to 7055 feet on Grayback Mountain. The Siskiyou Mountains in the vicinity of the California border rise above 5000 feet elevation and are snowbound during the winter months. The area is, for the most part, readily accessible by state highways and various county and forest roads.

Geology and Mineralization

Distribution of the gold deposits appears to be influenced by the elongate Wimer and Grayback diorite masses (figure 37) and other associated but smaller, scattered intrusive diorite bodies. The diorites intrude metamorphosed volcanic and sedimentary rocks of the Upper Triassic Applegate Group. The altered volcanics of the Applegate include lava flows, flow breccias, pyroclastics, and related intrusive rocks. The altered sediments consist of tuffaceous sandstone, volcanic wacke, argillite, quartzite, chert, marble, and some conglomerate. The Applegate Group has been tightly folded. Dips are seldom less than 30° and many beds are vertical or overturned. Recrystallization to schist and gneiss is common in contact aureoles surrounding the diorites. Many small bodies of serpentinite intrude the Applegate Group and are particularly numerous in the southern part of the area.

Calculations using data from a majority of the mines in the area show that about 45 percent of the gold deposits occur in metasedimentary rocks (for example, argillites and quartzites) of the Applegate Group; about 33 percent are in metavolcanic rocks or greenstone, also of the Applegate Group; 17 percent in diorite; and 5 percent in serpentinite.

Most of the veins in the area have rather steep dips and the predominant strike direction is northwesterly. About 42 percent strike between N. 20° W. and N. 70° W.; approximately 32 percent have a nearly east-west strike; 24 percent strike northeasterly; and about 2 percent strike north.

Ore minerals include gold, pyrite, chalcopyrite, some galena, pyrrhotite and arsenopyrite, and occasional sphalerite. Gangue minerals are mainly quartz and calcite. Secondary limonite and manganese oxides stain most of the veins near the surface. Veins in the area are characteristically small. Mineralization is occasionally found along shear zones without accompanying gangue minerals.

The rich, near-surface pocket deposits are poorly documented in the literature, due to the fact that very little digging was done to recover the values and when a pocket trace was found and worked out the hole was abandoned. It appears that most of the pockets were found in metasediments. Sooty manganese and iron oxides were usually present and gold often occurred in greater abundance than quartz.

A few areas of potentially large, low-grade gold mineralization are indicated in the literature. Among them is the Frog Pond mine (no. 60), where a 104-foot zone of low-grade mineralization is described. Another is the area of numerous small pockets referred to as the Mansfield placer (Department Bulletin 14-C [Jackson], 1943, p. 91) lying northeast of the Millionaire mine (no. 15). The Albright mine (no. 59) has two bodies of gossan, 80 by 900 feet and 20 by 300 feet, that have been explored by several adits and drilling. The underlying ore contains disseminated-to-massive pyrite with some chalcopyrite. Other such areas of large, low-grade mineralization are likely to be present.

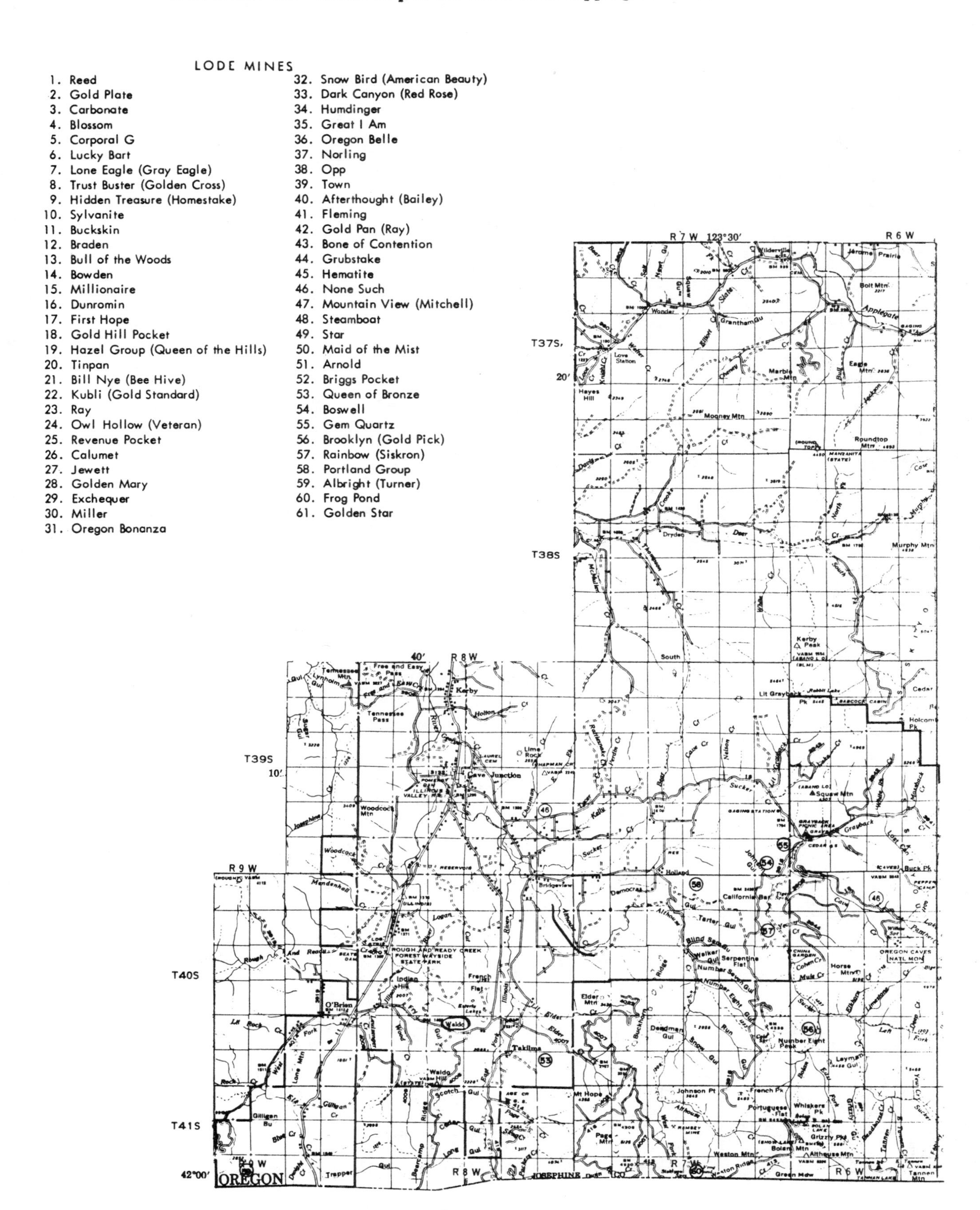

FIGURE 51. Index map of the Gold Hill-Applegate-Waldo area.

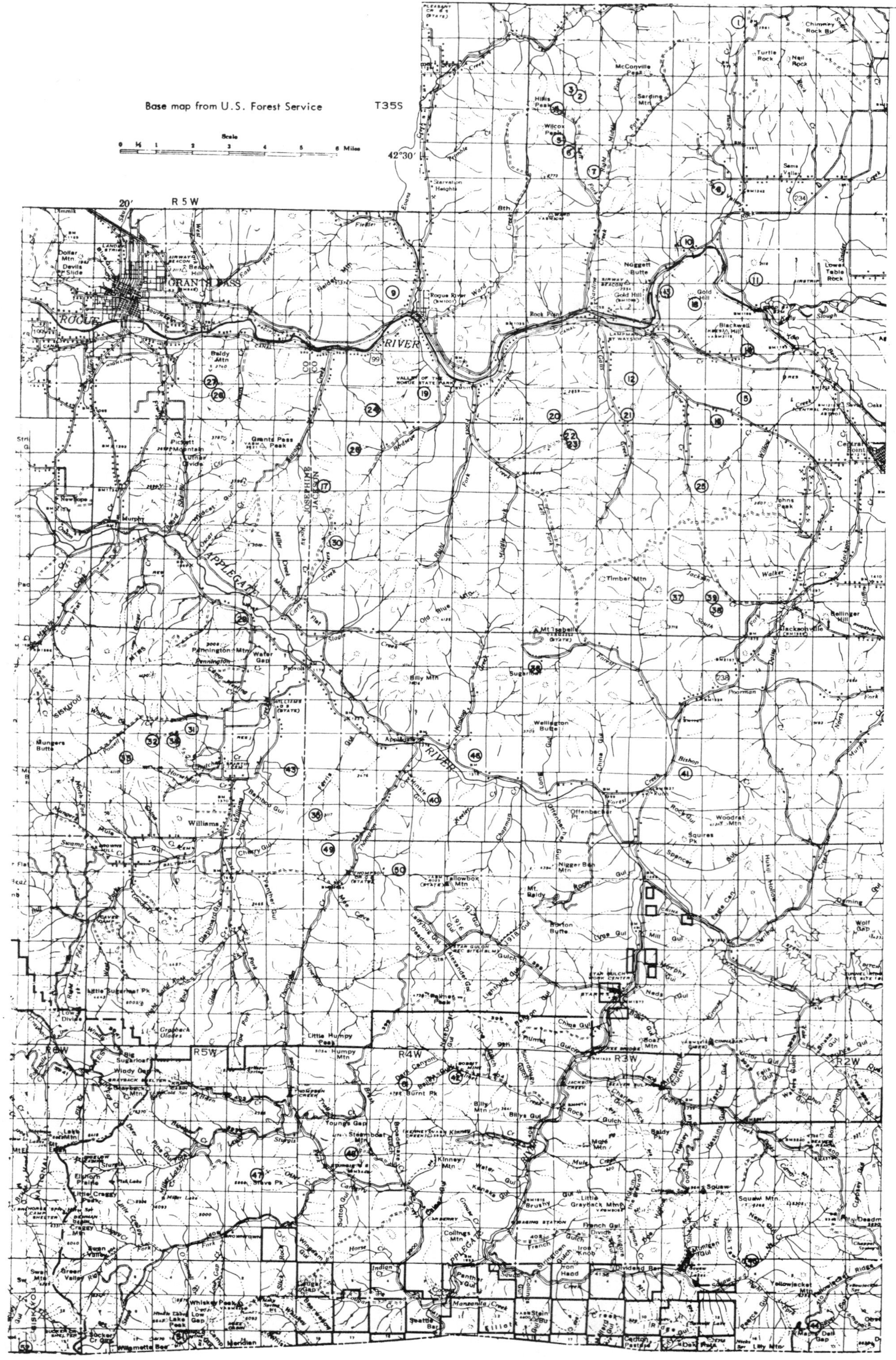

10′
123°00′
Base map from U.S. Forest Service
T35S
T36S
Scale
0 ½ 1 2 3 4 5 6 Miles
42°30′
20′
R5W
R6W
R4W
R3W
R2W
Chimney Rock Bu
Turtle Rock
Neil Rock
McConville Peak
Sardine Mtn
Hillis Peak
Wilcox Peak
Starvation Heights
Sams Valley
Dollar Mtn
Devils Slide
GRANTS PASS
ROGUE
RIVER
Rogue River Ward
Rock Point
Nuggett Butte
Gold Hill
Lower Table Rock
Blackwell Hill
Baldy Mtn
VALLEY OF THE ROGUE STATE PARK
Grants Pass Peak
Pickett Mountain
Murphy
New Hope
JOSEPHINE
JACKSON
APPLEGATE
Timber Mtn
Old Blue Mtn
Mt Isabelle
Jacksonville
Bellinger Hill
Central Point
Johns Peak
Pennington Mtn
Water Gap
Provolt
Billy Mtn
Sugarloaf
Wellington Butte
Applegate
Mungers Butte
SISKIYOU
Williams
Offenbacher Pt
Woodrat Mtn
Squires Pk
Yellowbox Mtn
Mt. Baldy
Burton Butte
Wolf Gap
Little Sugarloaf Pk
Low Divide
Grayback Glades
Little Humpy Peak
Humpy Mtn
Big Sugarloaf
Windy Gap
Grayback Mtn
Burnt Pk
Billy Mtn
Youngs Gap
Steamboat
Kinney Mtn
Stove Pk
Elkhorn Peaks
Little Craggy Peak
Craggy Mtn
Swan Mtn
Green Valley
Collings Mtn
Little Grayback Mtn
Squaw Pk
Squaw Mtn
Iron Knob
Iron Head
Dividend Bar
Yellowjacket Mtn
Whiskey Peak
Low Gap
Lake Peak
Seattle Bar
Manzanita Creek
Stein Bu
Dutchman Peak
Dell Gap

History and Production

Early history of the area centered around its placer-mining operations. Rich placer deposits were discovered at Jacksonville late in 1851, and for several years thereafter would-be miners and fortune hunters flooded into the region (Spreen, 1939). Soon after the Jacksonville discovery, highly productive diggings were found on Sterling Creek (Haines and Smith, 1964). Early in 1852 the famous deposits known as "Sailors Diggings" were found in the headwaters of the Illinois River. The resulting settlement at this site was later called Waldo. Shortly after this discovery was made, there was a rush to Althouse Creek where the stream bed, which soon became known for its large nuggets, was uniformly rich.

In the years immediately following the above discoveries, Jacksonville and Waldo became the most populous settlements in the area and the mining centers of Jackson and Josephine Counties, respectively. Jacksonville has survived as a small town, but scarcely a brick remains to remind us of the once bustling town of Waldo.

As in the other areas, the first work was done mostly with a pick, shovel, and pan. Rockers and long-toms were also used and streams were often diverted for short distances. A few years after the "cream" of the rich placers had been "skimmed" by concentrated hand work, large-scale hydraulic mining with ditches, pipes, and giants was developed.

The Sterling Creek placer south of Jacksonville, one of the largest hydraulic operations in the area, is reported to have produced about $3 million by 1914, and the Esterly placer near Waldo about half a million dollars. Production records are not available for the other placered areas, but it is estimated that the Foots Creek and Forest Creek dredged areas taken together probably produced well over one and a half million dollars, and that the output for the dredged area on Pleasant Creek was probably in excess of half a million dollars. Values recovered from dredged gravels on Forest Creek were reported to range from 6 cents to 36 cents per yard. During 1940 the Murphy Murray Dredging Co. produced 4253 ounces of gold and 616 ounces of silver from 627,261 cubic yards of Foots Creek gravel (Minerals Year Book for 1940, p. 433) which amounted to about 24 cents per yard recovery at $35 per ounce. In 1940, which was a peak gold-production year for Oregon, there were seven operating dredges in Jackson County and one in Josephine County.

Shenon (1933c) estimates that the rich, shallow gulch gravels in the vicinity of Takilma and Waldo, which were worked by the early miners beginning in 1852, paid as much as $2 per square yard of bedrock uncovered. He estimates the total minimum placer production for the Waldo-Takilma area at $4,000,000.

Lode mining in the Gold Hill-Applegate-Waldo area began in the 1860's after the richer and more easily obtainable placer deposits had been worked over. Exceptions were the rich pocket deposits discovered and quickly mined out in the early days. The area is particularly well known for its gold pockets, some of the more famous of which are described by Libbey (1963) as follows:

> "Hicks Lead: The first gold 'pocket,' also the first gold lode, discovered in Oregon was the so-called Hicks Lead found on the left fork of Jackson Creek above Farmers Flat in Jackson County. Sonora Hicks, the discoverer, working with his brother, took out $1,000 in two hours, according to the *Jacksonville Sentinel* of that time. Walling (1884, p. 328) relates that Hicks sold his claim to Maury, Davis, and Taylor, owners of the adjoining claim, who then built the first arrastra in Oregon in order to treat the Hicks ore. The yield from the Hicks claim was $2,000.
>
> "Gold Hill Pocket: The most famous of all was the astonishing Gold Hill Pocket (map no. 18) discovered in January, 1857 by Emigrant Graham and partners near the top of the hill 2 miles northeast of the town of Gold Hill in SW$\frac{1}{4}$NE$\frac{1}{4}$ sec. 14, T. 36 S., R. 3 W., Jackson County, at about 2,000 feet elevation. According to available records (Oregon Dept. Geology & Mineral Ind., 1943, p. 70), the outcropping rock was so full of gold that it could scarcely be broken by sledging. The crystallized quartz associated with the gold was not honeycombed as it generally is where sulfides have leached out of the rock, leaving sprays of gold in the cavity. The gold in this pocket went down only 15 feet and occurred in a fissure vein striking about N. 20° W., dipping about 80° E., with a vertical gash vein cutting the fissure nearly due east. The fissure vein averages 5 feet between the walls with 1 to 2 feet of gouge on the

footwall, which contains calcite and quartz mixed with a little pyrite, in spots containing free gold. A mass of granite, about 5 feet wide by 200 feet long, crops out in the footwall side of the fissure. The country rock is pyroxenite. It is said that this pocket produced at least $700,000.

"Revenue Pocket: Another large 'pocket' was named the Revenue (map no. 25). It was found and mined out (date unknown) by the Rhotan brothers 5 miles south of Gold Hill on Kane Creek in sec. 11, T. 37 S., R. 3 W., Jackson County, at an elevation of about 2570 feet. Reportedly it produced $100,000 (Parks and Swartley, 1916, p. 193) and was one of the larger pockets discovered by Rhotan brothers, who evidently were well-known pocket hunters.

"Steamboat Pocket: This important enrichment in a network of quartz veins in andesite was found in the Steamboat mine (map no. 48) about 1860. The location is on Brush Creek, a tributary of Carberry Creek, 2 miles west of Steamboat and 42 miles by road west of Medford. It is in sec. 20, T. 40 S., R. 4 W., Jackson County. The property has had several names and once was known as the Fowler mine, derived from the name of one of the owners of the Fowler and Keeler Trading Post on the Applegate River, 17 miles distant, and under this name was a litigant in long and costly law suits over title. The yield from the pocket (Parks and Swartley, 1916, p. 212) is reported to have been $350,000.

Figure 52. Photograph taken in 1904 of Briggs Pocket showing David Briggs, father of discoverer, holding a chunk of nearly pure gold. (Photograph courtesy of Grants Pass Courier.)

"Johnson and Bowden Pockets: Two pockets (map no. 39) in the Jacksonville locality are described under the name of Town Mine by Parks and Swartley (1916, p. 136). Date of discovery and extraction is not recorded. The deposits were discovered at points about 600 feet apart, approximately 2 miles west of the reservoir on Jackson Creek in sec. 25, T. 37 S., R. 3 W., Jackson County. The Johnson deposit yielded $30,000 and the Bowden $60,000.

"Roaring Gimlet Pocket: Diller (1914, p. 46) described a rich deposit known as the Roaring Gimlet pocket, discovered in 1893. It was found at the mouth of China Gulch, Jackson County, about 2½ miles south of the Gold Hill pocket. The high-grade ore was apparently liberated from oxidized sulfides, leaving very little quartz, and formed an enriched gouge seam from a quarter of an inch to 6 inches thick between a porphyry footwall and a slate hanging wall. At a depth of 40 feet the vein continued down between dioritic walls and contained some small kidneys of calcite and quartz with pyrite - a gangue looking very much like that of the Gold Hill pocket. Several small pockets were extracted just east of the large Gimlet pocket. The combined yield is said to have been $40,000."

Another dazzling, but short-lived, discovery was the Briggs pocket (map no. 52) found in 1904 at the head of Thompson Creek in Josephine County by Ray Briggs. According to the Grants Pass Courier for

July 28, 1904, the pocket produced $32,000 worth of gold in a two-week period from a narrow cut 10 feet long and 7 feet deep (figure 52). Large slabs of porous gold about 1 inch thick and 2 or 3 feet across were reportedly mined. The pocket is credited with an additional production of about $18,000 before it was cleaned out.

Other important producing lode mines in the Gold Hill-Applegate-Waldo area are the Sylvanite mine with $700,000; Oregon Belle, about $250,000; Lucky Bart and Opp mines with about $100,000 each; Boswell, $79,000; Rainbow, $46,000; and the Braden and Great I Am, $30,000 each. Other mines in the group that have had significant production but no reported records are the Humdinger and Jewett.

Placer Mines

The area has had a number of important hydraulic and dredge operations as well as extensive areas that were worked by various hand methods (see figure 35, in pocket). Some of the more important areas where placer mining was done and the types of equipment used are described below.

Rogue River and tributaries

The Rogue River and its tributaries, Kane Creek and lower Foots Creek, were the sites for some of the earliest gold dredging in the area. In later years dredges worked gravels in upper Foots Creek, Pleasant Creek, and Sardine Creek.

The first dredge was set up on the Rogue near Tolo, upstream from Gold Hill, in 1898. According to Winchell (1914), this was a short-lived operation. In 1903 the Champlin Electric Gold Dredging Co. purchased property on lower Foots Creek and constructed a steam-powered bucket-line dredge. Electric power from the Ray plant near Gold Hill was installed in 1905, thereby reducing the operating cost by one half (Diller, 1914, p. 107). The capacity of this dredge, which was equipped with 36 eight-foot buckets, was 2000 yards per day.

In 1908 the Electric Gold Dredging Co. worked a tributary of Kane Creek in the SW$\frac{1}{4}$ sec. 36, T. 36 S., R. 3 W. (Diller, 1914, p. 106). The operation made use of an electric power shovel which fed a washing plant at the rate of 500 cubic yards in 10 hours. Power for this dredge also came from the Ray dam on the Rogue River.

The area above the forks of Foots Creek, for a distance of about 2 miles on each fork, was dredged over a period of 7 years by the Rogue River Gold Mining Co. before its 1000-ton boat was moved to Grave Creek near Leland in September 1935 (see description of dredge in Greenback-Tri-County area). Dredging on Foots Creek apparently had several periods of inactivity and new starts. In late summer and fall, low water often necessitated the shutting down of operations until after rainfall again replenished the supply.

In January 1941, after dredge construction, the Murphy-Murray Dredging Co. (see photograph, fig. 12-a, p. 37) started digging on Middle Fork Foots Creek above the area dredged by Rogue River Gold Mining Co. and covered an area about 1$\frac{1}{2}$ miles up stream. In March 1941 the dredge was dismantled and moved to Ditch Creek, where digging began June 1941 and was discontinued in the same year and moved to eastern Oregon (Department Bulletin 14-C [Jackson], 1943, p. 97).

The Pleasant Creek Mining Corp. dredge, a steel-hulled, 3-cubic-foot connected bucket-line, diesel-powered, flume type, operated in secs. 22 and 27, T. 34 S., R. 4 W. during the period from 1939 to 1942. Testing reportedly showed a little better than 17 cents per yard values. This dredge was never reactivated after being shut down in 1942, and can still be seen on Pleasant Creek (1967).

A 1$\frac{1}{4}$-yard shovel and washer plant ("doodlebug"-type dredge) began operation on lower Sardine Creek in April 1939 and completed about a 1-3/4-mile stretch in September 1940. The operation was known as Gold Hill placers. The dredged area has been leveled and resoiled.

A number of other creeks which drain into the Rogue River should be mentioned for their placer-gold operations. From east to west they are Sams, Galls, Birdseye, Ward, Savage, Greens, and Bloody Run Creeks. Of these, Galls Creek, south of Gold Hill, was perhaps the most productive. The Blockert mine on Galls Creek was, until a few years before being reported on by Diller (1914, p. 106), the most important placer mine in the Gold Hill district. A few other placers were also worked on Galls Creek at that time.

Applegate River and tributaries

The Applegate River and its tributaries were very productive, and placers were extensively worked by hydraulic, dredging, and various nonmechanical methods. Some of the more important tributary streams placered were Sterling, Forest, Humbug, and Thompson Creeks.

The Sterling Creek placer was worked upstream from the mouth at Buncom, a distance of about 7 miles. The lower 3 miles of the Sterling channel is narrow, then it widens out considerably. Gravels being worked about 1908 were 20 to 40 feet thick, and gold was found across a width of nearly 200 feet.

The first hydraulic mining was done by the Sterling Mining Co., incorporated in 1877. The Sterling ditch from First Water Gulch on Little Applegate River is about 23 miles long and was completed in one year, 1877 (Haines and Smith, 1964, p. 49). After about 25 years of laborious seasonal hand mining along Sterling Creek by a large number of operators, during which time water was nearly always a precious commodity, the new ditch enabled a large-scale hydraulic operation to take over. The creek has been worked up to the level of the ditch terminus, about 2600 feet. Some power equipment was used to move the gravel in the later stages of operation during the 1930's. Diller (1914, p. 110) reported that value of the gravels was about 40 cents per cubic yard, at $20 per ounce, and that total production of the mine to that time was said to exceed $3,000,000.

Most of the area placered on Little Applegate River below Buncom belonged to the Sterling mine holdings and was served by the 6-mile China Ditch on the south side of the river. Part of the area is known as the Federal (Aurora) placer and was operated seasonally until recent years (about 1954). Bedrock is altered volcanic rock (greenstone) with some interbedded argillites of the Applegate Group. The main channel and bench gravels have been extensively mined. The recovered gold is coarse. Operation was seasonal, since water rights were shared with farmers. The giants were operated with a 200-foot head of water (Department Bulletin 14-C [Jackson], 1943, p. 160).

Forest Creek placers have been extensively worked by both hydraulic and dredging methods. Diller (1914, p. 110-111) describes three hydraulic placers on Forest Creek: the Spaulding mine in section 4; the Sturgis mine in section 10; and the Pearce mine in section 11. Diller reports values of from $7,000 to $12,000 per acre of recovery from Forest Creek gravels, which ranged in thickness from 10 to 60 feet and averaged about 25 feet. A number of other small placers were active on Forest Creek and the creek was among the most productive in the early days.

In 1940 and 1941 the B-H Company operated a 1½-yard dragline dredge and 150-yard-per-hour diesel-powered washing plant mounted on a steel frame and wooden pontoons on upper Forest Creek. The Hayfork dredge also operated for a brief period near Ruch in 1940-1941.

Total output of the Forest Creek area is not reported, but an estimate of more than $1,000,000 is suggested for the combined production of the hydraulic placers and dredges that operated on it.

Humbug Creek was another important placer-gold producer in the Applegate area. Several small placer operations, including some drift placer work, hydraulicking, and various other nonmechanized methods, have been active on Humbug Creek since the early days. Operations were seasonal due to limited water. Bedrock is metavolcanic. Gravel is generally small in size, with only a few large boulders and considerable clay. The ground is reported to run from 35 cents to $1.14 per cubic yard, at $35 per ounce, and an estimated 1,500,000 cubic yards are present (Department Bulletin 14-C [Jackson], 1943, p. 165-166).

The Hayfork dredge operation on Thompson Creek was a dragline "doodlebug"-type washer equipped with the usual trommel, stacker, and sluices. It was rated at 2500 yards per day. The operators began work on Thompson Creek in January 1940, dredged through the 3/4 mile of leased property, and in April 1940 moved to Forest Creek near Ruch (Mining World, September 1940, p. 19-20).

One of the larger hydraulic mines in the area, the Layton placer south of Provolt near the Jackson-Josephine County line, worked gravel deposits on Ferris Gulch as well as nearby Whiskey and Bamboo Gulches. A 21-mile upper ditch and an 18-mile lower ditch served the property. Extent of the area worked and total production are not reported. Parks and Swartley (1916) state that the mine had been a good producer for years. It was still being operated in 1940 (Department Bulletin 14-C [Jackson], 1943, p. 168-169.

A number of small placer operations were active in the early days along the upper Applegate River, especially on bench gravels adjacent to the principal channel. The area of placer activity extends on up the river into northern California. Carberry Creek, Squaw Creek, and especially Palmer Creek were all worked extensively.

On the lower river other creeks with significant placer-mining activity were Keeler, Williams, Powell, Slagle, Carris, Miners, Rocky, and Miller in the Missouri Flat area, Oscar, Board Shanty, Grays, and Murphy Creeks. The history and production records of these areas is scanty, but one can still see evidence of the early-day mining and a few small, seasonal operations continue in the area. A one-yard diesel shovel was operated on Oscar Creek in 1933 and gravel was transported about 1 mile from the shovel to a sluice by five-ton trucks.

Waldo area

In the southern part of the Gold Hill-Applegate-Waldo area the bulk of placer production has been along Sucker and Althouse Creeks and in the vicinity of Takilma and Waldo.

Diller (1914, p. 118) states:

> "From the gravels of Althouse and Sucker Creeks a large amount of gold was washed in the early days of placer mining in Oregon, but for several years the production has not been great, as the best ground was worked many years ago. During 1907 the production of the streams of this district probably did not exceed $6,000. There are no large mines, but numerous small ones..."

Placers in the Waldo area were along a number of small gulches which cut across old cemented bench gravels that are partly decomposed. These gravels, mapped by Shenon (1933c) as coarse "Tertiary conglomerate" and by Wells (1949) as early Pleistocene "auriferous gravels of the second cycle of erosion" are apparently the intermediate host rock for much of the placer gold in the area. Some of the more important early-day placers that were worked by a large number of individuals on closely spaced claims were on Scotch, Allen, Sailor, Waldo, and Fry Gulches, all within a two-mile radius of the town of Takilma.

Hydraulic mining: Large operations in the Waldo area that were mined by hydraulic methods after the ditch system was developed during the 1870's include the High Gravel mine, the Llano de Oro (Esterly) mine, and the Deep Gravel mine.

The High Gravel mine at the head of Allen Gulch in secs. 33 and 34, T. 40 S., R. 8 W. includes several pits covering an area of approximately 150,000 square yards. The mine, which operated to 1917, is estimated to have produced about $90,000, not including production from the old workings along the bottoms of Allen and Scotch Gulches. The gold is found in the old cemented and partly decomposed gravel deposit. The average value was estimated at about 3 cents per cubic yard, at $20 per ounce.

The largest mine in the Waldo area was the Llano de Oro or Esterly mine situated in secs. 8, 9, 10, 15, 16, 21, 22, and 27, T. 40 S., R. 8 W., which included more than 3000 acres of land. Mining was done on Carroll Slough at the head of Logan Cut in secs. 10, 15, and 16, and on French Flat in secs. 22 and 27. Hydraulic elevators were used to mine from pits below the water table on French Flat (Figure 36). These pits, which cover an area of more than 30 acres, are now called Esterly Lakes. The mine was operated by various groups up to 1945. Value of the gravels worked ranged from 12½ cents to 33½ cents per yard, at $20 per ounce. Total production to 1933 was estimated at about $500,000. Production since that time may have been as much as $100,000. The U.S. Bureau of Mines Minerals Yearbook review of 1939 (1940, p. 431) reports that during the year (operating season was generally from 4 to 9 months) 75,000 cubic yards of gravel were hydraulicked at the Esterly mine and 421 ounces of gold and 25 ounces of silver recovered. Some platinum-group metals are also recovered from the mine concentrates in this area. The ratio of platinum to gold in the Llano de Oro mine is estimated at about 1 to 50 (Shenon, 1933c, p. 187).

The Deep Gravel mine was on Butcher Gulch in secs. 16, 17, 20, and 21, T. 40 S., R. 8 W., just over the low ridge west of the Llano de Oro mine. In 1933 four deep pits and an aggregate of shallow pits covered an area of about 65 acres. The mine was operated from year of discovery in 1874 to 1933; the estimated total production was about $276,000. Recorded production between 1907 and 1933 was $26,316 (Shenon, 1933c, p. 188). Kay (1909, p. 74) reports that the average value of pay gravel over a period of five years was about 25 cents per yard, at $20 per ounce.

Dredging: Dredging in the Waldo area has been limited to a few short-lived operations on lower Althouse Creek, Sucker Creek, and along the East Fork of the Illinois near Takilma. A shovel and washing plant owned by Von der Hellen Brothers worked the Leonard placer in sec. 4, T. 40 S., R. 7 W. on lower Althouse Creek from 1936 to 1938. This area and down stream a short distance was also dredged by the Atlas Gold Dredging Corp. during 1940 and early 1941, using a 5-yard dragline that could handle 6000 yards daily and was the largest dragline washing plant in southwestern Oregon. Their work disclosed that much of the area had been drift-mined in the early days by Chinese (W. J. Cannon, oral communication, 1968).

In 1945 and 1946 B. H. Oregon, Ltd., ran a 3-yard dragline and washing plant on Sucker Creek a short distance above Grayback Creek.

A dragline and washing plant was operated intermittently by the Takilma Mining Co. along East Fork Illinois River just north of Takilma during the period of 1947 to 1950. Another dragline and washing plant was active at the Bailey mine on Fry Gulch west of Waldo during the same period. Production of these properties is not reported.

Principal Lode Mines

Sylvanite mine: The main workings of the Sylvanite mine (map no. 10) are on 80 acres of patented land about 3 miles northeast of Gold Hill. The property is under sales contract (1966) to Daniel Jones from the owner, George Tulare of Gold Hill. Libbey (1963) describes the mine as follows:

> "The discovery and early history of the mine are not of public record. Various published reports show that, beginning in 1916, owners and operators were, successively, E. T. Simons, with Stone and Avena, Denver, Colorado, lessees who found scheelite (tungsten ore) associated with the gold ore; Oregon-Pittsburg Co. in 1928; Discon Mining Co., A. D. Coulter, Manager, discoverer of the high-grade ore shoot along the Cox Lyman vein in 1930; Western United Gold Properties; Sylvanite Mining Co.; and finally Imperial Gold Mines, Inc., in 1939. This last company built a concentrating mill of 140 tons daily capacity and cleaned out underground workings to expose the openings where the rich ore shoot had been found.
>
> "The Sylvanite vein or shear zone occurs between metaigneous and metasedimentary (largely argillite) rocks. It shows intense shearing and alteration and is intruded in places by basic igneous dikes. It trends just east of north and dips southeasterly at about 45°. The Cox-Lyman shear zone strikes at right angles to the Sylvanite vein and stands nearly vertical. No certain sequence of faulting in the two shear zones has been established. Ore shoots are said to be from 5 to 12 feet thick and have averaged from $5 to $15 a ton. They have a gangue of quartz and calcite and carry galena, chalcopyrite, and pyrite. A fracture zone roughly parallel to the Sylvanite vein cuts the Cox-Lyman vein and at the intersection a rich ore shoot was found on the hanging wall, producing $1,000 per lineal foot of winze in sinking 600 feet. Discontinuous pockets of ore were found in the hanging wall of the shoot for 200 additional feet of depth. The winze reached 900 feet below the surface. This ore shoot was reported to have yielded about $700,000.
>
> "A total of more than 2,560 lineal feet of underground development work has been done. In addition, numerous surface pits and cuts, now caved, have been dug by pocket hunters.
>
> "Seemingly little effort has been made to explore the scheelite possibilities, although it is known that the Imperial Gold Mines Co. had such plans. They ran into difficulties underground because of caving ground, and presumably war-time conditions finally forced them to close down."

Lucky Bart mine: Lucky Bart Group (map no. 6) is about 6 miles northwest of Gold Hill, west of the left fork of Sardine Creek. Workings are between 2080 and 2900 feet elevation. There were 11 claims, and at least one in NE$\frac{1}{4}$ sec. 29 is patented. The mine was worked intermittently by various operators since discovery in 1890. Diller (1914, p. 38) reported on the group as follows:

> "Ore has been mined from five veins which run in a general direction a little south of west. These veins are on the average less than 2 feet wide. The country rock is metamorphosed sediment, mainly mica slates and micaceous quartzites. The general strike of these rocks in this vicinity is somewhat east of north; the dip is to the southeast and is in general at fairly high angles. The total amount of ore that has been milled exceeds 14,000 tons, which yielded from $4.80 to $100 a ton of free-milling ore. The ore from the Lucky Bart claim carried an average of 3 per cent of sulfides, which ran from 4 to 8 ounces of gold to the ton and a like amount of silver. Nine tons of ore from the deepest workings of this claim were shipped to the Tacoma smelter and gave returns of $130 to the ton. Practically all the ores from the group have been treated at a mill on Sardine Creek; the sulfides were shipped to the smelters at Tacoma, Wash., and Selby, Cal."

Extent of the workings on the Lucky Bart group is not described. Total production may have been as much as $200,000, but it has not been accurately reported.

Opp mine: The Opp mine (map no. 38) is situated about 2 miles west of Jacksonville. It contains workings that total about 7000 feet. It was discovered in the late 1800's and its major period of development was in the early 1900's. A total of 18 adits exposes three main veins which strike northwest to west and dip south. The ore shoots are reported to be where the veins are thickest. Thickness of the veins varies from a few inches to 12 feet. The country rock is siliceous argillite containing some chlorite and pyrite. The rocks belong to the Applegate Group. Their major trend is northerly, with steep dip east. Gangue minerals in the vein are quartz and calcite. Ore minerals are pyrite and gold. The ore was treated in a 20-stamp mill. Amalgamation, gravity concentration, and cyanidation were used. Total production is reported to be a little more than $100,000.

Braden mine: The Braden mine (map no. 12), situated 2 miles south of Gold Hill, was discovered about 1885. The first ore mined was ground in an arrastra. Total production of the mine is not known, but during the early 1900's it was equipped with a 10-stamp mill and for the year 1907 reported $30,000 production. In 1916 the mill was sold to owners of the Ashland mine. The workings total more than 3000 feet, but have reached less than 250 feet of depth. Winchell (1914) and Parks and Swartley (1916) report several quartz veins opened by six adits and an inclined shaft. The important veins strike about N 30° E., dip 25° SE., and have an average width of about 18 inches. The country rocks are mapped as metavolcanics of the Applegate Group. The mine report describes the rocks as interbedded sediments and andesites altered to a calcareous hornblende schist. The ore is described as mainly quartz with a little calcite, some pyrite, and minor amounts of arsenopyrite, chalcopyrite, and galena. About 65 percent of the gold and silver was recovered on amalgam plates and 25 percent as concentrates. Ore mined averaged about $8 to $10 per ton.

Oregon Belle mine: The Oregon Belle mine (map no. 36) is near the head of Forest Creek, a mile south of Mount Isabelle at about 3000 feet elevation. Development work, started in 1890, totals about 1750 feet of underground workings plus stopes. Most of the workings are caved and inaccessible. Total production is believed to have been in excess of $250,000. Two parallel veins mined are the Oregon Belle and the Roberts. The Oregon Belle vein strikes N. 70° to 75° E. and dips 50° to 60° N. It was extensively stoped and portions still exposed are from 2 to 4 feet thick, with an average value of $10 to $15. This vein characteristically pinches, swells, and changes direction of dip. The Roberts vein is about 100 feet south of the Oregon Belle. It has about the same thickness and it strikes N. 80° E. and dips 60° N.

Wall rocks are mostly volcanics of the Applegate Group with some interbedded argillite. The formation strikes in a north-northeast direction and dips at high engles. Transverse (northwest-striking) faults are common in the area and have offset and complicated the vein structure. A map of the accessible workings is available in Department mine files.

Jewett mine: The Jewett mine (map no. 27) is at about 2000 feet elevation on the south side of

Baldy Mountain, $2\frac{1}{2}$ miles southeast of Grants Pass. It was discovered by Thomas Jewett in 1860 and consists of seven patented claims. An eight-stamp mill used at the mine in 1863 was unsuccessful. It was equipped with a five-stamp mill when examined by Winchell (1914), but had been idle for some years. Winchell's profile of the mine workings and ore body shows more than 1500 feet of workings. It is known that the mine produced some gold, but the total amount is not recorded. Walling (1884, p. 330) reported a $40,000 pocket taken out near the surface by the Jewett brothers in 1860. Mineralization occurs in an irregular, sheared brecciated zone having a general strike of N. 20° W. and dip of 75° NE. Some quartz and a lesser amount of calcite fill the breccia. Ore minerals include gold, pyrite, pyrrhotite, and sylvanite (Winchell, 1914, p. 223; Diller and Kay, 1909, p. 61).

Humdinger mine: The Humdinger mine (map no. 34) is located 3 miles northwest of Williams. Shenon (1933b) described the property in detail and mapped 1200 feet of accessible workings. Ore was first discovered at the mine about 1900. Geology of the Humdinger is somewhat like that of several other mines in the area which occur in altered sedimentary and volcanic rocks of the Applegate Group flanking the elongate Grayback diorite body. Sedimentary rocks at the Humdinger are mainly thin-bedded argillite, which strikes northeast and dips steeply southeast. The volcanic rock is probably a metabasalt or meta-andesite (greenstone) and is a hard, massive, fine-grained, greenish-gray material in which no primary structures are readily visible. The greenstones are situated west of the argillite. Veins at the Humdinger strike northwest. They consist of multiple, semi-parallel and branching quartz veinlets which penetrate transverse shears in the argillite and greenstone. The main vein system developed in the greenstone dips about 60° NW. and varies from a little more than 3 feet thick down to a thin gouge. Gangue minerals reported are quartz, altered chloritic wall rocks, calcite, and apophyllite. Pyrite, arsenopyrite, and gold are the main ore minerals. A minor amount of galena has been found and tellurides have been reported.

Values of the ore are quite variable but generally increase with the amount of sulfides. A sample of oxidized ore cut across a 3-foot vein in the No. 2 raise of No. 1 drift assayed 0.40 ounce gold and 2.2 ounces silver per ton. A grab sample from the face of No. 1 drift consisting of quartz with pyrite and arsenopyrite assayed 1.12 ounces gold and 4.70 ounces silver per ton. Shenon (1933b, p. 51) gives results of mill tests run by the operator in which 25 tons oxidized ore from No. 2 tunnel averaged $24 a ton in gold and a test of 13 tons of oxidized ore from No. 3 level which contained more than $40 a ton. The same operator reported sampling the mine across mining widths and blocking out a considerable quantity of ore averaging $12 a ton in gold. No production records are available. A shipment of one carload of hand-sorted ore valued at $2700 was made in 1949. Most of the production was from milled ore in the early 1900's.

Boswell mine: The Boswell mine (map no. 54) is located on the west side of Sucker Creek near Johnson Gulch and consists of 244 acres of patented land. It was discovered by Robert Boswell and his son in 1914. About $75,000 gold was produced from an enriched pocket-type quartz-vein deposit in an area about 2 feet thick, 40 feet long, and 25 feet deep. The vein lies in greenstone near its contact with serpentine. The vein strikes N. 85° E. and dips 60° N. The serpentine-greenstone contact, which strikes northwest and dips east, terminates the vein to the west.

Rainbow mine: The Rainbow mine (map no. 57), formerly known as the Siskron, is located on the west side of Sucker Creek about 4 miles west of Oregon Caves National Monument. The mine was discovered by Mr. Siskron in 1915, and was operated by various parties between 1917 and 1937. The total reported production was $46,500. The main development work, consisting of about 1100 feet of drifts, crosscuts, winzes, and raises, is on the Siskron vein which strikes about N. 20° E. and dips west 35° into the hill. The vein is in metavolcanic rocks of the Applegate Group, and is actually a network of quartz veinlets which pinches and swells to a maximum of 8 feet. It is apparently offset or terminated by cross faults near both the north and south end of the workings. Ore values have a wide range. An average of samples about 1 foot wide taken at various places in the mine was $18.50 per ton (1938). Ore minerals are gold and sulfides including pyrite, arsenopyrite, and chalcopyrite in a gangue of quartz, calcite, and mixed wall rock.

Lode Mines of the Gold Hill-Applegate-Waldo Area

Afterthought (Bailey) mine — Gold Hill-Applegate-Waldo Area, 40

Location: Jackson County, SE¼ sec. 27, T. 38 S., R. 4 W., at 2300 feet elevation.

Development: About 500 feet in three tunnels, and 45-foot shaft and 50-foot winze.

Geology: Ore occurs in small, white-to-bluish quartz vein with some sulfides and rare calcite. The vein strikes N. 70° E., is nearly vertical, and occurs in argillite about 600 feet from a small granitic intrusion.

Production: Ore was milled in an arrastra. Production small; dates and amounts not reported.

References: Winchell, 1914:135; Parks and Swartley, 1916:7; Department Bulletin 14-C (Jackson), 1943:151; Department mine file report and map, 1960.

Albright (Turner) mine — Gold Hill-Applegate-Waldo Area, 59

Location: Josephine County, secs. 15 and 16, T. 41 S., R. 9 W., at about 3000 feet elevation.

Development: There are reported to be 14 short tunnels on the property.

Geology: Two bodies of gossan, 80 x 900 and 20 x 300 feet, occur in silicified greenstone adjacent to serpentine. Sulfide minerals are pyrite and chalcopyrite.

Production: The deposit was discovered in the late 1800's. Small gold production was obtained from the gossan. In 1940 cyanidation tests of gossan ore showed poor recovery. The deposit remains more promising as a copper prospect.

References: Shenon, 1933b:192; Department Bulletin 14-C (Josephine), 1942:177-178.

Arnold mine — Gold Hill-Applegate-Waldo Area, 51

Location: Josephine County, N½NE¼ sec. 16, T. 41 S., R. 5 W., at 5500 feet elevation.

Development: Workings examined include: 130-foot crosscut, 90 feet of drift, and 130-foot trench.

Geology: Average 20-inch-wide quartz vein striking northwest and dipping northeast in Applegate metasediments. Average tenor of ore $20. Minor sulfides (pyrite, galena, sphalerite[?], and chalcopyrite) occur in quartz.

Production: Small two-stamp mill with amalgamation plates. Production about 600 tons since 1915.

Reference: Department Bulletin 14-C (Josephine), 1942:180; Department mine file report, 1957.

Bill Nye (Bee Hive) mine — Gold Hill-Applegate-Waldo Area, 21

Location: Jackson County, SE$\frac{1}{4}$ sec. 33, T. 36 S., R. 3 W., elevation 1680 feet.

Development: Opened by several adits and a vertical shaft.

Geology: Northeast-striking, vertical quartz vein 2 feet wide, and larger northwest-striking vein in Applegate metasediments (impure quartzite).

Production: Records not complete. Was located about the turn of the century. Two shipments between 1907-1909 totaled $12,000 gold. A five-stamp mill was used.

References: Winchell, 1914:173-174; Parks and Swartley, 1916:26-27; Department Bulletin 14-C (Jackson), 1943:49.

Blossom mine — Gold Hill-Applegate-Waldo Area, 4

Location: Jackson County, NE$\frac{1}{4}$ sec. 19, T. 35 S., R. 3 W., elevation 2400 feet.

Development: More than 500 feet of underground workings, mostly caved. Upper and lower workings connected.

Geology: Vein of crushed rock with some quartz, calcite, sericite, and sulfides including pyrite, chalcopyrite, galena, pyrrhotite, and sphalerite(?), with some free gold. Strikes northwest and dips 55° NE. in greenstone.

Production: Early history not reported. Reopened 1928 and produced $2000 gold to 1936.

References: Winchell, 1914:179; Parks and Swartley, 1916:34; Department Bulletin 14-C (Jackson), 1943:51.

Bone of Contention mine — Gold Hill-Applegate-Waldo Area, 43

Location: Josephine County, on line of secs. 24 and 25, T. 38 S., R. 5 W., 1700 feet elevation.

Development: "Several thousand feet."

Geology: At west edge of Grayback diorite body and contact with argillite. Diorite cut by aplite dikes.

Production: History and production data lacking. A 15-stamp mill powered from water from ditch was at mine. Was worked in small way between 1916 and 1941.

References: Winchell, 1914:240; Parks and Swartley, 1916:39; Department Bulletin 14-C (Jackson), 1943:155.

Boswell mine — Gold Hill-Applegate-Waldo Area, 54

Location: Josephine County, 239 acres patented, sec. 36, T. 39 S., R. 7 W., at about 2700 feet elevation.

Development: About 600 feet total in four adits and shafts.

Geology: Quartz fissure vein or silicified shear 1 to 4 feet thick strikes N. 85° E. and dips 60° NW. in greenstone and is cut off by sheared serpentine contact to west. Rich shoot of gold was found in gossan zone adjacent to vein on north.

Production: Discovered 1914. Boswell produced about 4,200 ounces from 14-to-20-inch thick, 40-foot-long, 25-foot-deep zone. Total production possibly greater than $100,000.

References: Department Bulletin 14-C (Josephine), 1942: 182; Department mine file report, 1961.

Bowden claim

Gold Hill-Applegate-Waldo Area, 14

Location: Jackson County, SW¼ sec. 19, T. 36 S., R. 2 W., about 1480 feet elevation.

Development: About 700 feet adit and drifts and 198-foot shaft.

Geology: Quartz vein in diorite as much as 3 feet thick where stoped strikes N. 75° E., dips 85° N. Yielded free gold at 100-foot depth.

Production: No information.

References: Parks and Swartley, 1916:40; Department Bulletin 14-C (Jackson), 1943:52; Winchell, 1914:168-169.

Braden mine

Gold Hill-Applegate-Waldo Area, 12

Location: Jackson County, SW¼ sec. 27, SE¼ sec. 28, T. 36 S., R. 3 W., 1550 feet elevation.

Development: Six adits and an inclined shaft with several drifts total about 2500 feet.

Geology: Country rocks are interbedded metavolcanics and sediments of the Applegate Group. There are several quartz veins. Important ones strike northeast and dip 25° SE.; contain little calcite, some pyrite, arsenopyrite, chalcopyrite, and galena. Assays of a third to a half ounce gold reported.

Production: Mine discovered about 1885. Total production not reported. Production for one year, 1907, using a 10-stamp mill, was $30,000. Little work has been done since 1916.

References: Diller and Kay, 1909:56-58; Winchell, 1914:171-173, picture p. 179; Parks and Swartley, 1916:41; Department Bulletin 14-C (Jackson), 1943:53-54.

Briggs Pocket mine

Gold Hill-Applegate-Waldo Area, 52

Location: Josephine County, sec. 14, T. 41 S., R. 6 W., 4800(?) feet elevation.

Development: Shallow cuts only.

Geology: Metasediments and volcanics of Applegate Group intruded by serpentine, and about a mile distant from diorite stock are mapped in area. One-inch-thick slabs of porous supergene(?) gold in quartz vein on "contact of granite and porphyry."

Production: Found by Roy Briggs, age 18, in 1904. In two weeks the Briggs family recovered gold by hand mortaring valued at $32,000.

References: Oregon Observer, July 2 and 9, 1904; Rogue River Courier, July 7 and 28, 1904; Grants Pass Courier, April 3, 1935; Wells and others, 1940.

Brooklyn (Gold Pick) mine — Gold Hill-Applegate-Waldo Area, 56

Location: Josephine County, center(?) sec. 30, T. 40 S., R. 6 W., at about 3500 feet.

Development: Main adit 300 feet plus stopes.

Geology: Argillite intruded by amphibolite, diorite, and diorite aplite. Fissure vein 12 to 20 inches thick, with 2 to 12 inches of quartz, strikes N. 35° W. and dips 55° NE. Ore higher grade near surface.

Production: Adit connected to small mill on Bolan Creek 600 feet lower by tramway. Dates operated and production not reported.

References: Winchell, 1914:248; Parks and Swartley, 1916:44.

Buckskin (Curry Pocket, May Belle) mine — Gold Hill-Applegate-Waldo Area, 11

Location: Jackson County, south edge sec. 7, T. 36 S., R. 2 W., about 1700 feet elevation.

Development: Ninety-foot tunnel and surface cuts.

Geology: Snow-white to glassy quartz in metavolcanic rock of Applegate Group. Vein exposed for 150 feet.

Production: Coarse gold placered on hillside from Curry pocket. Small pocket mined. Dates and production not reported.

Reference: Department Bulletin 14-C (Jackson), 1943:57.

Bull-of-the-Woods mine — Gold Hill-Applegate-Waldo Area, 13

Location: Jackson County, NE$\frac{1}{4}$ sec. 15, T. 36 S., R. 3 W., 1200 feet elevation on east bank of Rogue River.

Development: One-hundred-foot shaft and 266 feet of drifts on three levels.

Geology: Country rock is dark diorite (gabbro). Fissure vein, resilicified shear zone 2 to 6 feet thick, strikes N. 55° W. and dips 75° to 80° SE., and contains some pyrite and free gold.

Production: Little is known of history prior to 1934. Produced $5,000 during 1934 and 1935. Total production not reported. A two-stamp mill was used.

Reference: Department Bulletin 14-C (Jackson), 1943:57-58.

Calumet mine

Gold Hill-Applegate-Waldo Area, 26

Location: Jackson County, east edge SE$\frac{1}{4}$SW$\frac{1}{4}$ sec. 5, T. 37 S., R. 4 W., elevation 2400 feet.

Development: About 500 feet in three tunnels.

Geology: Values occur in northwest-trending, 80° SW-dipping, 2-foot shear zone with 1- to 7-inch quartz veinlets. Country rocks are altered Applegate Group sediments and volcanics intruded by diorite dikes.

Production: Located in late 1800's. First production reported as about $700 from surface pocket. The amount of production between 1934 and 1937 from ore milled in an arrastra is not reported.

Reference: Department mine file report, 1963.

Carbonate mine

Gold Hill-Applegate-Waldo Area, 3

Location: Jackson County, W$\frac{1}{2}$ sec. 17, T. 35 S., R. 3 W., about 2100 feet in Murphy Gulch.

Development: About 250 feet in two tunnels which are now caved.

Geology: Tapering fissure vein from knife edge to as much as 4 feet thick, average 1 foot, in diorite strikes N. 20° W., dips 80° NE. Ore reportedly assays $20 to the ton. Country rock is altered diorite.

Production: Discovered in 1930, it was operated intermittently until about 1940. Was equipped with three-stamp mill using both amalgamation and cyanidation. Production not reported but believed small.

References: Department Bulletin 14-C (Jackson), 1943:58-59; Department mine file report and map, 1938.

Corporal G mine

Gold Hill-Applegate-Waldo Area, 5

Location: Jackson County, S$\frac{1}{2}$ sec. 19, T. 35 S., R. 3 W., at 2600 feet elevation.

Development: Three adits, each about 100 feet long with stopes.

Geology: A 3- to 12-inch vein striking N. 85° E., and dipping 60° N., occurs in Applegate Group micaceous slaty quartzites cut by diorite dikes. The ore contains quartz, calcite, pyrite, pyrrhotite, and a little chalcopyrite, bornite, sphalerite, galena, and rare free gold.

Production: Mine, discovered in 1904, was worked in a small way for several years. Production has not been reported.

References: Diller, 1914:37; Winchell, 1914:179; Parks and Swartley, 1916:81.

Dark Canyon (Red Rose) mine — Gold Hill-Applegate-Waldo Area, 33

Location: Josephine County, secs. 19 and 20, T. 38 S., R. 5 W., about 2900 feet elevation.

Development: About 700 feet of underground work in eight adits, plus surface cuts.

Geology: Country rocks are argillite, recrystallized metavolcanics, and serpentine. Four northwest-striking fissure veins average 2 to 4 feet thick, contain abundant pyrite, minor chalcopyrite, and rare galena, and some free gold.

Production: Located originally about 1900. Several pockets removed and some high-grade ore hauled out by pack animals. Seven claims were patented in 1959. A new mill was constructed in 1965 by owner, Dave Vallandigham, and some mining done. Total production not reported.

References: Department Bulletin 14-C (Josephine), 1942:172; Department mine file reports and maps, 1941 and 1944.

Dunromin mine — Gold Hill-Applegate-Waldo Area, 16

Location: Jackson County, SW$\frac{1}{4}$ sec. 36, T. 36 S., R. 3 W., 1500 feet elevation, 34 acres patented. On old Stage Road 5 miles from Gold Hill.

Development: A 25-foot caved shaft, 30-foot shaft with 16-foot drift, 8-foot winze, and small stope.

Geology: Values occur in N. 85° W. vertical quartz-filled fissure as much as 16 inches wide in quartz diorite. Rich ore containing pyrite, some galena, and gold formed at intersection of vein and N. 39° E. vertical fault.

Production: Mine reportedly produced $4,000 in 1897, $900 in 1935, and $200 in 1937.

Reference: Department Bulletin 14-C (Jackson), 1943:65.

Exchequer mine — Gold Hill-Applegate-Waldo Area, 29

Location: Josephine County, secs. 34 and 35, T. 37 S., R. 5 W., 1250 feet elevation.

Development: Underground workings total about 1200 feet on four levels and two shafts, plus stopes.

Geology: Country rocks are argillites and greenstones of the Applegate Group. The vein, which strikes N. 60° to 80° W. and dips 70° to 80° NE., contains highly fractured rock mixed with about a foot of limy fractured quartz and averages about 4 feet wide for a distance of 50 feet in the ore shoot. Ore minerals are pyrite and gold.

Production: One 11.7-acre claim has been patented. The 200-foot shaft was caved and flooded when reported by Winchell (1914). Other information on history and production is lacking. Estimated production is about $40,000.

References: Winchell, 1914:238; Parks and Swartley, 1916:92; Department mine file report, 1937; Department Bulletin 14-C (Josephine), 1942:155-156.

First Hope mine

Gold Hill-Applegate-Waldo Area, 17

Location: Jackson County, SW¼ sec. 7, T. 37 S., R. 4 W., elevation 3000 feet.

Development: Three tunnels have a little more than 300 feet of combined underground workings.

Geology: Northeast-trending quartz stringers sometimes carrying high-grade gold ore occur in manganese-stained, fractured, porphyritic andesite.

Production: Discovered in 1934 and produced a $1700 pocket. A $500 pocket was taken out in 1935 and about $100 since.

Reference: Department Bulletin 14-C (Jackson), 1943:66-67.

Fleming mine

Gold Hill-Applegate-Waldo Area, 41

Location: Jackson County, S½ sec. 23, T. 38 S., R. 3 W., about 1850 feet elevation.

Development: In 1941 there were 200 feet of crosscut, 400 feet of drift, and a 100-foot shaft connecting drift to surface.

Geology: Small quartz kidneys occur in fissure vein with slate footwall. Surrounding rocks are described as diorite. The area is mapped mainly as Applegate metavolcanics. Ore contains sulfides and some free gold.

Production: One 20-acre claim is patented. Information on production is lacking. Equipment includes a 10-ton Huntington mill. Mine has been operated in a small way for several years by Frank Gustis.

Reference: Department Bulletin 14-C (Jackson), 1943:160-161.

Frog Pond mine

Gold Hill-Applegate-Waldo Area, 60

Location: Josephine County, sec. 15, T. 41 S., R. 7 W., at the head of Johnson Gulch; elevation about 4400 feet. Part of a group of 17 claims extended into California.

Development: In 1940, accessible workings including two shallow shafts, three adits, and two drifts aggregated about 260 feet. There were in addition several small and one large open cut, two caved shafts, and old placered pits.

Geology: Country rocks are northeast-striking and southeast-dipping Applegate Group metasediments including argillite, quartzite, and limestone with some amphibole and chlorite schist, diorite, and serpentine. Assay samples indicate a mineralized zone at least 104 feet wide containing veinlets of quartz, calcite, and pyrite as well as disseminated pyrite, pyrrhotite, and small amounts of copper, arsenic, antimony, and mercury sulfides. Unweighted average of 123 gold assays taken by E. L. McNaughton in 1938 was about 0.19 oz./ton and for 74 silver assays 1.22 oz./ton. Three well-defined ore zones are reported. The 104-foot zone reportedly averages $4.20 per ton.

Production: Some early production was from ground sluicing. Dates and amounts are not reported. In 1938 the property was equipped with a small mill, buildings, compressor, and other mining equipment.

Reference: Department Bulletin 14-C (Josephine), 1942:199-200.

Gem Quartz mine

Gold Hill-Applegate-Waldo Area, 55

Location: Josephine County, $E\frac{1}{2}$ sec. 36, T. 39 S., R. 7 W., at 2000 feet elevation.

Development: There are two adits, the upper about 200 feet and the lower 150 feet long, as well as a number of surface cuts and small pits.

Geology: Pinching quartz veins 6 inches to 24 inches thick occur in high-angle shear zones in greenstone. Ore minerals are pyrite, chalcopyrite, and gold. Best values occur in swells.

Production: The property has 64 acres patented. Mr. Southerland worked the mine for several years prior to 1900 using an arrastra. Mine was reactivated in 1930 and a small home-made ball mill was used. No accurate records of the small production are reported.

References: Parks and Swartley, 1916:142; Department Bulletin 14-C (Josephine), 1942:200-201.

Golden Mary mine

Gold Hill-Applegate-Waldo Area, 28

Location: Josephine County, north edge sec. 34, T. 36 S., R. 5 W., at about 2500 feet elevation.

Development: About 600 feet of workings in two levels, and a shaft.

Geology: Highly fractured, light-colored siliceous country rock appears to be altered tuff or rhyolite. There are a few diorite dikes. Mineralization occurs in northeast-trending shear zones with some disseminated pyrite. Pocket-type enrichments of free gold with associated manganese and iron oxide have been reported.

Production: The latest work was done in the 1920's. Several rich pockets were reportedly mined but no record of total production is available.

Reference: Howard J. Black and Roy Stauch, Sr., Grants Pass, oral communication, 1966.

Golden Star mine

Gold Hill-Applegate-Waldo Area, 61

Location: Jackson County, secs. 3, 4, and 9, T. 40 S., R. 4 W., about 4000 feet elevation.

Development: Two shallow shafts 20 and 30 feet deep and cuts.

Geology: Nearly vertical quartz fissure vein as much as 2 feet wide in N. 70° E. shear zone in greenstone contains pyrite and some free gold.

Production: Discovered in 1935, produced $400, relocated in 1937.

Reference: Department Bulletin 14-C (Jackson), 1943:162.

Gold Hill Pocket mine — Gold Hill-Applegate-Waldo Area, 18

Location: Jackson County, SW$\frac{1}{4}$NE$\frac{1}{4}$ sec. 14, T. 36 S., R. 3 W., at 2000 feet elevation.

Development: Originally a small surface cut which was later enlarged.

Geology: Largely free gold with crystallized quartz, some pyrite, and calcite in a 5-foot, N. 20° W.-striking, 80° E.-dipping fissure vein which is cut by an east-west vertical gash vein. The country rock is metagabbro. Molybdenite has been found in the vein at greater depth.

Production: The rich surface pocket was discovered in 1857. It is reported that at least $700,000 was produced.

References: Diller, 1914:45-46; Parks and Swartley, 1916:109; Department mine file report, 1959.

Gold Pan (Ray) mine — Gold Hill-Applegate-Waldo Area, 42

Location: Jackson County, W$\frac{1}{2}$ sec. 2, T. 40 S., R. 4 W., at 2400 feet elevation, at mouth of Bailey Creek.

Development: Fifteen adits reportedly driven, only three were accessible.

Geology: Applegate Group host rocks largely greenstone, some argillite. Other information lacking.

Production: Located in 1919, production estimated at $5000. Equipment included three-stamp and 25-ton Forester rod mill with amalgamation plates.

Reference: Department Bulletin 14-C (Jackson), 1943:161-162.

Gold Plate mine — Gold Hill-Applegate-Waldo Area, 2

Location: Jackson County, sec. 17, T. 35 S., R. 3 W., at head of Murphy Gulch.

Development: Total of about 500 feet in two adits, including stopes.

Geology: The country rocks are metasediments of the Applegate Group. Ore occurred in the N. 60° W.-striking, 72° E.-dipping Gold Plate fissure vein in marble. The chief ore minerals are pyrite and gold. Stoped ore is said to have averaged $50.

Production: Claims were located in 1930. Estimated production to 1945 was about $20,000 from Gold Plate vein and $8000 from Iron Cube vein.

Reference: Department mine file report, 1945.

Great I Am mine — Gold Hill-Applegate-Waldo Area, 35

Location: Jackson County, NW$\frac{1}{4}$ sec. 31, T. 38 S., R. 4 W., near head of Farris Gulch, about 2600 feet elevation.

Development: About 1200 feet in three levels, including raises, shafts, and stoped areas.

Geology: Country rocks are highly sheared metamorphics of the Applegate Group, at least a portion being altered sediments. Ore shoots trend northwest and rake southeast. The upper level exposes a 6- to 10-inch brecciated quartz-filled vein.

Production: Original locator, Zeb Hyde (date ?), who sold to Harry Wilken, 1922. In 1940 Earle Young cyanided the dump. Total production amounted to about $30,000.

Reference: Department Bulletin 14-C (Jackson), 1943:162-163.

Grubstake mine

Gold Hill-Applegate-Waldo Area, 44

Location: Jackson County, $S\frac{1}{2}SW\frac{1}{4}$ sec. 9, T. 41 S., R. 2 W. at about 4400 feet elevation on the west bank of Silver Creek.

Development: Three hundred feet of tunnels, including 200 feet of drift on vein in 1916.

Geology: Country rocks are pre-Triassic schist with foliations striking about N. 70° E. and dipping 30° to 40° NW. A northwest-striking diorite dike penetrates the schist a short distance north of the mine. Ore minerals are mainly pyrite and gold in vein quartz and quartz mica schist with some talc and abundant chlorite. Assay values of about 1 oz./ton gold and 3 oz./ton silver are fairly common. Molybdenite has also been reported at the mine.

Production: Mine was operating in 1916, equipped with an arrastra, 32-foot overshot water wheel, and small cyanide plant. Other information lacking.

References: Parks and Swartley, 1916:115; Wells (1956) geologic map; Department assay records.

Hazel-Queen of the Hills Group

Gold Hill-Applegate-Waldo Area, 19

Location: Jackson County, $SW\frac{1}{4}$ sec. 27 and $NW\frac{1}{4}$ sec. 34, T. 36 S., R. 4 W., between 1200 and 1850 feet elevation.

Development: Numerous tunnels, cuts, and pits, mostly caved.

Geology: Country rocks are altered sediments including argillite and limestone of the Applegate Group which strike about N. 15° E. and dip 50° E. to vertical. Mineralization is associated with an andesite dike with quartz selvages.

Production: First mined in 1916. Production has been about $3000.

References: Department Bulletin 14-C (Jackson), 1943:76; Department mine file report, 1958.

Hematite mine

Gold Hill-Applegate-Waldo Area, 45

Location: Jackson County, $NW\frac{1}{4}$ sec. 6, T. 41 S., R. 2 W., and SW corner sec. 31, T. 40 S., R. 2 W., at 3400 to 3600 feet elevation.

Development: Two tunnels 110 and 150 feet, three shallow shafts, and various cuts expose the vein for 1500 feet.

Geology: Country rocks are argillite and meta-andesite of Applegate Group. Multiple quartz veins in a fissure as much as 4 feet thick strike N. 70° W. and dip 60° NE. A dolerite dike is exposed at various places adjacent to the vein. The contact of Paleozoic schists lies about 200 yards south of the vein. Ore minerals mainly free gold associated with limonite, some pyrite, malachite, and chalcopyrite.

Production: Originally located in early 1900's, called the Haskins and Traverso prospect. A small amount of high-grade ore was milled, but no record of value is reported.

References: Winchell, 1914:137; Department mine file reports, 1955 and 1963.

Hidden Treasure (Homestake) mine — Gold Hill-Applegate-Waldo Area, 9

Location: Jackson County, NW$\frac{1}{4}$ sec. 16, T. 36 W., R. 4 W., at 1600 feet elevation.

Development: In 1942 new work included a 31-foot shaft and 50-foot adit. Old workings (unsafe) contained about 500 feet.

Geology: Impure quartzites, argillite, and altered volcanics of Applegate Group. A 12- to 18-inch quartz vein with pyrite, some galena, sphalerite, chalcopyrite, and reported but not verified tellurides, strikes N. 35° W. and dips 35° NE.

Production: Was equipped with five-stamp mill in early 1900's, but production is not reported.

References: Winchell, 1914:178; Parks and Swartley, 1916:124; Department Bulletin 14-C (Jackson), 1943:76-77.

Humdinger mine — Gold Hill-Applegate-Waldo Area, 34

Location: Josephine County, N$\frac{1}{2}$ sec. 21, T. 38 S., R. 5 W., about 2400 feet elevation.

Development: About 1200 feet of underground development was mapped by Shenon, 1933b, plate 8.

Geology: A northwest-trending composite fissure vein from 6 inches to 9 feet wide, dipping about 60° NE. cuts northeast-striking, southeast-dipping greenstone and argillite of Applegate Group. Gangue minerals are quartz, chlorite, calcite, and apophyllite. Ore minerals are pyrite, arsenopyrite, gold, and minor galena with tellurides reported but not verified.

Production: Mine discovered about 1900. The main production was from milled ore in early 1900's. A $2700 shipment of lump ore was made in 1949. Total production is not recorded.

References: Winchell, 1914:239; Shenon, 1933b:48-51, and map.

Jewett mine — Gold Hill-Applegate-Waldo Area, 27

Location: Josephine County, SW$\frac{1}{4}$ sec. 27 and NW$\frac{1}{4}$ sec. 34, T. 36 S., R. 5 W. about

2500 feet elevation.

Development: Seven patented claims contain 104 acres. Total development about 1000 feet in two adits and shaft (1914).

Geology: Applegate metavolcanics are intruded by quartz diorite dikes. Ore body strikes N. 20° to 55° W. and dips 75° NE. Ore minerals include gold, pyrite, sylvanite, and pyrrhotite.

Production: Discovered about 1860 by Thomas Jewett. Was equipped with a five-stamp mill. Not operated since early 1900's and production not reported.

References: Diller and Kay, 1909:61; Winchell, 1914:222-223; Parks and Swartley, 1916:134; Department Bulletin 14-C (Josephine), 1942:82.

Kubli (Gold Standard) mine — Gold Hill-Applegate-Waldo Area, 22

Location: Jackson County, NW$\frac{1}{4}$ sec. 5, T. 37 S., R. 3 W., 2700 feet elevation (altimeter).

Development: There was about 650 feet of underground development in 1931.

Geology: Rich ore was mined from a narrow fissure vein as much as 18 inches wide, which strikes east and dips 60° N. Country rocks are metavolcanics and interbedded metasediments of Applegate Group, with nearby quartz-diorite intrusives.

Production: Development done in early 1900's. The production, probably small, was not reported. The tailings were cyanided in 1940.

References: Winchell, 1914:175; Parks and Swartley, 1916:107; Department Bulletin 14-C (Jackson), 1943:81-82.

Lone Eagle (Gray Eagle) mine — Gold Hill-Applegate-Waldo area, 7

Location: Jackson County, SE$\frac{1}{4}$ sec. 29, T. 35 S., R. 3 W., at about 1850 feet elevation.

Development: A 400-foot adit with 85-foot winze and raise to surface.

Geology: Quartz vein 9 to 35 feet thick strikes N. 70° E., dips 70° NW. Occurs in quartzite with associated andesite dike. Ore minerals are pyrite, chalcopyrite, galena, and gold.

Production: Development was prior to 1911, followed by inactivity until small operation just prior to 1942. Equipped with aerial tram to 10-stamp mill on Sardine Creek. Production not reported.

References: Winchell, 1914:180-181; Parks and Swartley, 1916:111 (with map); Department Bulletin 14-C (Jackson), 1943:87.

Lucky Bart mine

Gold Hill-Applegate-Waldo Area, 6

Location: Jackson County, northwest corner sec. 29, T. 35 S., R. 3 W., about 2100 feet elevation.

Development: Not described in reports. Several tunnels known.

Geology: Five veins, all striking nearly east-west, occur in argillite and quartzite of Applegate Group. Gangue quartz and calcite; ore minerals pyrite, minor chalcopyrite, galena, and gold.

Production: Discovered about 1890. Portion of group in section 29 patented. Worked intermittently since 1916. Was equipped with five-stamp mill. Mined ore was mainly from oxidized zone. Production possibly about $200,000.

References: Diller, 1914: 38; Winchell, 1914:179-180; Parks and Swartley, 1916:144; and Department Bulletin 14-C (Jackson), 1943:88-89.

Maid of the Mist mine

Gold Hill-Applegate-Waldo Area, 50

Location: Jackson County, SE$\frac{1}{4}$ sec. 4, T. 39 S., R. 4 W., at 2300 feet elevation.

Development: In 1914 there was a 200-foot shaft and about 500 feet of other workings.

Geology: Several quartz veins in greenstone. Most important strike east and dip 55° S. The quartz has minor calcite, some arsenopyrite, and pyrite.

Production: Information lacking in report.

References: Diller and Kay, 1909:61; Winchell, 1914:135; Parks and Swartley, 1916:148; Department Bulletin 14-C (Jackson), 1943:172.

Miller mine

Gold Hill-Applegate-Waldo Area, 30

Location: Jackson County, NW$\frac{1}{4}$NE$\frac{1}{4}$ sec. 19, T. 37 S., R. 4 W., about 2100 feet elevation.

Development: There is about 300 feet of workings in one adit, plus stopes.

Geology: Small fractured quartz vein with disseminated pyrite in argillite and greenstone strikes east. Near surface, oxidized ore was quite rich in places.

Production: Mine has been worked in a small way for several years by Sidney Miller and his father. Equipment includes a small rod mill and concentrating table. Total production is not known.

Reference: Sidney L. Miller, Grants Pass, oral communication, 1966.

Millionaire mine

Gold Hill-Applegate-Waldo Area, 15

Location: Jackson County, NE corner sec. 36, NW corner sec. 31, SW$\frac{1}{4}$ sec. 30, T. 36 S., Rs. 2 and 3 W., between 1600 and 1800 feet elevation.

Development: There are 4000 feet of drifts and crosscuts; three shafts 400 feet, 262 feet, and an incline 200 feet deep; all are under water.

Geology: Country rocks are sheared argillite of the Applegate Group, including small lenses of limestone. Four parallel quartz veins that strike east and dip steeply north contain pyrite and rare galena and chalcopyrite. Two other veins which strike north and dip east contain calcite, quartz, pyrite, and sylvanite(?).

Production: Mine is on 720 acres of patented homestead and claims. Was equipped with a two-stamp mill, amalgamation plates, and a table. Last development work was done in early 1920's. Production is not reported.

References: Winchell, 1914:169; Parks and Swartley, 1916:153; Department Bulletin 14-C (Jackson), 1943:93-94.

Mountain View (Mitchell) mine

Gold Hill-Applegate-Waldo Area, 47

Location: Josephine County, sections 23 and 24, T. 40 S., R. 5 W., about 4800 feet elevation.

Development: A 150-foot drift and 76-foot shaft are on the upper workings on No. 3 claim. No.1 claim has 115-foot tunnel, two 50-foot cuts, and a 35-foot shaft. No. 2 claim has a 75-foot tunnel, an open cut, and a 30-foot shaft. No. 1 and 2 claims' workings are caved.

Geology: Gold and associated pyrite occur in a tapering, west-striking, steep, north-dipping fissure vein which averages 18 inches thick on No. 3 claim. Country rocks are altered sediments and volcanics of the Applegate Group lying a short distance west of serpentine. The vein on No. 1 claim also strikes west, dips 80° north, and is from 1 to 5 feet thick.

Production: Worked intermittently since early 1900's. Produced about $2500 between 1937 and 1942.

References: Department Bulletin 14-C (Josephine), 1942:209-210; Department mine file report, 1960.

None Such mine

Gold Hill-Applegate-Waldo Area, 46

Location: Jackson County, center section 23, T. 38 S., R. 4 W., about 1700 feet elevation.

Development: A 215-foot adit, plus older caved workings and stope to surface.

Geology: Country rock is sheared greenstone cut by a quartz vein striking N. 65° W. and dipping 75° SW. A second vein was encountered in crosscut.

Production: Ore was being treated in a water-powered three-stamp mill on the Applegate River in 1913. Production was not reported.

References: Winchell, 1914:134-135; Parks and Swartley, 1916:162; Department Bulletin 14-C (Jackson), 1943:175.

Norling mine

Gold Hill-Applegate-Waldo Area, 37

Location: Jackson County, SW$\frac{1}{4}$SW$\frac{1}{4}$ sec. 26, T. 37 S., R. 3 W., about 3200 feet elevation.

Development: There are about 500 feet in two adits.

Geology: Country rocks are altered lavas. The main vein strikes west and dips 75° S. It is 8 to 18 inches wide and mainly quartz. Gold occurs chiefly in the quartz, although pyrite is abundant in surrounding rock.

Production: During 1905-1907, 120 tons of ore worth $6400 were produced. Further development work was done in 1913 and 1916. Later production is not reported.

References: Winchell, 1914:148-149; Parks and Swartley, 1916:162-163; Department Bulletin 14-C (Jackson), 1943:133.

Opp mine

Gold Hill-Applegate-Waldo Area, 38

Location: Jackson County; includes 360 acres patented land in NW$\frac{1}{4}$ sec. 36, T. 37 S., R. 3 W., about 1850 to 2850 feet elevation.

Development: Workings total about 7000 feet in 18 adits.

Geology: Country rocks are siliceous argillite with some chlorite and pyrite. Workings are on three veins which strike northwest to west and dip south. Widths of a few inches to 12 feet contain quartz and calcite gangue and pyrite and gold.

Production: Discovered late 1800's. Principal development in early 1900's. Total production a little more than $100,000. Ore was ground in a 20-stamp mill. The mine was also worked later in 1931-1935, and last worked in 1939 and 1941.

References: Winchell, 1914:149-153; Parks and Swartley, 1916:168-170; Department Bulletin 14-C (Jackson), 1943:133-135.

Oregon Belle Mine

Gold Hill-Applegate-Waldo Area, 36

Location: Jackson County, S$\frac{1}{2}$ sec. 6, T. 38 S., R. 3 W., about 3000 feet elevation.

Development: There are about 1750 feet of underground workings, plus stopes.

Geology: Country rocks are mostly greenstone with some argillite. Workings are on two semi-parallel quartz veins which strike N. 70° to 80° E. and dip about 60° N. The veins are about 100 feet apart and vary from 2 to 8 feet thick.

Production: Development work started in 1890. Total production is believed to exceed $250,000.

References: Winchell, 1914:134; Parks and Swartley, 1916:170; Department Bulletin 14-C (Jackson), 1943:176-177.

Oregon Bonanza mine

Gold Hill-Applegate-Waldo Area, 31

Location: Josephine County; 80 acres patented in SE¼ sec. 16, T. 38 S., R. 5 W., about 2100 feet elevation.

Development: Roughly 1300 feet of underground workings, including shafts.

Geology: Country rocks are mainly metasediments of the Applegate Group including argillite, limestone, and impure quartzite with some altered volcanic rock - greenstone. Mineralization occurs along sheared footwall contact of a northwest-striking 30-foot-wide diorite dike which dips 60° to 70° NE. Gangue minerals in the 3- to 4-foot vein include quartz, calcite, chlorite, epidote, and so forth. Ore minerals are pyrite, gold, and small amounts of galena. Some of the ore was unusually high grade.

Production: Early history is unreported. Workings were caved and buildings in ruins, according to 1916 report by Parks and Swartley. The mine was operated more or less continuously from 1936 to 1940. Production is unreported except for smelter returns for 1936 to 1938 of $6500.

References: Winchell, 1914:238; Parks and Swartley, 1916:170; Department Bulletin 14-C (Josephine), 1942:167-171.

Owl Hollow (Veteran) mine

Gold Hill-Applegate-Waldo Area, 24

Location: Jackson County, SW¼ sec. 32, T. 36 S., R. 4 W., about 2600 feet elevation.

Development: There are about 500 feet in two adits and some surface cuts.

Geology: Country rock is greenstone. A northwest-striking, nearly vertical vein of broken quartz and sheared greenstone was 3 to 4 feet wide where stoped. Some pyrite in greenstone.

Production: Discovery date unknown. Some development was done before 1914. A $14,000 pocket was reportedly mined near surface. Production from stopes in upper tunnel is not known. Was operated in a small way by Peter Bozich during 1940's using a small, home-made drag mill.

References: Diller, 1914:45; Department mine file report, 1963.

Portland Group

Gold Hill - Applegate-Waldo Area, 58

Location: Josephine County, sec. 3, T. 40 S., R. 7 W., elevation about 2600 feet, 1½ miles southeast of Holland.

Development: About 570 feet including shaft plus stoped area.

Geology: Country rocks are mapped as Applegate Group metasediments, largely argillite. Ore occurs in northeast-striking quartz vein in a silicified bleached and altered shear zone. The average stope width is 2½ feet.

Production: Dates of operation not reported. Considerable stoping has been done from adit level.

Manager reported $3 to $4 gold recovered per ton using 20-ton Gibson mill. Total production unreported.

Reference: Parks and Swartley, 1916:182.

Queen of Bronze mine — Gold Hill-Applegate-Waldo Area, 53

Location: Josephine County, NW$\frac{1}{4}$ sec. 36, T. 40 S., R. 8 W., about 2400 feet elevation.

Development: More than 7000 feet of drifts and crosscuts, numerous stopes, and an open pit.

Geology: Mineralization occurs as massive to disseminated sulfide (pyrite, chalcopyrite, pyrrhotite, sphalerite) bodies in greenstone ranging in size from mere stringers to 10,000 tons.

Production: The mine was located in 1862. It is a copper mine, but an estimated $120,000 of gold has been produced as a by-product.

References: Shenon, 1933c:163-167; Diller, 1914:81-83; Winchell, 1914:251-253; Parks and Swartley, 1916:184-186.

Rainbow (Siskron) mine — Gold Hill-Applegate-Waldo Area, 57

Location: Josephine County, NW$\frac{1}{4}$ sec. 12, T. 40 S., R. 7 W., about 2400 feet elevation.

Development: About 1200 feet of underground in the main workings, with numerous surface cuts and short tunnels.

Geology: The country rock is mainly greenstone. The main vein, a network of several small quartz veinlets, strikes about N. 20° E. and dips 35° W. It pinches and swells to a maximum of 5 feet. Ore minerals are pyrite, a small amount of chalcopyrite and gold.

Production: Discovered in 1915 by Mr. Siskron. Reported production to 1937 was $46,500.

References: Winchell, 1914:247; Department Bulletin 14-C (Josephine), 1942:216; Department mine file reports and maps, 1941.

Ray mine (Fair View claim) — Gold Hill-Applegate-Waldo Area, 23

Location: Jackson County, S$\frac{1}{2}$NW$\frac{1}{4}$ sec. 5, T. 37 S., R. 3 W., about 3000 feet elevation.

Development: Some drifting and winze sunk.

Geology: High grade found near surface in a 1- to 2$\frac{1}{2}$-foot, northwest-striking, quartz vein with calcite, pyrite, and galena. Country rock is greenstone.

Production: Discovery date unknown. Deeded land measures 40.52 acres. Has been worked intermittently since early days. About $9,000 produced from pocket in 1934-5.

References: Winchell, 1914:175-176; Department mine file report, 1944.

Reed mine

Gold Hill-Applegate-Waldo Area, 1

Location: Jackson County, SE$\frac{1}{4}$NE$\frac{1}{4}$ and NE$\frac{1}{4}$SE$\frac{1}{4}$ sec. 1, T. 35 S., R. 3 W., about 2450 feet elevation.

Development: Two crosscut tunnels, length not reported.

Geology: Country rocks in section 1 mapped by Diller, 1924, as quartz diorite and greenstone. Samples taken by J. E. Morrison in 1937 on 2- to 5-foot quartz veins indicate values from trace to $20.

Production: Discovered about 1900. Eighty acres patented. Reed produced about $1000 between 1922 and 1937 with 3$\frac{1}{2}$-foot Huntington mill. Total production not reported.

References: Department mine file report, 1937; Department Bulletin 14-C (Jackson), 1943:104.

Revenue Pocket mine

Gold Hill-Applegate-Waldo Area, 25

Location: Jackson County, SE$\frac{1}{4}$ sec. 11, T. 37 S., R. 3 W., at about 2750 feet elevation.

Development: Relatively small.

Geology: A 2-foot quartz vein about 100 feet east of limestone outcrop in argillite of Applegate Group. Reported 2-inch vein of solid gold with some mixed quartz (hearsay).

Production: One of the more famous, almost legendary, extremely rich, surface "pockets" mined by the Rhotan brothers in the early 1900's; it is credited with $100,000 production.

References: Parks and Swartley, 1916:193; Department Bulletin 14-C (Jackson), 1943:104-5.

Snow Bird (American Beauty) mine

Gold Hill-Applegate-Waldo Area, 32

Location: Josephine County, NE$\frac{1}{4}$ sec. 20, T. 38 S., R. 5 W., about 3200 feet elevation.

Development: About 550 feet in three adits plus surface cuts.

Geology: Gold occurs in a N. 35° W.-striking, steep northeast-dipping composite quartz fissure vein from 3$\frac{1}{2}$ to 12 feet thick in a zone of diorite-impregnated Applegate metasediments. Ore minerals are mainly pyrite with minor chalcopyrite, galena, and gold. Calcite is an associated gangue mineral.

Production: Original location in early 1900's. Some ore was reportedly milled at the Bone of Contention mill across the valley. No record of the small production is reported. Values are spotty. Recent development work includes extending the lower drift and testing ore with a small ball mill.

References: Department mine file reports, 1946 and 1959.

Star mine

Gold Hill-Applegate-Waldo Area, 49

Location: Jackson County, NE¼ sec. 6, T. 39 S., R. 4 W., about 2600 feet elevation.

Development: There are at least 2 adits, an open quarry, and a shaft.

Geology: Country rocks are argillite and metavolcanics of Applegate Group. Workings are on small limy quartz veins.

Production: Located 1896 by J. J. Kunutzen. About 800 tons of $2.00 to $4.00 rock were milled by 1914 from the open pit.

References: Diller and Kay, 1909:60; Diller, 1914:44; Wells and others, 1940; Lowell Downing, oral communication, 1958.

Steamboat mine

Gold Hill-Applegate-Waldo Area, 48

Location: Jackson County, secs. 17 and 20, T. 40 S., R. 4 W., about 3200 feet elevation.

Development: There were eight lode claims and one placer claim in the group. Most development is on three claims: the Rich Mortar and Blue Jay in sec. 17 and the Fowler in sec. 20, with an approximate total of 1200 feet.

Geology: The old Steamboat pocket came from shallow surface workings in greenstone in north edge NE¼ sec. 20. Numerous faults and small quartz veins present a complicated geologic setting. The pocket was located on a N. 10° W.-striking, 45° E.-dipping, 10-inch quartz vein. A west-striking, south-dipping vein from 0 to 3 feet thick is exposed on both the Rich Mortar and Blue Jay claims. Some other veins in the area trend northerly. Ore minerals are mainly pyrite and gold. Other sulfides, including arsenopyrite, are probably present but not reported.

Production: Discovered about 1860. A $350,000 pocket was mined out before 1869. Since then, the mine has been worked intermittently in a small way. During the 1930's a few hundred to $1000 were produced annually. A two-stamp mill was built in 1912.

References: Winchell, 1914:136; Parks and Swartley, 1916:212-213; Department Bulletin 14-C (Jackson), 1943:188-189; Department mine file maps, 1938.

Sylvanite mine

Gold Hill-Applegate-Waldo Area, 10

Location: Jackson County, S½ sec. 2, T. 36 S., R. 3 W., from 1300 to 1600 feet elevation.

Development: Contains a total of more than 2660 lineal feet of workings.

Geology: Mineralization occurs in shear zones in Applegate Group metavolcanics and argillite. Gangue minerals are quartz and calcite and carry galena, chalcopyrite, and pyrite. Some scheelite has also been noted.

Production: Discovery date not reported, but probably late 1800's. Worked by various operators to about 1940, with small amount of work since. Total production not reported, but main rich shoot reported to have produced about $700,000. A 140-ton float mill was at the mine in 1940.

References: Winchell, 1914:166; Parks and Swartley, 1916:219-220; Department Bulletin 14-C (Jackson), 1943:110-112; Libbey, 1963.

Tinpan mine

Gold Hill-Applegate-Waldo Area, 20

Location: Jackson County, SE$\frac{1}{4}$ sec. 31, T. 36 S., R. 3 W., at 2800 feet elevation.

Development: By 1916, had more than 1200 feet of drifts, shafts, etc.

Geology: Country rocks are mainly metasedimentary rocks including argillite, quartzite, and limestone with some metaigneous or metavolcanic rock. The quartz fissure vein varies from about 1$\frac{1}{2}$ feet to more than 6 feet in width, is vertical and strikes between northeast and east. Minor pyrite and galena occur in the vein.

Production: Located in late 1800's. In 1908, was prospected extensively by Pacific American Gold Mining Co. without finding any large body of good ore. Production, if any, not reported.

References: Diller and Kay, 1909:60; Winchell, 1914:174-175; Parks and Swartley, 1916:222; and Department Bulletin 14-C (Jackson), 1943:113.

Town mine

Gold Hill-Applegate-Waldo Area, 39

Location: Jackson County, S$\frac{1}{2}$ sec. 25, T. 37 S., R. 3 W., on ridge 800 feet west of Jackson Creek reservoir at about 2200 feet elevation.

Development: About 700 feet in two adits reported in 1914.

Geology: Semi-parallel quartz veins in argillite strike about west and are nearly vertical. One vein is associated with a basic hornblende-rich dike.

Production: Two rich near-surface pockets. The Johnson pocket reported to have yielded $30,000 and the Bowden pocket $60,000 are the only production reported. No dates reported.

References: Winchell, 1914:146-147; Parks and Swartley, 1916:222-223; Department Bulletin 14-C (Jackson), 1943:136-137.

Trust Buster (Golden Cross) mine

Gold Hill-Applegate-Waldo Area, 8

Location: Jackson County; three claims in northwest corner sec. 36, T. 35 S., R. 3 W. at about 1700 feet elevation, adjacent to and associated with Golden Cross (seven claims) and Golden Wedge (two claims) mines in sec. 35, T. 35 S., R. 3 W.

Development: A 100-foot tunnel with raise to surface and a 20-foot shaft on Golden Wedge. Length of adit on Trust Buster not reported.

Geology: Country rocks are Applegate metasediments and diorite. Junction on two veins in diorite formed small ore shoot that was mined to surface on Trust Buster. One quartz vein 18 to 20 feet wide contains values in sulfides which are reported to average $6.00 per ton on Golden Cross mine.

Production: No record.

References: Winchell, 1914:166; Parks and Swartley, 1916:224; and Department Bulletin 14-C (Jackson), 1943:68-69 and 115.

Paddle-wheel arrastra (top) **working in lower Rogue River. Note sluice boxes on right side of picture**. —Curry County Historical Society. **A portion of concrete arrastra** (lower) **abandoned near upper end of Solitude Bar on the river. The arrastra is a large grinding wheel used to crush ore so valuable minerals can be removed.** —*Bulletin, Museum of Natural History*, No. 22. Univ. of Oregon. **A stamp mill (page 67), expensive to buy then to haul to a mining site, did the same thing. Arrastras were often built by hand by the miners.**

MYRTLE CREEK AREA

Location

The Myrtle Creek area lies east and northeast of the town of Myrtle Creek, in the drainage areas of North and South Myrtle Creeks (figure 53). Known areas of mineralization are in T. 28 S., R. 4 W., and T. 29 S., R. 3 W. The area is generally mountainous, with a well-developed drainage pattern. Most of the streams flow in either a southerly or a southwesterly direction. Elevations range from about 900 feet in the stream valleys to a little more than 3000 feet on the higher ridges and buttes. White Rock, a prominent feature composed of Western Cascades volcanic rock lying just east of the area, is 4019 feet high.

Access roads follow valleys of the principal streams, and spur logging roads extend into the hills.

Geology and Mineralization

The geology of the Myrtle Creek area was originally mapped by Diller (1898), and was reinterpreted on the basis of subsequent work in this and adjacent regions by Diller and Kay (1924) and by Wells and Peck (1961). The oldest rocks are sediments and volcanics of the Dothan, Rogue, and Galice Formations. These rocks, mapped as "metagabbro" in the Roseburg quadrangle by Diller (1898), are in part altered to gneiss and have been complexly intruded by diorites, gabbros, and dacites. They occupy the drainage areas of North and South Myrtle Creeks.

Rocks of the Myrtle Group, principally conglomerates and sandstones of the Riddle Formation, underlie the area around the town of Myrtle Creek and extend to the northeast about 5 miles.

A narrow northeast-trending band of serpentinite from a quarter to a mile wide lies along the northwestern boundary of the area in the vicinity of Dodson and Brushy Buttes. Other small bodies of serpentinite are situated to the northeast along this same general trend.

Wells and Peck (1961) map three small bodies of diorite and related intrusives in the Myrtle Creek area: one in the vicinity of Buck Fork, one near Frozen Creek, and the third on a ridge between the branches of Lewis Creek. The Buck Fork and Lewis Creek bodies were mapped by Diller as "dacitic rocks (generally conspicuously porphyritic)."

Mineralization in the northern portion of the area is in part localized in shear zones within and around the serpentinite bodies near Brushy and Dodson Buttes. These zones contain some chalcopyrite, pyrite, and secondary copper carbonates with limonite. Some placer gold has probably been derived by weathering and erosion of these deposits.

The bulk of the area in which mineralization occurs is underlain by the rocks mapped as "metagabbro" by Diller, who described them as having a texture like granite, with an original mineral composition of plagioclase feldspar and lime-soda pyroxene, which is largely altered to hornblende or chlorite. The feldspar is altered in part to an aggregate of quartz, muscovite, epidote, and kaolin. Mineralization at the Chieftain and Continental lode mines on South Myrtle Creek occurs in a quartz vein in the "metagabbro."

Diller (1898) believed the bulk of placer gold mined on tributaries of North Myrtle Creek was derived from numerous small quartz veins in the "metagabbro." He reports that nuggets show little rounding and are generally attached to pieces of vein quartz. The gold occurs both in gravels of the stream valleys and on slopes of hills in the decomposed rock.

History and Production

Diller (1898) reported that placer mining had been active during the rainy seasons for many years,

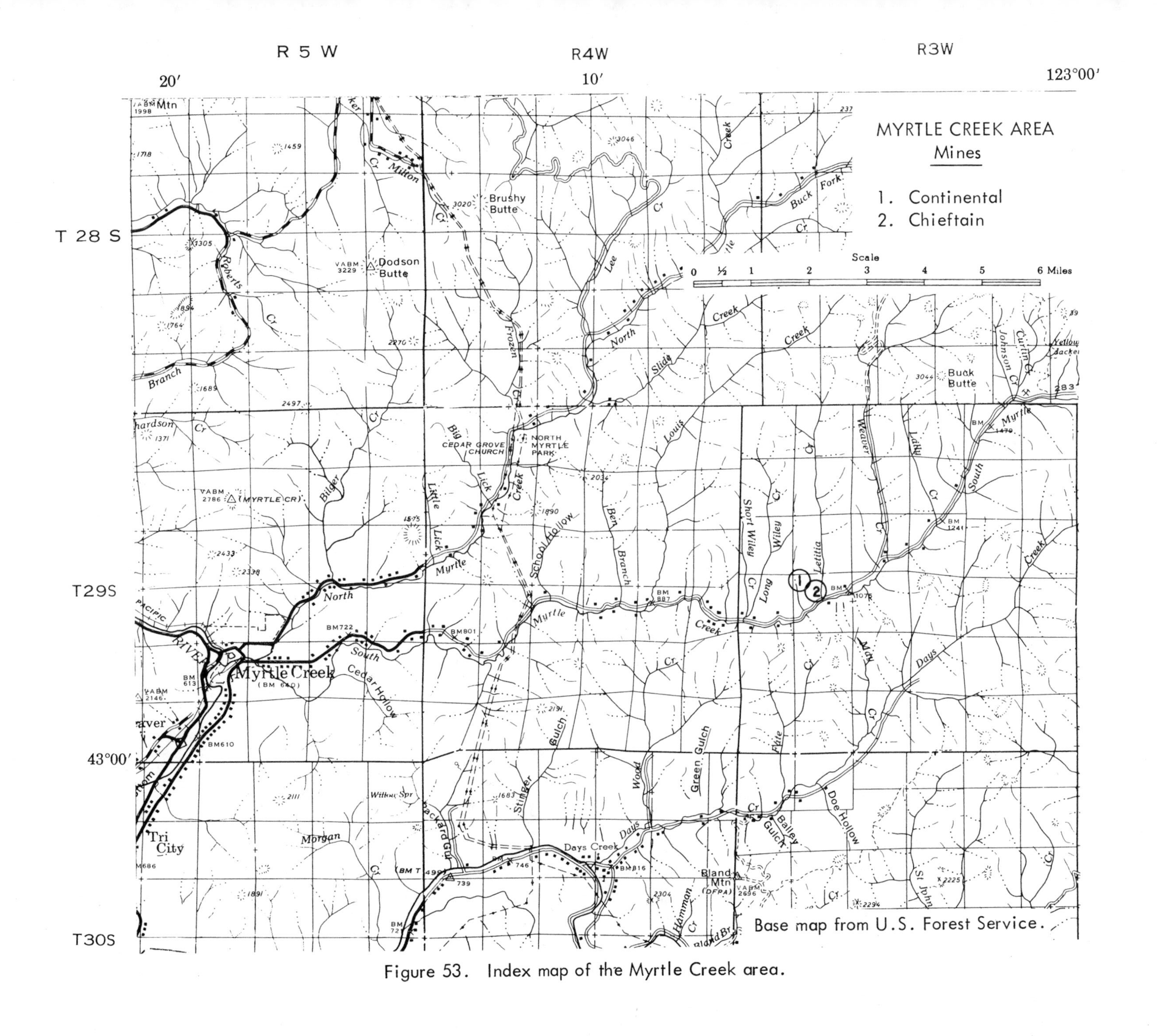

Figure 53. Index map of the Myrtle Creek area.

principally on Lee Creek and Buck Fork of upper North Myrtle Creek, and he estimated placer production to 1898 at $150,000. Some seasonal placer mining continued to about 1942. The two principal lode mines of the area (Chieftain and Continental) were apparently discovered shortly after Diller's work in the area. They were being worked during the early 1900's. Each of these mines is believed to have had a production of about $100,000.

Principal Lode Mines

Chieftain and Continental mines: The Chieftain and Continental mines are located near Letitia Creek, a tributary of South Myrtle Creek. Both mines are situated on the same fissure vein that strikes N. 80° E. and dips 60° to 75° N. Wells (1933) reports that the vein has been traced by disconnected outcrops for a distance of $1\frac{1}{4}$ miles. It consists of lenses and discontinuous stringers of quartz in a shear zone bounded by slickensided walls from less than a foot to about 4 feet apart. The vein has been cut by a set of steeply dipping, north- to northeast-striking fractures and faults of small displacement. Wells describes the mineralization at the Chieftain mine, which is situated on the eastern portion of the vein, as follows:

> "....Irregular grains, patches, and streaks of sulfides in places form as much as 10 percent of the vein. Coarsely crystalline pyrite is the predominant sulfide. Chalcopyrite and sphalerite occur in subsidiary amounts. The pyrite is mostly bright, though in part dull and dirty, probably owing to granulation. Chalcopyrite forms small patches near the pyrite but is rarely associated directly with it. Sphalerite is likewise commonly associated with the chalcopyrite.
>
> "Under the microscope the sphalerite is seen to contain blebs and veinlets of chalcopyrite. Sylvanite and petzite (tellurides of gold and silver) occur as small irregular patches or threads in both the chalcopyrite and sphalerite and here and there by themselves in the quartz. Neither was found, however, in the pyrite. Petzite contains a smaller amount of tellurium than sylvanite, and the silver content of both is variable; in the specimens from the Chieftain mine it is low, probably less than 25 percent. No free gold was seen. During the period of mineralization the deposition of quartz was continuous. Pyrite is the oldest sulfide. Sphalerite was deposited next and was succeeded by chalcopyrite. Sylvanite and petzite were deposited last. The tellurides are almost exclusively associated with chalcopyrite and sphalerite, and the abundance of these sulfides, which are readily seen, is therefore some indication of the value of the ore."

He further states on page 61:

> "...Though the sulfide minerals that carry the gold are abundant in spots they are not concentrated in definite shoots but are distributed irregularly throughout the vein. Much of the quartz now showing carries considerable sulfide, and the vein on the main level beyond the fault is well mineralized. From these facts it is reasonable to assume that the ore continues in depth and that within the limits imposed by the size and tenor of the vein a considerable tonnage can be mined."

Geology and mineralization of the Continental mine which is situated on the western extension of the vein are much the same as the Chieftain.

Production of the Chieftain and Continental mines is not definitely known. The deposits were discovered about 1898 and production occurred mainly prior to the 1930's. Wells (1933) reports that evidence presented in a law suit claimed the Continental mine had produced ore worth $168,000.

Lode Mines of the Myrtle Creek Area

Chieftain and Continental mines — Myrtle Creek Area, 1 and 2

Location: Douglas County, NW$\frac{1}{4}$ sec. 20, T. 29 S., R. 3 W., 1200 feet elevation.

Development: About 1500 feet in Chieftain and more than 2000 feet in Continental.

Geology: Both mines are on the same fissure vein, which strikes about N. 80° E., dips 60° to 75° N., and is from less than 1 foot to 4 feet wide. Country rock is "metagabbro." Ore minerals are pyrite, chalcopyrite, sphalerite, sylvanite, and petzite.

Production: Deposits discovered about 1898, and the bulk of production was prior to 1930. Records are incomplete. Production was probably greater than $100,000 for each mine.

Reference: Wells, 1933:57-61.

Hydraulic mining brought a terrific force of water at pressures of hundreds of pounds per square inch through giant swivel-mounted nozzles. The jet of water broke the soil which it struck, the soil then washed through lines of sluices. Operating costs were so low that working even poor ground was generally profitable. Eventually, the governments outlawed hydraulic mining.

—Curry County Historical Society

ASHLAND AREA

Location

The Ashland area lies west and south of the city of Ashland (figure 54). It includes a small group of mines, the most productive of which has been the Ashland mine, lying mainly in T. 39 S., Rs. 1 E. and 1 W.

The area is mountainous, with hills rising rapidly from about 1700 feet in Bear Creek Valley to more than 7000 feet on Wagner Butte, 7 miles to the south. The region is drained by Wagner and Ashland Creeks.

Geology and Mineralization

The area is underlain by the northern part of the Ashland diorite batholith (figure 37), which intrudes metamorphosed sedimentary and volcanic rocks of the Applegate Group. The mines and prospects occur in quartz veins in both the metamorphic and dioritic rocks. Small patches of Late Cretaceous Hornbrook Formation, occurring along the northeastern margin of the metamorphic and dioritic rocks, include conglomerates which locally contain minor amounts of gold eroded from the older rocks. The geology of the area has been mapped and described by Wells (1956).

Two prominent strike directions of veins are apparent in the area. One set of veins strikes north to N. 20° E., parallel to the fold axes of the Applegate Group; the other set strikes N. 10° to 50° W. across this structural trend. The latter veins appear to be the more numerous. The lowest angle of dip reported in either set of veins is 40°, and the dominant direction of dip in both is easterly.

History and Production

The Ashland mine is credited with a production of $1,500,000. Second in importance as a producer in the area has been the Shorty Hope with a production of $30,000. Burch (1942, p. 105-128) reported the date of discovery of the Ashland mine as 1886, and Parks and Swartley (1916) report that in 1898 and 1899 ore from the mine was treated in a five-stamp water-powered mill located in the city of Ashland. Shortly afterward, a 10-stamp mill was erected near the mine.

Placer Mines

During the early days of mining in the Ashland area placers operated on Bear and Anderson Creeks; on Wagner Creek and its tributaries, Yankee Gulch, Arrastra Creek, and Horn Gulch; and on Ashland Creek. Some evidence of this early mining can still be seen and perhaps a few "old timers" in the area have some knowledge of its extent; but except for a brief report on the Forty-nine diggings placer (Diller, 1914, p. 90-93) there is no reported information on placer mining in the area.

The Forty-nine diggings group, northwest of Ashland near Phoenix, operated regularly for a few months during the wet season from about 1860 to about 1900, and produced from 60 to 150 ounces of gold annually (Diller, 1914, p. 90). The source of this gold was Cretaceous conglomerate of the Hornbrook Formation.

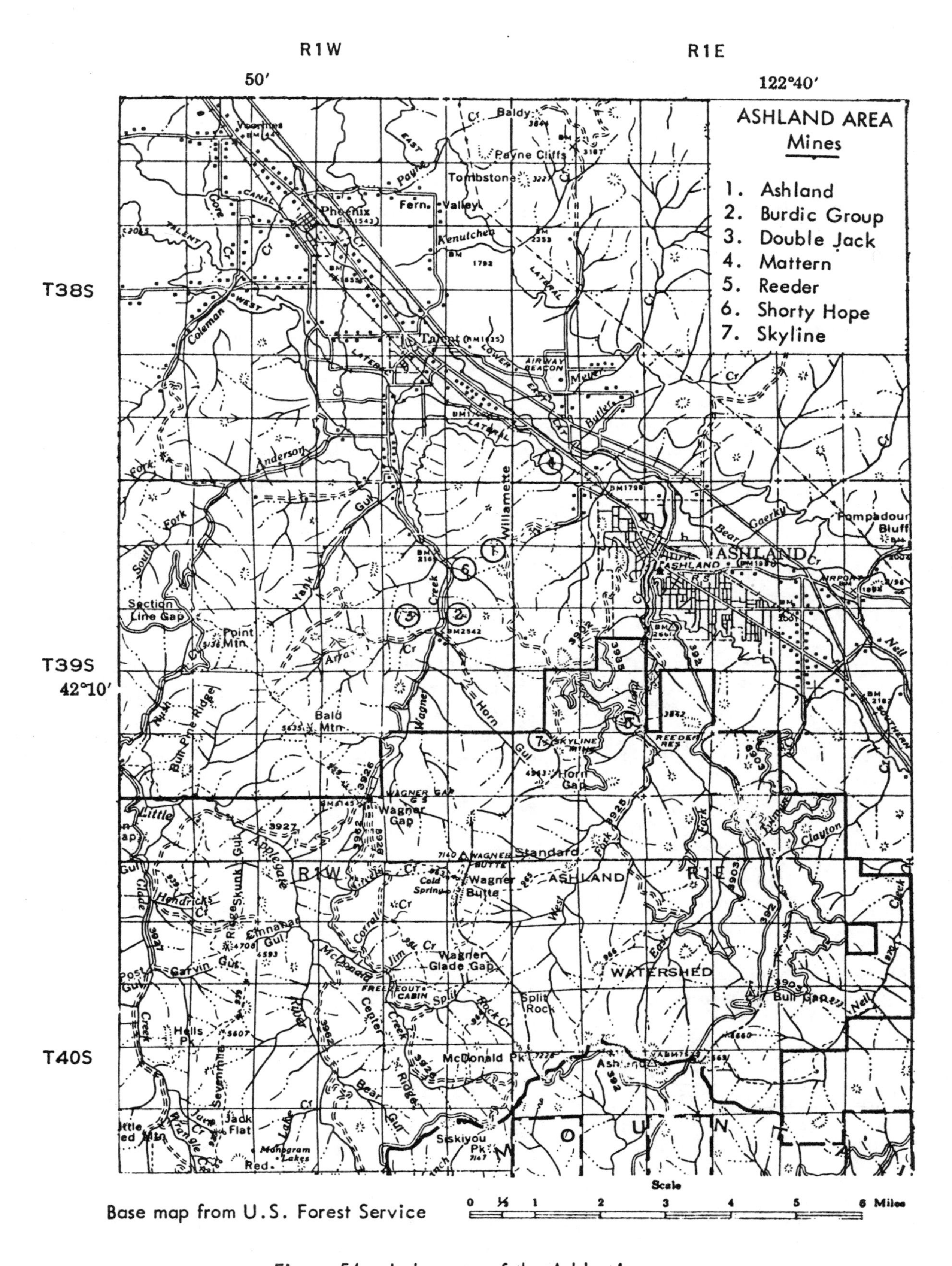

Figure 54. Index map of the Ashland area.

Principal Lode Mines

Ashland mine: The Ashland vein is one of the most persistent in southwestern Oregon. It has been traced for more than a mile along its strike and to a depth of about 1200 feet down dip, and it extends beyond the limits of the workings. The vein is a quartz fissure from 2 to 12 feet wide; it strikes about N. 20° E. and dips 45° SE. The enclosing rocks are highly altered sediments of the Applegate Group that are in large part replaced by diorite. Quartzite, quartz mica schist, and hornblende-biotite-quartz diorite are found alongside the vein. Ore minerals are gold, pyrite, and minor chalcopyrite.

Underground workings at the Ashland mine total 11,000 feet including shafts and raises (plate 2, in pocket). Ore mined varied considerably from \$3 to \$30 per ton in the milling ore and the shipping ore averaged \$100 (at the old price of \$20.67, according to Parks and Swartley, 1916).

Shorty Hope mine: The Shorty Hope vein strikes northwest and dips steeply. The mine workings, which total 3500 feet, lie one mile west of the Ashland mine. At one time a tunnel was proposed to connect the two mines so that the ore could be processed at one mill. Ore at the Shorty Hope is similar to but somewhat lower grade than that at the Ashland mine. The average assay of mill heads at the Shorty Hope was reported to be \$18.35 per ton. Its total production, including concentrates from a 10-stamp mill and some high-grade shipping ore, was about \$30,000, all prior to 1939. The vein is reported to have an estimated 50,000 tons of ore reserves above the main working level.

Lode Mines of the Ashland Area

Ashland mine

Ashland Area, 1

Location: Jackson County; the workings straddle the Willamette Meridian in the SW$\frac{1}{4}$ sec. 6, NW$\frac{1}{4}$ sec. 7, T. 39 S., R. 1 E., and the E$\frac{1}{2}$ sec. 12, T. 39 S., R. 1 W. Elevations range from about 2500 to 3450 feet.

Development: Total 11,000 feet of tunnels, shafts, and raises explore vein to 1200 feet on dip.

Geology: Quartz vein in Applegate metasediments and diorite is 2 to 12 feet wide, strikes N. 20° E., and dips 45° SE. Ore minerals are gold, pyrite, and minor chalcopyrite.

Production: Discovered 1886. The first mill was a five-stamp in Ashland. Later a 10-stamp mill operated at the mine. Gross production was \$1,500,000.

References: Winchell, 1914:114-117; Parks and Swartley, 1916:16-18; Burch, 1942:107; and Department Bulletin 14-C (Jackson), 1943:23-26.

"Burdic Group" (includes the Ruth, Little Pittsburgh, and Growler mines)

Ashland Area, 2

Location: Jackson County, sec. 13, T. 39 S., R. 1 W. These are separate properties, and not previously written up as a group.

Development: Growler (NE$\frac{1}{4}$ sec. 13) contained about 190 feet in one adit, including a 40-foot drift plus a small stope, and a 40-foot inclined shaft south of the tunnel (1937). The Burdic (center, sec. 13) contained more than 300 feet in two adits (1914). The Ruth (NW$\frac{1}{4}$ sec.13), about 500 feet east of Wagner Creek, contained about 150 feet in one adit (1914). The Little Pittsburgh, about 700 feet east of the Ruth, contained a 150-foot adit caved at the portal, a 50-foot inclined shaft, and a few shallow cuts (1914).

Geology: The group lies in the contact area of highly metamorphosed Applegate sediments and the Ashland diorite batholith. The Burdic vein follows a 4- to 10-foot altered diorite dike which strikes N. 80° W. and dips 85° S. The Growler vein also strikes about east and contains 0 to 2 feet of quartz with some pyrite and chalcopyrite. The Ruth and Little Pittsburgh veins both strike about north and dip 70° to 80° east, and contain some calcite and pyrolusite.

Production: Information mostly lacking. Growler, located in 1937, produced about $1200 the first year in a home-made one-stamp mill.

References: Winchell, 1914:119-121; Department Bulletin 14-C (Jackson), 1943:33.

Double Jack mine

Ashland Area, 3

Location: Jackson County, NW¼ sec. 14, T. 39 S., R. 1 W., at 3300 feet elevation.

Development: Shallow cuts, and about 500 feet in shafts, crosscuts, and raises.

Geology: Country rocks are metasediments of Applegate Group. A northwest-striking, northeast-dipping quartz-fissure vein from a few inches to 3 feet thick carries occasional pocket-type enrichments with visible galena, pyrite, chalcopyrite, and free gold.

Production: Mine was first worked in 1870 and has history of small, intermittent operation to 1963. Total production is estimated to be about $7000. Ore milled in an arrastra.

Reference: Department mine report file, 1963.

Mattern mine

Ashland Area, 4

Location: Jackson County, SE¼ sec. 31, T. 38 S., R. 1 E., at 1760 feet elevation on Southern Pacific Railroad.

Development: One adit 325 feet long and a winze.

Geology: North-trending, 40° E.-dipping, quartz-calcite vein occurs in sheared diorite, thought to be north extension of the Ashland vein.

Production: Not reported.

References: Winchell, 1914:117; Parks and Swartley, 1916:150.

Reeder mine

Ashland Area, 5

Location: Jackson County, SE¼ sec. 20, T. 39 S., R. 1 E., about 3000 feet elevation.

Development: Three adits contain about 790 feet of workings.

Geology: Quartz vein about 4 feet wide strikes northwest and dips about 80° NE. in sheared quartz diorite and with pegmatitic aplite dike wall. Vein is exposed over 300 yards along strike. Values are spotty and in some areas especially rich.

Production: Not reported.

References: Winchell, 1914:117-118; Parks and Swartley, 1916:192.

Shorty Hope mine

Ashland Area, 6

Location: Jackson County, NW¼ sec. 12, T. 39 S., R. 1 W., from 2400 to 3500 feet elevation. It lies on 107 acres of patented land.

Development: The main adit at 2450 feet elevation is 1480 feet long and follows a northwest-striking vein. Other adits and shaft bring total workings to about 3500 feet plus extensive stopes.

Geology: Country rocks are Applegate Group recrystallized metavolcanics and sediments, complexly intruded by dikes and masses of the Ashland diorite batholith. The 4- to 16-feet-wide vein strikes N. 55° W., dips very steeply southwest. Ore minerals are pyrite, some galena and gold.

Production: Location date not reported -- probably late 1800's. Ore was milled in stamp mill with ten 1000-pound stamps. Total production was about $30,000.

References: Winchell, 1914:118-119; Parks and Swartley, 1916:201-202; Wells and others, 1956; Department Bulletin 14-C (Jackson), 1943:39-40.

Skyline mine

Ashland Area, 7

Location: Jackson County; NW¼ sec. 30, T. 39 S., R. 1 E., at 4750 feet elevation.

Development: A shaft is shown on Ashland 15-minute quadrangle topographic map. Four shafts are reported to have been dug: one is about 70 feet and one 109 feet deep. Some drifting has been done from the shafts. Pumping is required to keep shafts de-watered.

Geology: Area is mapped as quartz diorite by Wells (1956). The vein is reported to strike northeast and dip southeast. It is small but rich in places.

Production: The mine was worked from time to time since the early 1900's by various owners. A small mill with amalgamation was used to recover the gold. Total production is not reported.

References: Wells, 1956; E. G. Taylor, oral communication, January 10, 1968.

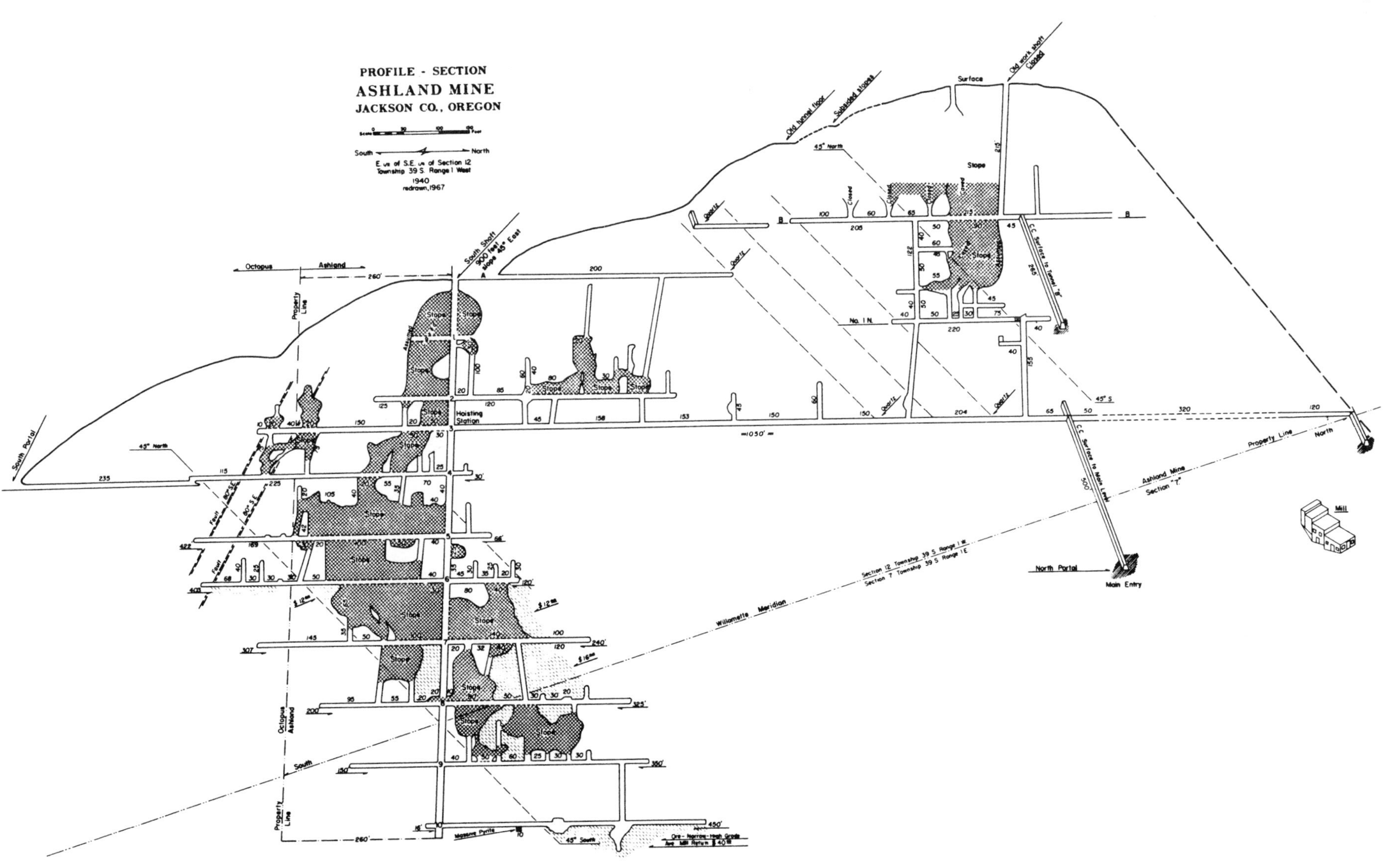
PROFILE - SECTION
ASHLAND MINE
JACKSON CO., OREGON
South
North
E. ½ of S.E. ¼ of Section 12
Township 39 S. Range 1 West
1940
redrawn, 1967
Surface
Old work shaft
Closed
Old tunnel floor
Subsided stopes
Stope
Octopus
Ashland
Property Line
South Shaft
Hoisting Station
South Portal
North Portal
Main Entry
Mill
C.C. Surface to Main Level
C.C. Surface to Tunnel "B"
Ashland Mine
Section "7"
Section 12 Township 39 S Range 1 W
Section 7 Township 39 S Range 1 E
Willamette Meridian
Massive Pyrite
1050'

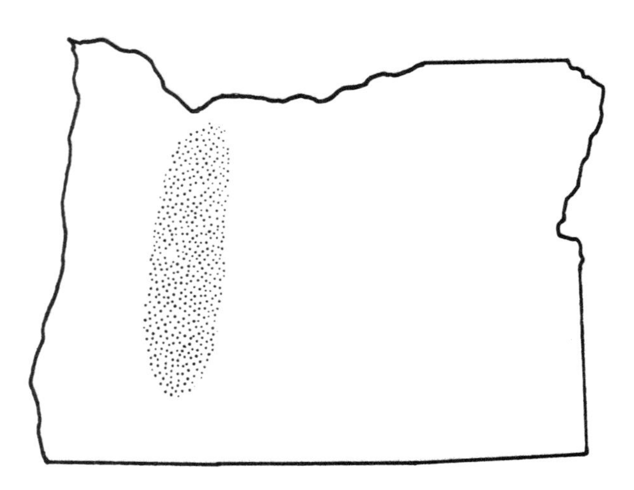

Part III-B Deposits In Western Oregon The Western Cascades

Most miners' camps were built of tents, lean-to's or shacks. This cabin, built of prime logs was in Curry County. The extended roof helped keep out rain. Note ladder at corner. It gave access to the roof (probably sheet iron) to knock the soot out of the stove pipe. —Curry County Historical Society

THE WESTERN CASCADES

Table of Contents

THE WESTERN CASCADES

INTRODUCTION

Mineralized areas in the Western Cascades of Oregon have produced a total of approximately $1,500,000 in gold and silver since 1858, when gold was first discovered in that region. Most of the production took place prior to 1920; the periods of mining activity were intermittent. Production since 1920 has been about half a million dollars in total value of gold, silver, copper, lead, and zinc.

The gold-producing areas are fairly evenly spaced along a north-south alignment extending from Clackamas County on the north to Jackson County on the south (figure 55). The deposits are in Tertiary volcanic rocks and are associated with small Tertiary intrusive bodies.

GEOGRAPHY

The Cascade Range is a broad, mountainous region extending almost due north from northern California through Oregon and Washington and into British Columbia. In Oregon, the range is 250 miles long and about 50 miles wide. It is made up of two sections (see figure 8, page 30) -- the western belt, called the Western Cascades, which is a maturely dissected region of older volcanic rocks; and the eastern belt, termed the High Cascades, a less eroded region underlain by younger volcanic flows and cones. Since the gold-mining areas lie in the Western Cascades, the discussion is confined largely to that part of the mountain range.

Drainage in the Western Cascades is westward and a dendritic pattern is well developed. To the east the stream canyons are narrow, have relatively steep gradients, and are separated by long, sharp ridges whose tops rise 3000 to 5000 feet above sea level; to the west, valleys broaden and ridges are lower and more rounded.

The climate is generally mild and wet during winter months and dry in the summer. The average annual rainfall is from 40 to 50 inches. Timber productivity is excellent and dense stands of Douglas fir and hemlock are common.

PREVIOUS WORKERS

One of the most comprehensive reports on the geology and mineral deposits of the Western Cascades was prepared by Callaghan and Buddington (1938). Earlier workers who describe some or all of the mines include Diller (1900, 1914); Stafford (1904); MacDonald (1909); Swartley (1914); Parks and Swartley (1916), and Stowell (1921). Smith and Ruff (1938) review the geology and mineral resources of Lane County. Thayer (1939) describes the geology of the North Santiam River basin. Taber (1949) summarizes information on the mines and prospects of the Bohemia district, and Lutton (1962) describes in detail the geology of the Bohemia district. Peck and others (1964) present a description of the geology of the central and northern parts of the Western Cascades, and Wells (1956) covers the geology of the southern portion.

The compilation of mines reports, geology, and mineral-resource data by the Oregon Department of Geology and Mineral Industries (1951) in its Bulletin 14-D is the source of most of the information presented here. In addition, data on recent developments in the Bohemia district were provided by H. E. L. Barton, mining engineer of Eugene, Oregon.

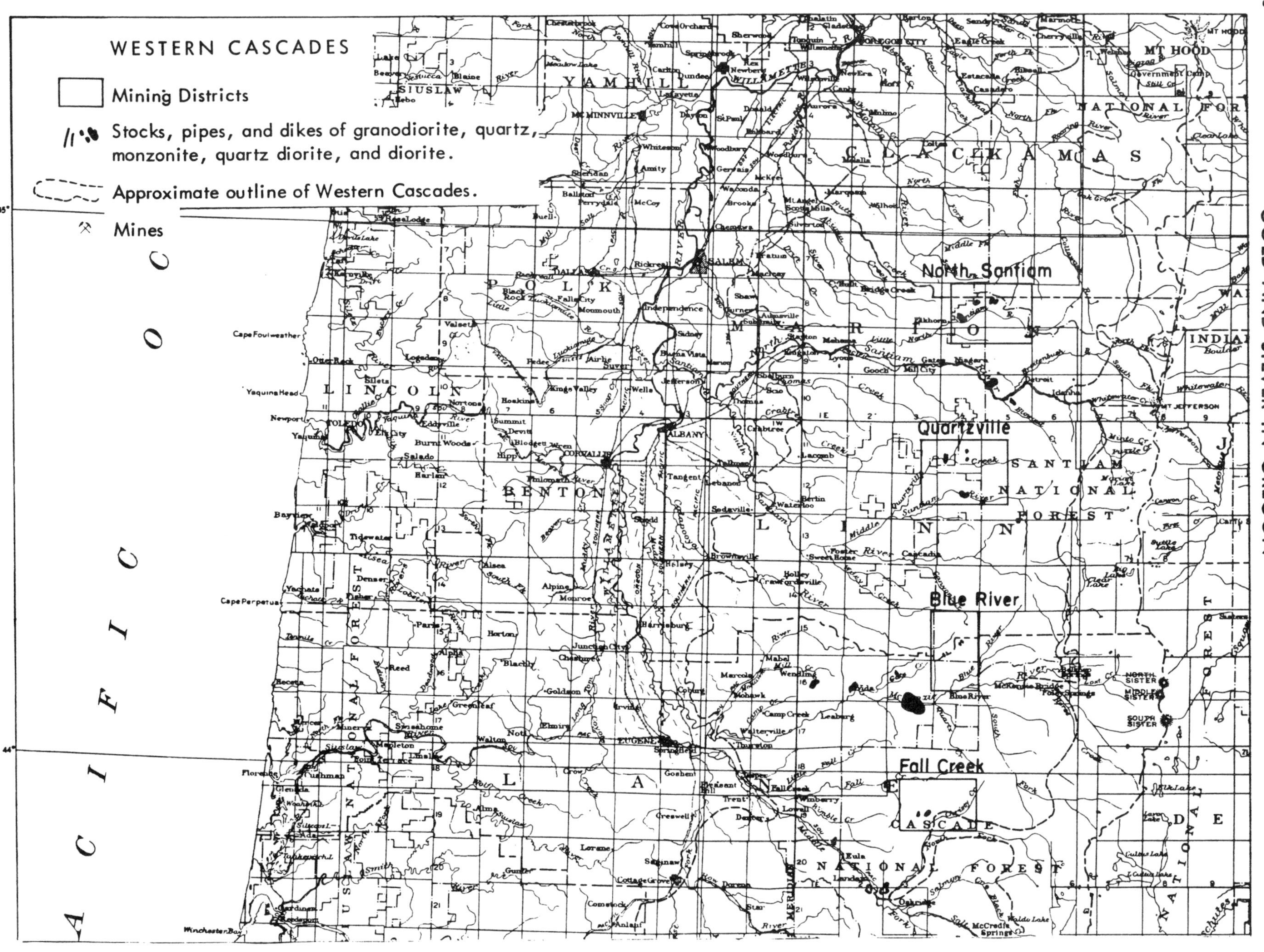
WESTERN CASCADES
Mining Districts
Stocks, pipes, and dikes of granodiorite, quartz, monzonite, quartz diorite, and diorite.
Approximate outline of Western Cascades.
Mines
North Santiam
Quartzville
Blue River
Fall Creek
MT HOOD
YAMHILL
CLACKAMAS
POLK
MARION
LINCOLN
BENTON
LINN
LANE
SALEM
ALBANY
CORVALLIS
EUGENE
OREGON CITY
McMINNVILLE
SANTIAM NATIONAL FOREST
CASCADE NATIONAL FOREST
SIUSLAW NATIONAL FOREST
NORTH SISTER
MIDDLE SISTER
SOUTH SISTER
MT JEFFERSON
PACIFIC OCEAN

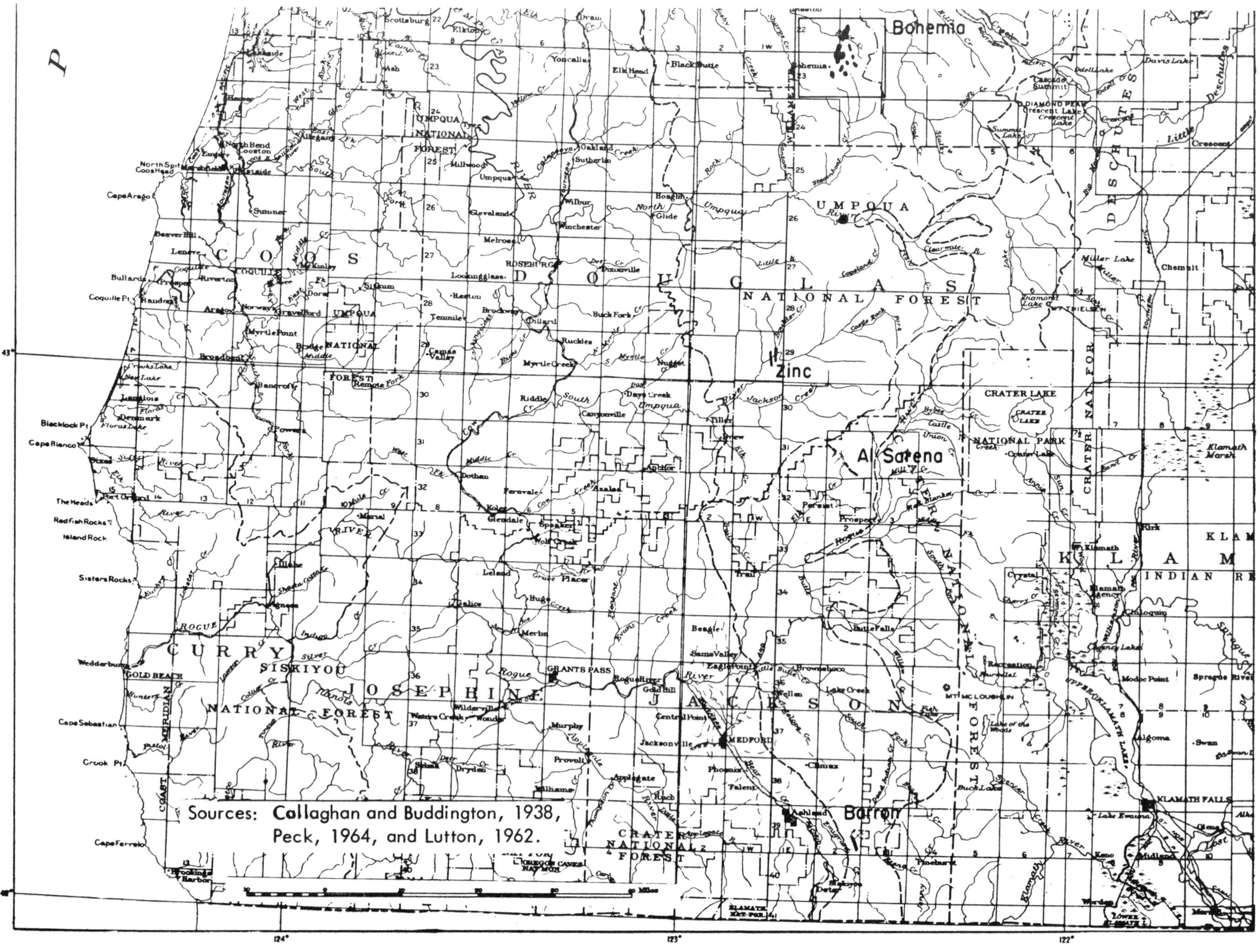

Figure 55. Index map of the Western Cascades showing location of the mining districts and intrusive bodies.

HISTORY AND PRODUCTION

The discovery of placer gold near Jacksonville in 1851 stimulated interest in the prospecting of all streams in Oregon shortly thereafter. Small amounts of gold panned in tributaries of the Row, Molalla, Santiam, and McKenzie Rivers led to discoveries at Bohemia in 1858; in the Molalla and North Santiam valleys in 1860, and at Blue River in 1863. The town of Quartzville in Linn County was established in 1864. Little production was realized until about the turn of the century. The Lawler mine at Quartzville was active in the 1890's as were the Musick, Champion, and Noonday mines of the Bohemia district. The major producing mine in the Blue River district, the Lucky Boy, was active in the early 1900's.

The nature of the ore, the technology, and the economy were all important factors in the short productiveperiods of these mines. Mills equipped with stamps and amalgamation plates were able to recover only the free gold found in near-surface oxidized ores, and were unsuited to recover values in complex primary sulfides. Selective flotation had not been developed and gravity concentration, where tried produced a complex concentrate of copper, lead, and zinc not desired by smelters. The result was only intermittent activity from about 1910 until the price of gold increased in 1934. A small annual production was maintained from 1934 to World War II.

Total production for the Western Cascades has been approximately $1.5 million, the bulk of it from the Bohemia district.

GEOLOGY

The Western Cascades region is underlain by a thick section of Tertiary volcanic rocks composed mainly of andesitic flows with some basalts and rhyolites, extensive tuff beds, and lenses of lacustrine and fluviatile sediments. These rocks range in age from late Eocene to late Miocene (see correlation chart on following text page). To the west the volcanics overlie or interfinger with Tertiary marine beds. The volcanic units have been folded locally and faulted. In general they have a low dip to the east or northeast. Undeformed Pliocene and Pleistocene fluviatile deposits and flows of andesite and basalt are locally present. Small diorite intrusives of late Miocene age occur in mineralized areas and are believed to be the source of the mineralizing solutions.

Stratigraphy

The following brief summary of the stratigraphy of the Western Cascades is derived largely from the work of Peck and others (1964).

The oldest unit, the Colestin Formation, crops out in the southern part of the region. It is composed of tuffs, tuff breccia, volcanic sandstone and conglomerate, and intercalated lava flows, all of andesitic composition. The most abundant rock type is a massive pumice lapilli vitric tuff. In addition to the volcanic rocks, the formation includes local deposits of nonvolcanic conglomerate and coal. The Colestin contains an upper Eocene flora and it overlies or interfingers with Eocene marine sediments of the Spencer Formation to the north and the Umpqua Formation of Wells (1956) in the Medford area. Average thickness of the formation is about 1500 feet.

Overlying the Colestin Formation is the Little Butte Volcanic Series, consisting of 3000 to 15,000 feet of pyroclastic rocks and flows of Oligocene and early Miocene age. These rocks make up the bulk of the Western Cascades south of the McKenzie River, and they crop out in the foothills and in axes of anticlines to the north. Massive beds of andesitic and dacitic tuff are the most common unit in the series. Other components include basalt and andesite flows and breccia, welded tuff, and small intrusives of dacite and rhyodacite. The series includes other previously mapped formations such as the Mehama Volcanics. Fossil floras ranging from middle Oligocene to early Miocene age occur in beds of water-laid tuff within the series. To the west, the series interfingers with marine beds of middle Oligocene to early Miocene age.

Columbia River Basalt unconformably overlies the Oligocene and lower Miocene rocks in the northern part of the Western Cascades. The formation consists of columnar-jointed and hackly flows of basalt and basaltic andesite. Maximum thickness is about 1500 feet.

The Sardine Formation of middle and late Miocene age overlies older formations in much of the Western Cascades region, particularly in the northern portion. The Sardine Formation consists of andesitic lavas, tuff, and breccia and has an average thickness of about 3000 feet. It conformably overlies Columbia River Basalt, and it includes the Fern Ridge Tuffs, the Rhododendron Formation, and the Boring Agglomerate. Middle and late Miocene fossil plants are the basis for dating the Sardine Formation.

Undeformed Troutdale Formation containing Pliocene flora unconformably overlies the Sardine and older formations in the northwestern corner of the Western Cascades. The Troutdale consists of about 1000 feet of mudstone, sandstone, and conglomerate of fluvial origin.

Pliocene to early Pleistocene Boring Lava, which makes up much of the High Cascades, also occurs in the Western Cascades as andesitic and basaltic flows and breccia. These lavas overlie the Troutdale and part of the older Tertiary rocks in the Western Cascades. They occur as intracanyon flows and isolated shield volcanoes.

Quaternary deposits in the Western Cascades consist of gravel, sand, and silt and local glacial till in the major stream valleys. Some of these deposits are gold bearing, but none have a history of significant placer production.

Intrusive rocks

Small bodies of diorite and other intrusive rocks are fairly numerous in the Western Cascades. They are related to the volcanic units and range in age from late Eocene to late Miocene. Most of the dioritic rocks (including quartz diorite, granodiorite, and quartz monzonite) associated with the mineralized areas are probably of Miocene age, as indicated by the correlation chart below. Figure 55 shows the distribution of dioritic intrusives in the Western Cascades province.

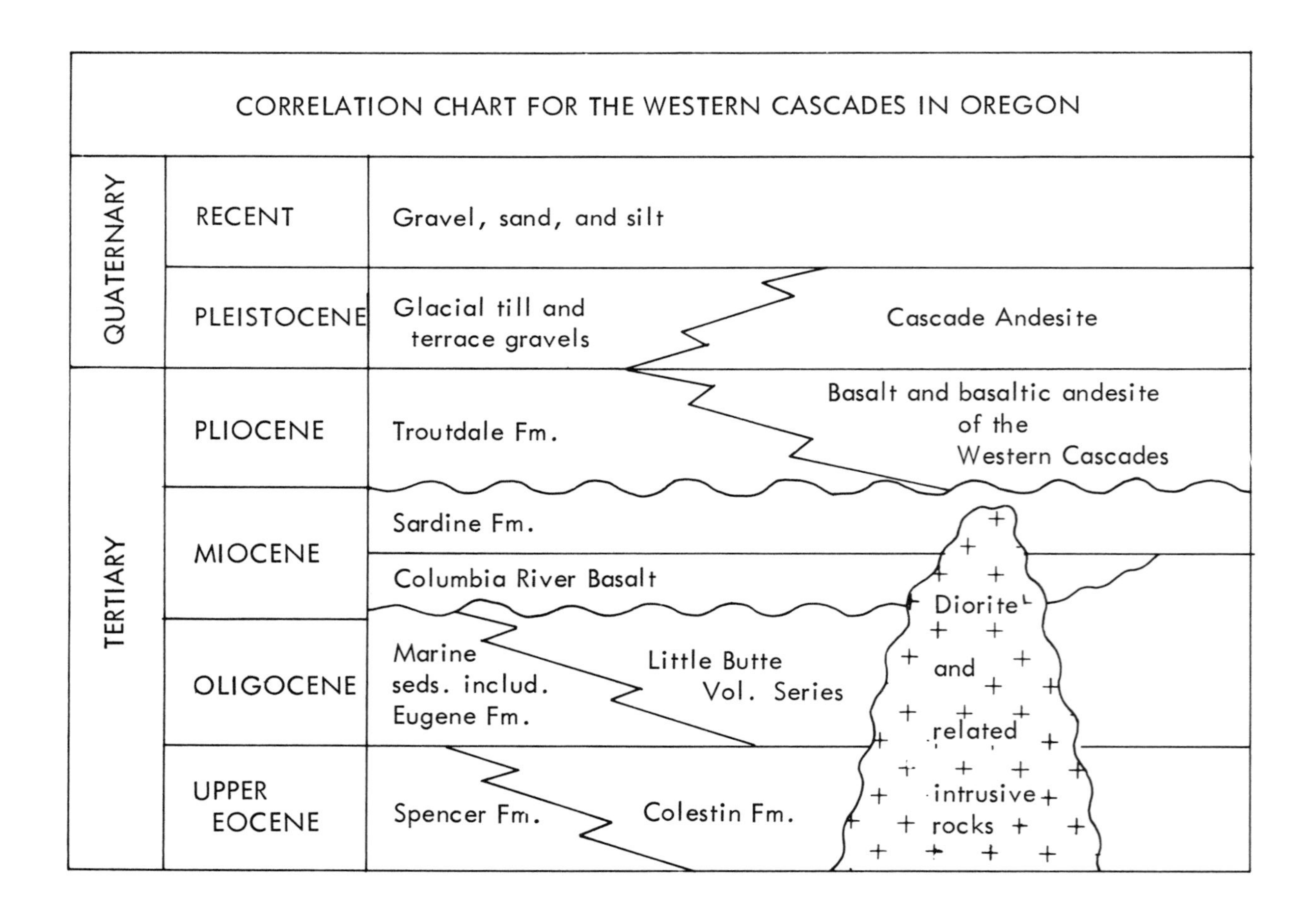

ORE DEPOSITS

Host rocks

The gold and silver deposits of the Western Cascades are associated with areas of propylitic alteration surrounding small stocks, dikes, and plugs. The intruded and altered host rocks consist of a variety of volcanic materials ranging in age from Eocene to Miocene, which are referred to by Peck and others (1964) and by Wells (1956) as Colestin Formation, Little Butte Volcanic Series, and Sardine Formation.

Mineralization

Mineralization in the Western Cascades is described by Callaghan and Buddington (1938, p. 2-3) as follows:

> "The mineral deposits occur in fissure veins of probable upper Miocene age and are believed to be genetically related to the dioritic intrusive bodies. They are of low-temperature and shallow-depth, or epithermal type, except those in the Bohemia district, which bear evidence of an initial high-temperature stage. Typical vein matter consists of altered brecciated country rock cemented with quartz that contains sphalerite, galena, chalcopyrite, and pyrite. Very minor amounts of tetrahedrite occur in some of the veins, bournonite was seen in one vein, arsenopyrite in two veins, and stibnite in several veins. Gold and silver occur in most of the sulfide ore, commonly in small amounts, but visible gold in dendrites and "wires" is present in weathered vein matter. Quartz and altered rock are the principal materials accompanying the sulfides, but other minerals occur, including calcite, dolomite, mesitite, ankerite, adularia, johannsenite, epidote, sericite, chlorite, clay minerals, barite, and both specular and red hematite. The variation in the proportions of some of the minerals permits distinction of several types of veins. Complex sulfide veins in which sphalerite predominates are the most abundant. In a few veins chalcopyrite is the dominant sulfide. A few quartz veins contain free gold without appreciable sulfides. Another group is characterized by a gangue of carbonate, chiefly calcite. Veins with stibnite in cherty quartz occur in the southwestern part of the Bohemia district. Wall rocks of the veins are altered to an aggregate of quartz, carbonate, and chlorite, to quartz, carbonate, and sericite, or to quartz and clay minerals. Pyrite occurs in almost all the altered rock.
>
> "Veins exposed to weathering and readily permeated by water are characteristically leached and iron stained, and secondary sulfides in them are negligible. In some of the veins gold remains as minute flakes, "wires," and dendrites. No secondary zinc minerals remain, and galena is represented by only very minor amounts of anglesite and cerusite, except in one vein. Secondary copper minerals include covellite, which forms films on chalcopyrite, chalcocite, which remains as a powder in some veins, a very little azurite, malachite, and chrysocolla. Pyrite is leached or changed to limonite.
>
> "Some of the veins are nearly a mile long, and a few of these contain several ore shoots that consist of brecciated country rock cemented with quartz and sulfides or of sulfide streaks in quartz. No definite evidence of change in the primary mineral composition with depth was found, but evidence of areal zoning was seen in the North Santiam and Bohemia districts. In the North Santiam district a central zone of chalcopyrite veins is surrounded by a zone of complex sulfide veins, which in turn is surrounded by carbonate veins. In the Bohemia district a central zone of large veins with abundant sulfides is surrounded by a zone of veins containing minor amounts of sulfides, and this zone in turn is surrounded by one with stibnite veins."

Discussing the genesis of Western Cascades mineralization, Callaghan and Buddington (1938, p. 36) suggest a deep-seated, cooling magma source for the solutions that penetrated newly formed fissures. The resulting veins cut both the exposed diorite and older volcanic rocks. The majority of the veins trend northwest to west and dip steeply. Mineral content and texture of veins suggest a wide range of temperature of ore-bearing solutions. Some ores in the Bohemia district indicate an initial high-temperature stage. However, many veins in the area contain fine-grained cherty quartz, some colloform with shrinkage cracks, black with disseminated pyrite dust, and in places having microconcretionary structures. Such veins must have been formed at somewhat lower temperatures.

The Western Cascades contain many local areas of a mile or so in diameter where the volcanic rocks have been intensively altered, silicified, and impregnated throughout with pyrite. It is possible that these zones are surface indications of buried intrusive bodies and that some may have mineral potential at depth. Bleached and altered rocks surrounding or adjacent to such places as Quartz Mountain, the Zinc mine, and the Cove Sulfur prospect in southeastern Douglas County, and Abbott Butte and the Al Sarena mine in northeastern Jackson County exemplify the type of localities in the Western Cascades where exploration might be warranted.

Placer deposits

A small amount of placer mining has been done in the upper reaches of Steamboat Creek south of the Bohemia area, along Quartzville Creek, and on a limited scale in several other creeks (Callaghan and Buddington, 1938). The deposits are reported to be generally low grade and most of the gold occurs as very fine particles that are difficult to recover.

The general paucity of placer deposits in the Western Cascades as compared to those of northeastern and southwestern Oregon may be due in part to the fact that lode deposits in the Western Cascades are much younger than those in the older rocks and consequently have been subjected to less uplift and erosion. Perhaps only the tops of the deposits have been exposed to erosion and little gold has been transported by streams to form placers.

GOLD MINING AREAS OF THE WESTERN CASCADES

The gold deposits of the Western Cascades fall naturally into separate areas, each of which is associated with small, dioritic intrusive bodies of Tertiary age (figure 55). The areas or mining districts are distributed along a north-south zone through the center of the Western Cascades province. The five main mineralized districts, from north to south, are:

1. North Santiam district in Clackamas and Marion Counties.

2. Quartzville district in Linn County.

3. Blue River district in Linn and Lane Counties.

4. Fall Creek district in Lane County.

5. Bohemia district in Lane County.

6. Other deposits: Zinc mine in Douglas County and the Al Sarena and Barron mines in Jackson County.

NORTH SANTIAM DISTRICT

Location

The North Santiam district lies about 37 miles east of Salem in T. 8 S., Rs. 4 and 5 E., along the Little North Santiam River (figure 56). A few prospects lie outside this area in T. 7 S., R. 4 E. and in T. 9 S., Rs. 3 and 6 E. The area is 23 miles east of Mehama, and is reached by way of the Little North Santiam road. It is partly on the Mill City and partly on the Battle Ax 15-minute quadrangles. The terrain is steep, and elevations range from about 1400 feet on the Little North Santiam to 5547 feet on Battle Ax Mountain near the east edge of the area.

Geology and Mineralization

Light-colored andesites and rhyolites belonging to the Oligocene-Miocene Sardine Series of Thayer (1939) (Little Butte Volcanic Series and Sardine Formation by Peck and others, 1964) predominate over fragmental volcanic rocks in the district. These rocks are intruded by small dacite-porphyry dikes and plugs. One quartz-diorite body is exposed at the Crown mine. Large areas of contact metamorphic alteration and zoning of mineralization indicate the possibility of larger unexposed intrusive bodies. Zoning of mineralization appears to be better developed in the North Santiam district than elsewhere in the Western Cascades mineralized areas. Chalcopyrite-bearing veins from the Crown mine to and along the river form a central high-temperature zone. This is succeeded in the section by pyrite-bearing veins on Gold Creek that, in turn, give way to complex sulfide veins in the Blende Oro and, farther east, in the Ruth mine. These complex sulfide veins carry sphalerite with variable amounts of galena and chalcopyrite. An outer low-temperature zone is represented by the calcite vein on lower Elkhorn Creek and at the Ogle Mountain mine.

History and Production

Mineralization was first discovered in the district about 1877, and by 1903 most of the properties had been located. The district's total recorded production for the period 1896 to 1947 is about $25,000 divided as follows: gold, 454 ounces; silver, 1,412 ounces; copper, 41,172 pounds; lead, 40,700 pounds; and zinc, 110,063 pounds.

The most intensively developed mines are the Ogle Mountain, Ruth, and Santiam Copper. Because of lack of information and apparent lack of significant values and development, only 8 of the 26 known occurrences are described in the alphabetical list at the end of this section. A more complete coverage is given in Department Bulletin 14-D (1951, p. 115-132).

Principal Mines

Ogle Mountain mine: Ogle Mountain mine on upper Ogle Creek, a tributary of the Molalla River, was first worked in 1903. Machinery was brought to the mine in 1905. It was operated until 1914 and again under lease in 1918-1919. The owner reported a total production of about $10,000.

Equipment included a 10-stamp mill and a 5 x 20-foot pebble-filled tube mill that were operated by steam engines. The workings in a number of scattered adits and levels totaled about 3000 feet. The longest reported is a lower crosscut 1460 feet long. Drifts and stopes were fairly extensive. Oxidized upper ore bodies were stoped to a width of 6 feet. They contained free gold and some visible wire gold.

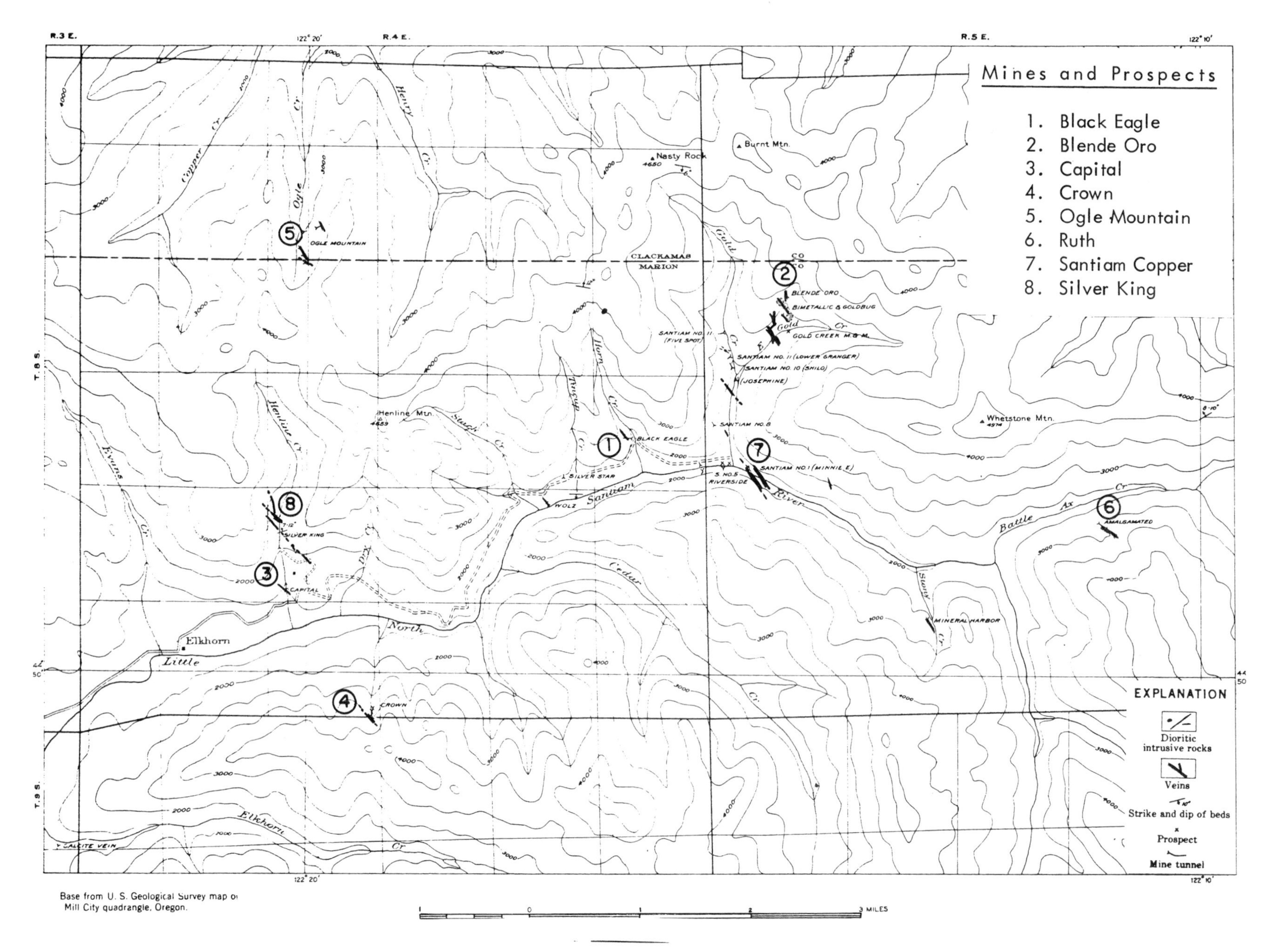

Figure 56. Index map of the North Santiam district (from U.S.G.S. Bull. 893).

Ruth mine: The Ruth mine was formerly known as the Lewis & Clark Mining & Milling Co. This company located five claims south of Battle Ax Creek, a tributary of Little North Santiam River, some time before 1902. By this date they had opened several hundred feet of tunnel on two levels. In 1920 the Amalgamated Mining & Milling Co. took over the original claims and located 18 more. During 1929-1934 a combined effort of Amalgamated and Columbia Mines Development Co. constructed a road to the mine, erected several buildings and a mill, and shipped nine carloads of crude ore and mill concentrates during 1931 and 1932. In 1939 the mine was purchased by the Pacific Smelting & Refining Co. Total production and ore values are not reported.

A total development of more than 4000 feet is reported, mostly on the No. 4 and No. 5 levels. About 200,000 tons of ore reserves, containing about 6 percent zinc, are said to have been blocked out. Various engineers' reports on the property estimate reserves of from 200,000 to 800,000 tons that will average from 4.33 to 11 percent zinc in blocks of ore with an average thickness of about 10 feet. Gold and silver assays are usually low.

Santiam Copper mine: The Santiam Copper mine, situated on both sides of the Little North Santiam and Gold Creek, was first worked about 1900. It has been known at various times as: Freeland Consolidated (dissolved 1914); Electric Mining & Smelting Co. (dissolved 1914); Consolidated Copper Mining & Power Co. (1914-1925?); Lotz and Larsen mine (1916-1925); Northwest Copper Co. (1926-1930); and Rainbow mine (lessees 1941).

During the years 1923 to 1925, 138 tons of ore and some concentrates were shipped. In 1928, 3 tons of sorted ore were shipped. In 1940, 14 tons of ore and 73 tons of concentrates were shipped. These shipments averaged about 10 percent copper, 3 ounces silver, and 0.03 ounces gold.

The bulk of production has been from the main 1000-foot drift on the Minnie E. or Northwestern vein on the south side of the Little North Santiam River.

Chalcopyrite is the principal ore mineral and occurs in places as 1- to 4-inch seams of pure mineral. Four fairly distinct narrow ore shoots exposed in the main workings are described by Callaghan and Buddington (1938, p. 96). The shoot in the north drift is 100 feet long and about 18 inches thick. The ore is reported to average about 4.47 percent copper and 1.22 ounces per ton silver for a width of 6 feet. A winze, now full of water, is reported to have 14 inches of chalcopyrite 96 feet below the drift. A 200-foot-long, 5-foot-thick ore shoot, largest of three mined in the south drift, averaged 2.41 percent copper, 0.75 ounce silver, and trace gold. A small mill at the mouth of Gold Creek was used to concentrate some of the ore, but the type of mill equipment is not reported.

Lode Mines of the North Santiam District

Black Eagle mine

North Santiam District, 1

Location: Marion County, W½ sec. 24, T. 8 S., R. 4 E., at about 2000 feet elevation.

Development: About 1000 feet in main tunnel plus several short tunnels and open cuts in 1916.

Geology: Country rock is andesite. Main portal is in gravels. Vein is quartz-cemented andesite breccia with chalcopyrite, very little galena, and secondary limonite, malachite, azurite, and chrysocolla. Several parallel veins are exposed in a 20-foot stope. Main vein is 2½ feet thick, with an additional 7 feet of low-grade material exposed in an 85-foot shoot.

Production: Owned by Black Eagle Mining & Milling Co. in 1916. Had a small concentrating mill. Amount of production is not reported.

References: Parks and Swartley (1916:33-34); Callaghan and Buddington (1938:89-90); Department Bulletin 14-D (1951:116).

Blende Oro mine

North Santiam District, 2

Location: Marion County, NE¼ sec. 18, T. 8 S., R. 5 E., at about 2700 feet elevation on north branch of East Fork Gold Creek.

Development: Two drifts; upper 40 feet and lower 215 feet in 1941.

Geology: Country rock is highly altered tuff consisting of sericite, cherty quartz, and disseminated pyrite. Bunches or bands of sulfides from 1 to 5½ feet thick form an ore shoot more than 100 feet long in a vertical zone. Sphalerite, pyrite, galena, and chalcopyrite occur as well-defined crystals lining vugs. Some light, resinous sphalerite crystals more than an inch in diameter are reported.

Production: Located about 1900 by the Hart brothers. Production is not reported.

References: Callaghan and Buddington, 1938:90; Department Bulletin 14-D, 1951:116-117.

Capital claims

North Santiam District, 3

Location: Marion County, SW¼ sec. 28, T. 8 S., R. 4 E., at about 1500 feet elevation. Two patented claims.

Development: About 400 feet in one tunnel including a winze, raise, and stope; a shaft reported elsewhere.

Geology: Branching veins 15 to 30 inches thick consist of silicified andesite breccia with sericite, iron-magnesium carbonate, and quartz veinlets with sphalerite, minor galena, and chalcopyrite. Tunnel cuts an old gravel channel of Henline Creek.

Production: Claims patented about 1872-1878 were among first in this part of the state. A shipment of ore is said to have been made to Swansea, Wales. Production not reported.

References: Callaghan and Buddington, 1938:91; Department Bulletin 14-D, 1951:118.

Crown mine

North Santiam District, 4

Location: Marion County, SE¼ sec. 33, T. 8 S., R. 4 E., and NE¼ sec. 4, T. 9 S., R. 4 E., at about 3000 feet elevation.

Development: About 1000 feet crosscuts and drifts in main tunnel, plus short tunnels and surface cuts.

Geology: Altered andesite, tuff, and rhyolite are cut by diorite and andesite dikes near the contact of an intrusive mass of quartz diorite. Workings cut several northwest-striking veins and gouge zones (refer to map Callaghan and Buddington, 1938, fig. 3, p. 92) varying from a few inches to 10 feet thick. Sulfides reported include pyrite, chalcopyrite, and a little sphalerite. Some secondary malachite and chrysocolla are also reported.

Assays reported by Callaghan and Buddington (1938, p. 93)

		1	2	3	4	5	6
Width	inches	stringers	selected	18.00	6.00	-	12.00
Gold	ounces to the ton	0.10	0.06	0.03	0.22	0.08	0.08
Silver	ounces to the ton	2.76	0.90	0.68	1.80	0.90	1.60
Copper	ounces to the ton	1.11	1.93	1.93	4.24	2.31	6.26

Note: Samples 1 and 2 are from Blind vein, 3 from Blackwall vein, 4 from Thirteen-foot vein, and 5 and 6 from Winze vein.

Production: Most of the development was done prior to 1927. Production is not reported.

References: Parks and Swartley, 1916:83-84; Callaghan and Buddington, 1938:91-93; Department Bulletin 14-D, 1951:118-119.

Ogle Mountain mine — North Santiam District, 5

Location: Clackamas-Marion County line, sections 9 and 16, T. 8 S., R. 4 E., between 3000 and 3500 feet elevation near the head of Ogle Creek.

Development: Probably in excess of 3000 feet of workings, mainly from two adits but with several other tunnels described.

Geology: Glacial gravels cover porphyritic andesite and tuff breccia. Main vein strikes N. 40° W., dips 45° to 60° E., and averages 5 feet wide. Cherty quartz gangue is cut by carbonate veins and vuggy calcite. Other vein minerals reported are ankerite, sphalerite, pyrite, galena, chalcopyrite, and adularia. Free gold visible in oxidized ore has been reported in the upper stopes, with average recovery of 0.24 ounce gold per ton.

Production: Work was started in 1903, and production of $10,000 to 1919 was reported by the owner. The mine had a ten-stamp mill, tube mill, cyanide tanks, a table, and various other equipment.

References: Parks and Swartley, 1916:166; Callaghan and Buddington, 1938:94-95; Department Bulletin 14-D, 1951:123-124.

Ruth (Amalgamated) mine — North Santiam District, 6

Location: Marion County, sec. 27, T. 8 S., R. 5 E., at about 3000 feet elevation.

Development: More than 4000 feet of workings in No. 4 and No. 5 levels on Ruth vein and 340 feet in Blue Jay vein, plus several small tunnels and cuts.

Geology: Country rock is andesite cut by a porphyritic rhyolite dike in No. 4 level. Mineralization occurs in a fault zone from 1 to more than 60 feet thick, trending N. 55° W., and dipping 55° to 70° NE. It contains crushed rock, clay gouge, chlorite, sericite, calcite, quartz, sphalerite, galena, chalcopyrite, and pyrite. Lenses of high-grade sulfide (mainly sphalerite) dip more steeply than the enclosing fault zone. Thickness of 15 ore blocks is from 5 to 15 feet. Ore reserve estimates reported from 200,000 to 800,000 tons, with from 4 to 11 percent zinc and about 1 percent lead. Gold, silver, and copper values are generally low.

Production: Worked by several operators from about 1900 to 1948. Nine carloads of ore and concentrates were shipped in 1931-1932. Total production is not reported.

References: Parks and Swartley, 1916:140; Callaghan and Buddington, 1938:87-89; Rosenberg, F. J., 1941, unpublished report; Department Bulletin 14-D, 1951:126-127.

Santiam Copper mine

North Santiam District, 7

Location: Marion County, sec. 19, T. 8 S., R. 5 E., at about 2000 feet elevation on the Little North Santiam River.

Development: Most work is on Minnie E. or Northwestern vein which crosses the river. On south side of the river there is 1000 feet of drift, with 6 raises in 3 stopes, and on the north side workings include 300 feet of drift and a 96-foot winze (1941).

Geology: The vein strikes northwest, dips about 70° E. in andesite country rocks. Ore minerals occur as seams of chalcopyrite with subordinate pyrite and sphalerite in breccia and altered rock with quartz veinlets. Average thickness is about 5 feet. Ores average 2 to 4½ percent copper, 0.75 to 1.22 oz./ton silver, and trace gold.

Production: Work began about 1900. Smelter shipments from 1923 through 1941 amounted to 222.71 tons of ore and concentrates averaging (weighted) about 10 percent copper, 3 ounces silver, and .03 ounces gold per ton.

References: Stafford, 1904; Parks and Swartley, 1916:69; Callaghan and Buddington, 1938: 95-97 including map; Department Bulletin 14-D, 1951:128-130.

Silver King group

North Santiam District, 8

Location: Marion County, NW¼ sec. 28, T. 8 S., R. 4 E., between 2000 and 2500 feet elevation.

Development: The main crosscut is reported to be about 1700 feet long. A 50-foot eastern drift tunnel on the east bank of Henline Creek, a west drift on the west side of the creek (length not reported), and an 80-foot shaft south of the crosscut portal.

Geology: The crosscut tunnel cuts several northwest-striking veins. The principal vein is the Queen of the West. It strikes northwest, dips 40° to 45° SW., and averages 18 to 24 inches thick. Vuggy quartz veins contain altered andesite fragments as well as sphalerite, some galena, a little chalcopyrite, and pyrite. Footwall rock is andesite; a dacite-porphyry dike forms the hanging wall on the east drift and forms both walls on the west drift. Assays indicate gold from a trace to 0.07 oz./ton, silver 1.6 to 40.8 oz./ton, lead 0.1 to 4.7 percent, and zinc 1.4 to 4.2 percent.

Production: Information lacking.

References: Parks and Swartley, 1916:202-203; Callaghan and Buddington, 1938:97-98; Department Bulletin 14-D, 1951:130-131.

QUARTZVILLE DISTRICT

Location

The Quartzville district lies in Linn County in Ts. 11 and 12 S., R. 4 E., in the drainage area of Quartzville Creek, a major tributary of the Middle Santiam River (figure 57). The district is reached from Foster on U.S. Highway 20 by way of the Green Peter dam and reservoir road, a total distance of 27 miles.

The region is steep and heavily timbered; elevations range from 1500 feet on Quartzville Creek to about 3500 feet on the ridges. Gold Peak and Galena Mountain are about 4000 and 5000 feet in elevation, respectively. The area is covered by the 15-minute Quartzville topographic map.

Geology and Mineralization

The district is underlain by volcanic rocks assigned to the Sardine Formation by Peck and others (1964). The following summary is quoted from Department Bulletin 14-D (1951, p. 96):

> "As in other Cascade mining districts, rocks of the area are made up of andesite and rhyolite flows and interbedded tuffs and breccias in about equal amounts. There are scattered dikes and plugs of dacite porphyry with narrow contact aureoles of hornfels. Vein systems are markedly uniform in strike, averaging N. 40° W., with steep dips in either direction. Although individual ore shoots are of rather small dimensions, the shear zones of the system to which they belong usually have considerable lateral extent, several of them being traceable for several thousand feet, with widths of 50 feet or more.
>
> "Much of the underground workings are now inaccessible, and those that are open are usually in the upper weathered portion of the veins, so that knowledge of the sulfide zones is scanty. Depth of oxidation varies greatly and in places unaltered rock containing sulfides crops out at the surface. Most of the veins contain mixed sulfides with sphalerite and pyrite common. Bournonite and tetrahedrite have been recognized. Pyritization is widespread, and pyrite is found streaking some vein walls. Concentrations of metallic gold in the oxidized zone at or near the surface have been the incentive for much of the prospecting, so that development work, except in a few cases, has been principally in following small ore shoots. Almost no systematic crosscutting which would show the potentialities of the large shear zones has been done. Many exceptional specimens of wire gold have been found in pockets."

History and Production

Mineralization was discovered in 1863 and the area was organized as a mining district in 1864. The Lawler and Albany mines were operated during the 1890's. The district has been relatively inactive since that time and less than 200 ounces of gold has been produced since 1896. Total production for the district reported in Department Bulletin 14-D (1951, p. 96) is $181,255, which included 8557 ounces of gold and 2920 ounces of silver.

Some small-scale lode and placer mining was done in the district during the 1930's.

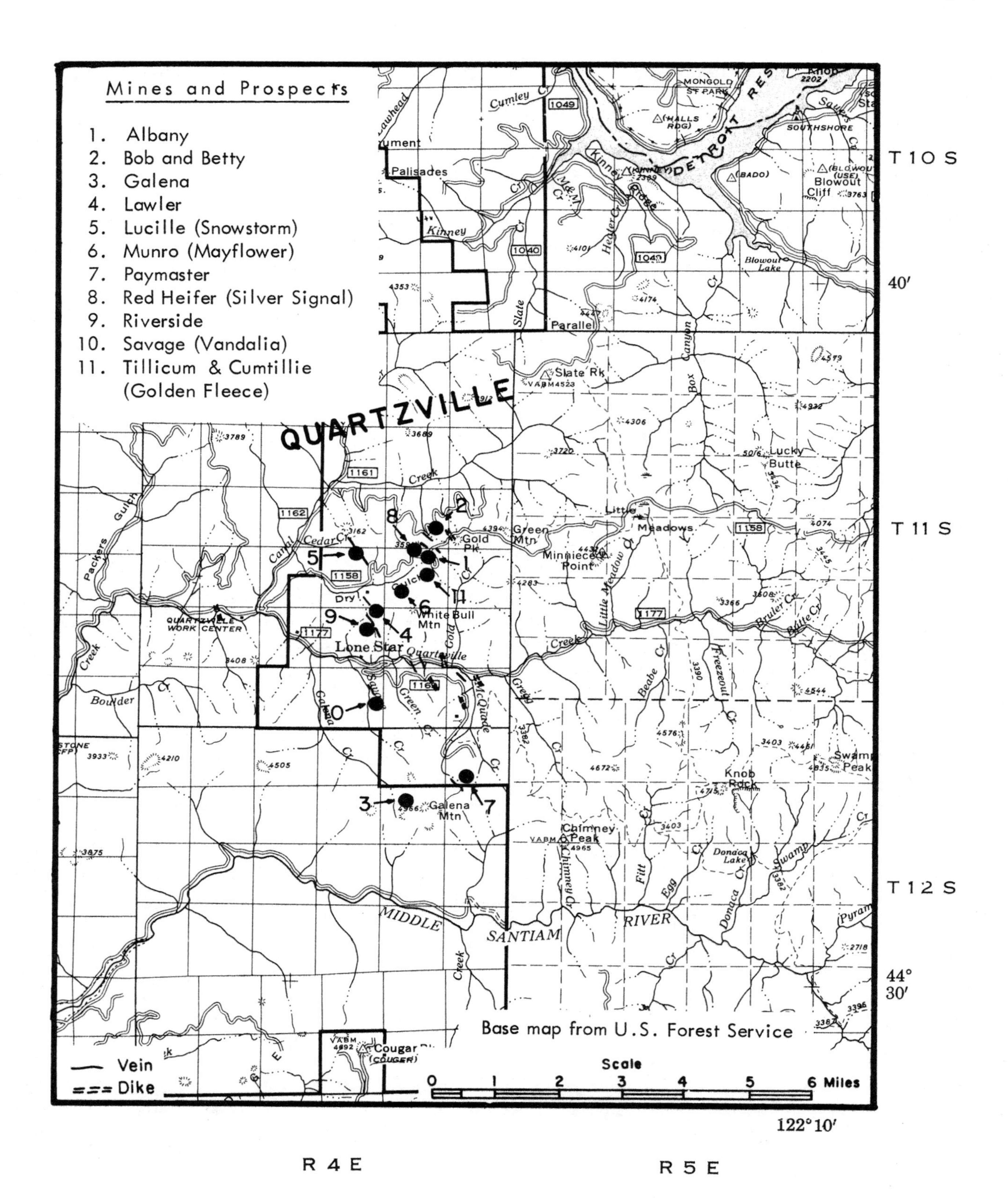

Figure 57. Index map of the Quartzville district.

Principal Mines

Lawler mine: The Lawler mine, located on White Bull Mountain and Dry Gulch, was the original discovery in the district; it was said to have been found by Dr. E. O. Smith in 1861. A company was organized and it optioned the property to W. B. Lawler, who interested English capital. A 20-stamp mill was built and considerable development work done. The mill is reported to have ceased operations in 1898. Total production was about $100,000.

Albany mine: The Albany mine, near the head of Dry Gulch in sec. 23, T. 11 S., R. 4 E., was first prospected in 1888. In 1892 a Tremain two-stamp mill was built and used to mill good oxide ore from two small shoots. Recovery was poor. Later a 10-stamp mill was installed by the Lincoln Mines Co., and U.S. Mint records for 1890, 1892, and 1893 credit the property with 653 ounces of gold. In 1925 a Lane mill was installed and ore brought down to it by a cable tramway.

Lode Mines of the Quartzville District

Albany mine — Quartzville District, 1

Location: Linn County, sec. 23, T. 11 S., R. 4 E. Workings are between about 2570 and 3070 feet elevation. Three patented claims.

Development: In 1931 a total of 1090 feet of tunnels was accessible. In addition, there are several cuts and shallow shafts. (Maps of workings in Callaghan and Buddington, 1938, plate 19.)

Geology: Country rocks are andesite, rhyolite, and tuffs. Mineralization occurs along four main, northwest-trending, brecciated and altered silicified zones containing some clay gouge and drusy and cherty quartz veinlets. The most abundant sulfide is pyrite with some sphalerite, galena, and chalcopyrite.

Production: The property was first prospected in 1888. Ore was processed in three mills: The first was a 2-stamp; a 10-stamp mill operated by steam in 1892; and a chilean mill was installed in 1925. Total production is not reported, but may have been about $50,000.

References: Parks and Swartley, 1916:141; Stowell, 1921; Callaghan and Buddington, 1938: 103-105; Department Bulletin 14-D, 1951:97.

Bob and Betty (Smith & McCleary) mine — Quartzville District, 2

Location: Linn County, SE$\frac{1}{4}$ sec. 14, T. 11 S., R. 4 E., about 3100 feet elevation.

Development: Workings include about 1650 lineal feet of tunnel, plus a large open cut and 100-foot shaft (caved).

Geology: Mineralization occurs along a fairly wide and sheared and altered zone in andesite and tuff. The zone strikes N. 55° to 80° W. and dips 70° S. Clay gouge, reticulating quartz veinlets, calcite, and spots of sulfides occur in the zone. Sphalerite, galena, pyrite, and minor chalcopyrite are found. A sample of hand-sorted sulfides assayed 0.20 oz./ton gold, 1.0 oz./ton silver, trace copper, 2.8

percent lead, and 2.95 percent zinc.

Production: Discovered in 1881. In 1885, 100 tons were milled in a steam-operated 10-stamp mill. Recovery was poor and production small.

References: Stowell, 1921; Callaghan and Buddington, 1938:105-106; Department Bulletin 14-D, 1951:98.

Galena mine

Quartzville District, 3

Location: Linn County, NW¼ sec. 11, T. 12 S., R. 4 E., between 3400 and 3700 feet elevation.

Development: Two crosscut tunnels with some drifting total 725 feet plus 2 small open cuts.

Geology: Seams of quartz and disseminated sulfides occur in a northwest-trending, brecciated shear zone in andesite and tuff. The following assays were reported by Stowell (1921, p. 34-36):

Location	Width	Ounces per ton Gold	Ounces per ton Silver	Percent Copper	Percent Lead	Percent Zinc
North drift	18 inches	0.56	2.44	0.65	0.20	4.25
6' South drift	12 inches	0.02	3.68	0.15	8.85	2.4
Open cut 3500' elevation	4 feet	0.10	0.4	0.35	4.65	1.6
Face south drift lower adit	16 inches	0.03	0.31	0.45	0.20	2.0
100'x50' outcrop 650' north of upper cut elev. 3750'	4 feet	0.10	0.4	0.05	2.6	3.7

Production: Some development work was being done in 1921. Production not reported.

References: Stowell, 1921; Callaghan and Buddington, 1938:106.

Lawler mine

Quartzville District, 4

Location: Linn County, secs. 21, 22, 23, 26, and 27, T. 11 S., R. 4 E. from about 1600 to 3280 feet elevation. Two large placer and 11 lode claims patented.

Development: By 1903 a total of 2000 feet of tunnel had been driven. There are four principal adit levels. Stopes extend from the main level 155 feet to the surface. There are numerous open cuts as well.

Geology: The vein strikes about N. 50° W. and dips steeply NE. It is a gouge and quartz-cemented breccia of rhyolite, andesite, and tuff containing some sphalerite, galena, chalcopyrite, and pyrite. Caved stopes indicate a width of 8 feet. Most of the ore mined was oxidized and showed no sulfides.

Production: First discovery in the district in 1861. Reported production is about $100,000. Ore was ground in a 20-stamp mill which shut down in 1898.

References: Stafford, 1904:58; Callaghan and Buddington, 1938:107-108; Department Bulletin 14-D, 1951:101.

Lucille (Snowstrom & Bell, formerly Edson Group) mine — Quartzville District, 5

Location: Linn County, NE$\frac{1}{4}$ sec. 22 and south edge sec. 15, T. 11 S., R. 4 E. between about 2480 and 3400 feet elevation.

Development: About 650 lineal feet of tunnels and several cuts. One cut 40 feet long, 12 feet wide, and 15 to 20 feet deep at the face.

Geology: Rhyolite, tuff, and andesite are cut by a northwest-trending shear zone with silicified, iron and manganese oxide-stained rock and gouge. The zone is as much as 70 feet wide with values in narrow seams.

Production: Discovered by Edson in 1897. Most of the development work was done prior to 1920. A $2000 pocket was reportedly mined.

References: Stowell, 1921; Callaghan and Buddington, 1938:111-112; Department Bulletin 14-D, 1951:102, 106-107.

Munro (Mayflower) group — Quartzville District, 6

Location: Linn County, SW$\frac{1}{4}$SW$\frac{1}{4}$ sec. 23, T. 11 S., R. 4 E., at about 2800 feet elevation.

Development: Main adit tunnel had 263 feet of crosscut and 217 feet of drift, along with several other tunnels with short drifts, winzes, and so forth, on both sides of the ravine, as described by Callaghan and Buddington (1938, p. 109).

Geology: Country rocks are altered, iron-stained andesite and rhyolite. Two or three narrow veins or mineralized seams trending N. 20° to 40° W. are described. Vein matter is partly silicified, brecciated, and leached rock with very little quartz and no sulfides other than pyrite.

Production: Discovered about 1890. Owner reported production of 72.56 ounces of gold. In 1931 owner was using a small prospector's mill; an old, dismantled arrastra was on the property.

References: Callaghan and Buddington, 1938:109; Department Bulletin 14-D, 1951:102-103.

Paymaster claim — Quartzville District, 7

Location: Linn County, SW$\frac{1}{4}$ sec. 1 and NW$\frac{1}{4}$ sec. 12, T. 12 S., R. 4 E., about 3600 feet elevation.

Development: In 1916 a drift 130 feet long was reported.

Geology: Mineralized zones in tuff and andesite strike northwest. They contain altered silicified rock with sulfide-bearing quartz veins a few inches wide. A sample of the sulfide-bearing quartz taken by Stowell (1921) about 8 inches wide assayed 0.04 oz./ton gold, 0.86 oz./ton silver, 0.5 percent copper, 3.2 percent lead, and 4.0 percent zinc. A one-foot sample from the vein in the main tunnel assayed 0.76 oz./ton gold, 1.94 oz./ton silver, 0.10 percent copper, 10.1 percent lead, and 8.4 percent zinc.

Production: Developed by a capital-stock promotion about 1913 called the Paymaster Mining & Milling Co., dissolved by proclamation in 1917. Production not reported.

References: Parks and Swartley, 1916:178; Stowell, 1921; Callaghan and Buddington, 1938: 109-110; Department Bulletin 14-D, 1951:103.

Red Heifer (Silver Signal) claim

Quartzville District, 8

Location: Linn County, near line of secs. 14 and 23, T. 11 S., R. 4 E., about 3300 feet elevation. On north side of ridge.

Development: Two tunnels and several cuts. Lower adit is caved. Upper has a 60-foot crosscut, short drift southeast, and 30-foot raise.

Geology: A sheared and silicified breccia zone trending N. 50° E. in andesite contains comb quartz, iron and manganese oxide-stained gouge, and residual sulfide cores chiefly of sphalerite, a little pyrite, chalcopyrite, and minor galena. A selected sample of sulfides assayed 0.32 oz./ton gold and 4.90 oz./ton silver. A sample of leached ore assayed 0.92 oz./ton gold and 6.2 oz./ton silver. Vein is about 3 feet thick.

Production: In 1921 the vein was reported to have produced a high-grade pocket. In 1938 the owner had installed a small mill and was working the property. Total production is not reported.

References: Callaghan and Buddington, 1938:110-111; Department Bulletin 14-D, 1951:104.

Riverside group

Quartzville District, 9

Location: Linn County, E½ sec. 27, T. 11 S., R. 4 E. Workings are from about 2500 to 3000+ feet elevation.

Development: About 500 lineal feet in 9 tunnels and several open cuts.

Geology: Brecciated, altered andesite with iron-stained gouge and bunches of quartz with sphalerite, pyrite, and galena strikes about N. 30° W. and is vertical. A 4-foot sample of vein taken by Stowell (1921) assayed 1.32 oz./ton gold and 0.98 oz./ton silver.

Production: First located in 1912 and operated in small way for many years. Ore was ground in a Gibson prospector's mill. Amount of production is not reported.

References: Stowell, 1921; Callaghan and Buddington, 1938:110; Department Bulletin 14-D, 1951:105.

Savage (Vandalia) group

Quartzville District, 10

Location: Linn County, SE¼ sec. 34, T. 11 S., R. 4 E. between 2700 and 3100 feet elevation.

Development: There is development of about 600 lineal feet in 6 tunnels, a caved shaft 80 feet deep, and several open cuts.

Geology: Country rocks are andesites and tuffs, with a mineralized shear zone about 50 feet wide with branching quartz veinlets, disseminated pyrite, and small bunches of other sulfides. The zone strikes northwest and dips steeply southwest. A $2\frac{1}{2}$-foot sample across the vein in east drift of upper tunnel assayed 0.16 oz./ton gold and 0.74 oz./ton silver with traces of lead, zinc, and copper. A $3\frac{1}{2}$-foot sample across the vein in Golden West tunnel assayed 0.36 oz./ton gold and 0.34 oz./ton silver.

Production: Discovered in 1900. Combined production of the Vandalia of two claims and the Golden West group of five claims in 1921 was about $7000. The mine was being worked in a small way during the 1930's.

References: Stowell, 1921; Callaghan and Buddington, 1938:110-111 (includes map); Department Bulletin 14-D, 1951:105-106.

Tillicum and Cumtillie (Golden Fleece) claims — Quartzville District, 11

Location: Linn County, SW$\frac{1}{4}$ sec. 23, T. 11 S., R. 4 E., about 3000 feet elevation.

Development: Early development in oxidized ore is inaccessible. Later workings 100 feet lower total about 300 feet.

Geology: Workings follow both walls of a dark-colored diorite dike trending N. 40° W. in altered andesite. The soft, pyrite-bearing gouge seam (about 20 inches wide) on the southwest side of the dike carries some gold.

Production: Property was owned by the Advance Mining & Milling Co. in 1916. A later owner worked the property with a small, two-stamp, water-powered mill during the 1930's. Production is not reported.

References: Parks and Swartley, 1916:7; Callaghan and Buddington, 1938:106; Department Bulletin 14-D, 1951:107.

BLUE RIVER DISTRICT

Location

The Blue River mining district (figure 58) is situated in southern Linn County and northern Lane County about 5 miles north of the small town of Blue River, which is on U.S. Highway 126 on the north side of the McKenzie River about 43 miles east of Eugene. The area is mainly in Ts. 15 and 16 S., R. 4 E., on both sides of the drainage divide between Calapooya and McKenzie Rivers. The principal mine, Lucky Boy, is reached by a road up Quartz Creek from Blue River. Other mines and prospects in the area are accessible from various trails and logging roads. Elevations range from slightly more than 4800 feet on the county line ridge $2\frac{1}{2}$ miles northeast of Gold Hill to about 2000 feet at the lower prospects. The slopes are generally steep and well timbered. The area is covered by the Blue River 15-minute quadrangle topographic map.

Geology and Mineralization

The district is underlain by volcanic rocks assigned to the Sardine Formation by Peck and others (1964). The following summary is given in Department Bulletin 14-D (1951, p. 45):

> "Rock and mineral associations are in general the same as in other Cascade deposits. The proportion of flows to fragmental rocks (tuffs and volcanic breccias) appears to be about equal. The nearly horizontal lavas are predominantly andesitic; some rhyolite, however, appears on the north side of the divide on the headwaters of Uncle Sam and Badger Creeks. Two groups of dioritic dikes and plugs with narrow contact aureoles of hornfels occur, one northeast of Gold Hill and the other on the south fork of Tid Bits Creek.
>
> "Most of the past mining was from the oxidized portions of the veins, which are fewer and less developed than in the Bohemia district to the south. The Lucky Boy vein contains the usual Cascade assemblage of pyrite, sphalerite, galena, and chalcopyrite, with quartz the dominant gangue material. It also contains some small grains of tetrahedrite. Calcite is predominant in a few of the veins in the district, especially the Great Northern, Higgins, and Cinderella. Adularia is more common here than in other districts. The secondary minerals are largely leached out, but occur in small amounts."

Veins in the area strike in a north-to-northwest direction and generally dip steeply.

Callaghan and Buddington (1938, p. 116-117) describe the veins of the area as follows:

> "Only the weathered parts of many of the veins are revealed in accessible workings, but there are several workings in which the nature of the primary vein matter is apparent. One vein, the Rowena, is characterized by chalcopyrite but contains minor amounts of sphalerite and pyrite and possibly a little galena. The Great Northern vein is characterized by massive calcite and minor quantities of the sulfides. Calcite is exposed in the Higgins workings, and its former existence in the Cinderella vein is indicated by a brown powder of manganese oxide. The other veins are of the usual type, with varying amounts of the sulfides, chiefly sphalerite. Vein matter from the Lucky Boy contains small grains of tetrahedrite and considerable pyrite in addition to sphalerite, galena, and chalcopyrite. Quartz is the dominant gangue mineral, occurring as coarse crystals or as cryptocrystalline aggregates with colloform structure, as at the Durango prospect. Adularia is more abundant in the Blue River

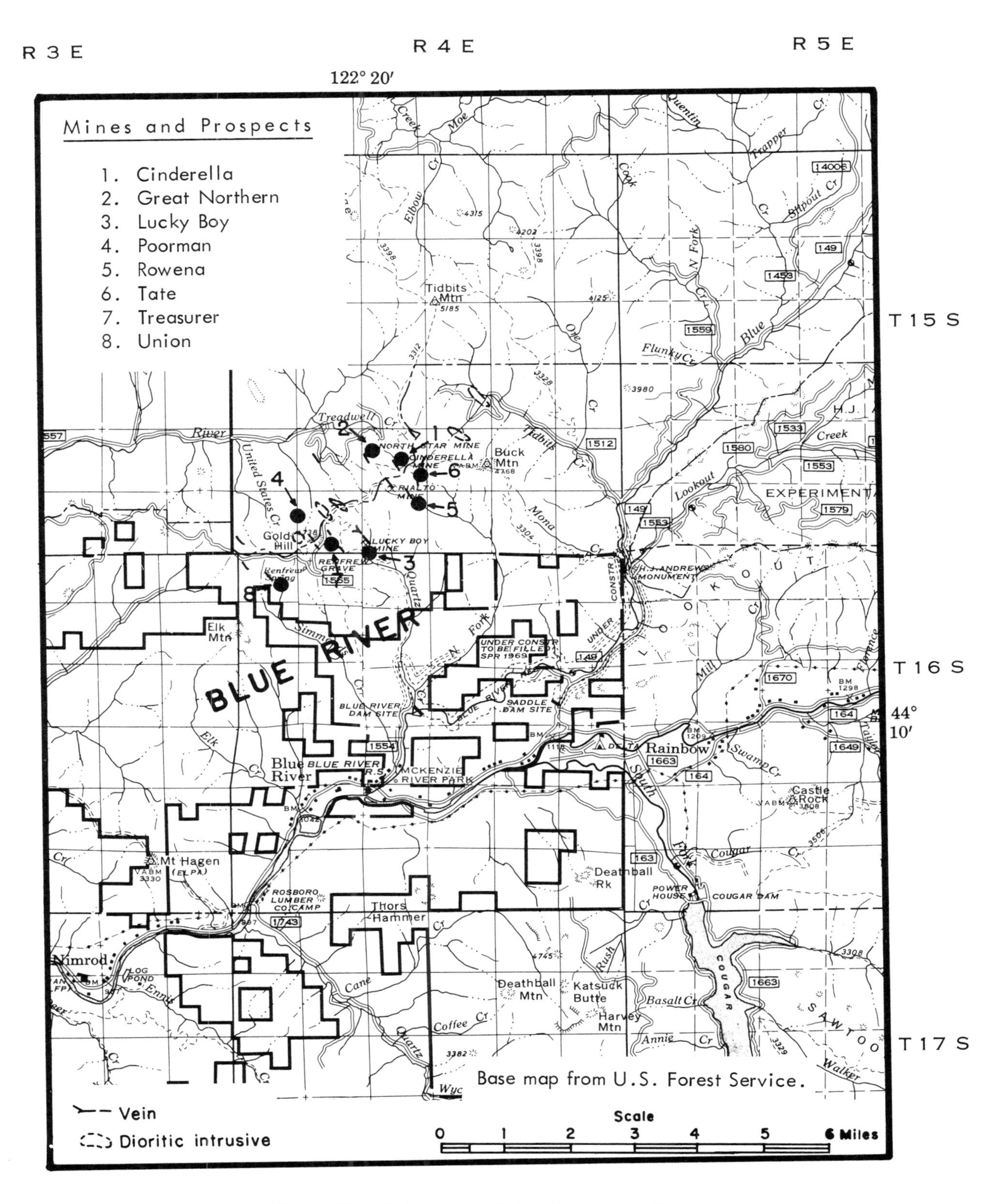

Figure 58. Index map of the Blue River district.

district than in any of the other districts, and at the Tate property it makes up nearly half of the vein matter in the main drift. A little barite occurs with calcite in vein matter from the Treasure mine. Vein matter in the weathered zones is leached, but in some places films of chalcocite and covellite remain, and there are very small amounts of chrysocolla, malachite, cerusite, and anglesite and larger amounts of limonite. A brown powder of oxides of iron and manganese occurs in the weathered parts of the carbonate veins.

"The veins in the Blue River district are smaller and less persitent than those in the Bohemia district. The only large ore shoot was that at the intersection of the Lucky Boy and Daisy Creek veins, a favorable location for weathering. The sulfides appear to have a low content of the precious metals. The Lucky Boy vein has been explored by drifts for more than 1100 feet, and the drift on the Union vein is 700 feet long. According to Parks and Swartley (1916, p. 224) the main drift in the Treasure mine followed the vein for 1800 feet.

"Possibly some pockets or small shoots of gold ore from the weathered parts of the veins may be found, though considerable prospecting was done in the early days of the camp. Probably moderate quantities of sulfides with a low gold content remain in the Lucky Boy mine and might be extracted when prices of base metals become sufficiently high. The discovery of large ore bodies is not anticipated, and any newly developed ore should be blocked out prior to selection and installation of milling equipment."

History and Production

Production in the district has come mainly from the Lucky Boy mine, which was discovered in 1887. The Great Northern mine had a small output and was second in the district to the Lucky Boy. Production at the Treasure mine was small, and in 1902 the mine was the most extensively developed of any in the district.

Most of the properties were patented and little mining work has been done in the area since about 1913 when the Lucky Boy mine was closed. Callaghan and Buddington (1938, p. 115-116) tabulated a total production of 7,727.89 ounces of gold, 17,162 ounces of silver, and 257 pounds of copper for the district between 1896 and 1924, and state that if actual production records prior to 1902 were known they would possibly increase the total by $50,000 to $100,000. Some exploration work was done at the Lucky Boy mine in 1964 by a California group.

Callaghan and Buddington (1938, plate 20, p. 112) list 21 mines and prospects in the district. Owing principally to lack of pertinent information, only eight are included and described in the alphabetical list of mines.

Principal Mine

Lucky Boy mine: A year after its discovery in 1897, a 15-stamp mill was built at the Lucky Boy mine and in 1903 a 40-stamp mill was erected and a power plant installed on the McKenzie River, with a transmission line to the mine. Production figures from 1902 to the 1912 mine-closure date reported in Department Bulletin 14-D (1951, p. 48) are $174,000 itemized as follows: gold, 7,737 ounces; silver, 12,844 ounces; copper, 4,257 pounds; and lead, 1,051 pounds. The production between 1897 and 1902 is not accurately known but is estimated to have been between $50,000 and $100,000, which would make a total production for the mine in excess of $200,000.

Lode Mines of the Blue River District

Cinderella mine

Blue River District, 1

Location: Linn County, E½ sec. 28, T. 15 S., R. 4 E., about 4100 feet elevation.

Development: About 400 feet in two tunnels and a 50-foot shaft.

Geology: The vein is described as a northwest-striking, nearly vertical, highly altered and leached band as much as 2 feet thick of quartz and brownish-black powder, largely manganese oxide, occurring in altered, iron-stained andesitic breccia.

Production: In 1902 workings included a 50-foot shaft and a 240-foot tunnel. In 1916 a 3-stamp mill was on the property. Production was small and values were not reported. A truckload of high-grade ore was shipped about 1961 (H. E. L. Barton, oral communication, 1967).

References: Stafford, 1904; Parks and Swartley, 1916:55; Callaghan and Buddington, 1938: 117; Department Bulletin 14-D, 1951:45; H. E. L. Barton, oral communication, 1967.

Great Northern mine

Blue River District, 2

Location: Linn County, W½ sec. 28, T. 15 S., R. 4 E., about 4000 feet elevation.

Development: Several hundred lineal feet of tunnels and raises reported.

Geology: Parallel veins about 30 feet apart and brecciated, altered andesite with calcite, quartz, pyrite, and iron and manganese oxides. A low-grade oreshoot 75 feet long and 10 feet wide is reported.

Production: The mine operated in the early days, reopened in 1917, and was worked again in the 1920's. Equipment included a four-stamp mill and aerial tramway. Production was small but was second in the district to that of the Lucky Boy. H. E. L. Barton (oral communication, 1967) reports that the mine had a total production of about $25,000.

References: Parks and Swartley, 1916:112; Callaghan and Buddington, 1938:118; Department Bulletin 14-D, 1951:47; H.E.L. Barton, oral communication, 1967.

Lucky Boy mine

Blue River District, 3

Location: Lane County, 14 patented claims in secs. 32 and 33, T. 15 S., R. 4 E., and secs. 4 and 5, T. 16 S., R. 4 E., about 3000 feet elevation.

Development: Workings on seven levels total about 5000 feet.

Geology: The country rocks are tuffs with some andesite dikes. Five veins are described. The main ore shoot was at the intersection of the Daisy Creek vein, which strikes N. 45° W. to due west, and the Lucky Boy vein, which strikes N. 33° W. and dips 80° NE. The veins contain gouge, silicified breccia, and sulfides in quartz consisting of chalcopyrite, galena, pyrite, and sphalerite. The vein on No. 6

level is reported to be as much as 25 feet thick. Selected samples of ore assayed:

Gold (ounces)	Silver (ounces)	Lead (percent)	Zinc (percent)
0.03	19.77	71.52	0.8
0.07	18.50	65.10	0.9

Production: Most of the ore was milled in an electrically operated 40-stamp mill erected in 1903. Production from 1896 to 1912 was more than $200,000.

References: Stafford, 1904; Parks and Swartley, 1916:69-70; Callaghan and Buddington, 1938:119-121; Department Bulletin 14-D, 1951:48-49.

Poorman group

Blue River District, 4

Location: Linn County, five patented claims in secs. 31 and 32, T. 15 S., R. 4 E., at about 4200 feet elevation.

Development: One 600-foot drift, caved.

Geology: Workings follow a vein striking N. 70° W. and dipping 70° to 80° S. that consists of silicified tuff and massive quartz. Assays from the tunnel are reported to run $3 to $24 per ton.

Production: A two-stamp Tremain mill was on the property in 1916 and the operators were Calapooya & Blue River Mill & Mining Co. Production not reported, probably small.

References: Parks and Swartley, 1916:48; Callaghan and Buddington, 1938:121; Department Bulletin 14-D, 1951:49-50.

Rowena mine

Blue River District, 5

Location: Lane County (claims extend into Linn County), secs. 28 and 33, T. 15 S., R. 4 E. at about 4000 feet elevation.

Development: About 750 feet total workings in several adits.

Geology: A nearly vertical vein trending N. 10° to 35° W. consists of brecciated, altered tuff with quartz stringers and partly leached sulfides, mainly chalcopyrite in a zone 1 to 6 feet wide. A 6½-foot sample reported by Callaghan and Buddington assayed 0.03 oz./ton gold, 9.6 oz./ton silver, 7.6 percent copper, and 4.1 percent zinc. Another sample assayed 0.8 oz./ton gold, 7 oz./ton silver, 5.2 percent copper, and 3 percent zinc.

Production: May be the same as Ravena Group reported by Stafford (1904). No production reported.

References: Stafford, 1904; Callaghan and Buddington, 1938:122; Department Bulletin 14-D, 1951:50.

Tate mine

Blue River District, 6

Location: Lane County, SE$\frac{1}{4}$ sec. 28, T. 15 S., R. 4 E., probably between 4000 and 4400 feet elevation.

Development: There are open cuts and three tunnels containing a total of about 435 feet and small stopes. The main adit is 276 feet long and contains a short drift.

Geology: The veins strike about north and the dip is nearly vertical. They are from 1 to 5 feet wide and consist of altered tuff and andesite breccia with lenses of quartz, adularia, limonite, and manganese oxide fracture coatings. Residual sulfides are mainly sphalerite, with some pyrite, chalcopyrite, and galena.

Production: Not reported.

References: Callaghan and Buddington, 1938:123; Department Bulletin 14-D, 1951:51.

Treasure mine

Blue River District, 7

Location: Lane County, three patented claims in sec. 32, T. 15 S., R. 4 E., about 3300 feet elevation.

Development: Lower tunnel 1800 feet, upper tunnel 500 feet, 30-foot shaft. Total development on five levels including raises and crosscuts is about 4000 feet.

Geology: Country rocks are tuff breccia and andesite. Northwest-striking, steep, southwest-dipping veins are iron-stained, altered rock with drusy quartz and some disseminated pyrite; contain low values in gold and silver over widths from 2 to 15 feet.

Production: A 15-stamp mill was in ruins in 1930. There was small production prior to 1902, but no accurate record.

References: Stafford, 1904; McDaniel and Marshall, 1911; Parks and Swartley, 1916:224; Callaghan and Buddington, 1938:123; Department Bulletin 14-D, 1951:51-52.

Union mine

Blue River District, 8

Location: Lane County, sec. 6, T. 16 S., R. 4 E., at about 3800 feet elevation.

Development: Drifts and crosscuts total about 1200 feet on one level plus stopes.

Geology: The main vein, in andesite, is a few inches to 5 feet wide, strikes N. 43° W., and consists of brecciated altered rock cemented by quartz, with minor disseminated sulfides mainly oxidized and leached.

Production: There was a small production prior to 1915. Ore was ground in a small stamp mill.

References: Callaghan and Buddington, 1938:125-126 including map of workings; Department Bulletin 14-D, 1951:52-53.

FALL CREEK DISTRICT

Location

The Fall Creek district is situated in eastern Lane County between Fall and Christy Creeks north of Oakridge (figure 59). Most of the prospects in the area are in T. 19 S., Rs. 3 and 4 E., near Christy Creek. They are reached from Oakridge, on Oregon Highway 58, by following the road up the North Fork of the Middle Fork of the Willamette River from Westfir to Christy Creek a distance of about 18 miles.

Alpine Ridge forms the divide between Fall Creek and Christy Creek, and Sinker Mountain, a high point on the ridge, is 4764 feet above sea level. The elevation on Fall Creek at the mouth of Gold Creek is about 1500 feet; and on North Fork of the Middle Fork of the Willamette River at the mouth of Christy Creek it is about 1640 feet. The area is covered by the Sardine Butte 15-minute topographic map.

Geology and Mineralization

The area is underlain by a thick section of Tertiary volcanic rocks, mainly tuffs with andesitic to basaltic lavas which are assigned to the Little Butte Volcanic Series by Peck and others (1964). These rocks have been intruded by a few augite-diorite and dacite-porphyry plugs and dikes and show some contact alteration. The youngest rocks in the area are Quaternary intra-canyon basalts which underlie Christy Flats and High Prairie on the south side of Christy Creek and on the southeast side of North Fork of the Middle Fork of the Willamette River.

Callaghan and Buddington (1938, p. 127 and 128) describe and map two large areas and several smaller areas of rock alteration. One of the large areas, which includes most of the prospects, is on the east side of Alpine Ridge between Perdue and Billy Creeks. The other is on the west side of Alpine Ridge along Portland and Logan Creeks. On the surface the altered rocks are bleached or iron-stained and contain disseminated pyrite in unweathered portions. Callaghan and Buddington (p. 128) summarize the mineral deposits as follows:

> "The mineral deposits in this district are of low grade and consist (1) of zones without definite veinlike appearance in weathered altered rock which, according to prospectors, yields a little gold on panning; (2) of silicified zones in altered rock that apparently do not yield any appreciable gold; and (3) of veins in altered rock with stringers of quartz in comb or cockade structure. Only leached vein matter was found on the dumps of caved workings, and pyrite was the only sulfide seen. No appreciable production is expected in this area."

History and Production

Gold was discovered in the Fall Creek district in 1901, and the area was actively prospected in 1903 (Stafford, 1904, p. 61). Prospects in the district have only a small amount of development work. The Ironside mine was worked for several years and the Blanket property was prospected as late as 1931. Production records are not reported.

None of the prospects described in the Fall Creek district appear to be important enough to be given special attention here. Brief descriptions are included in the alphabetical list of mines and prospects which follows.

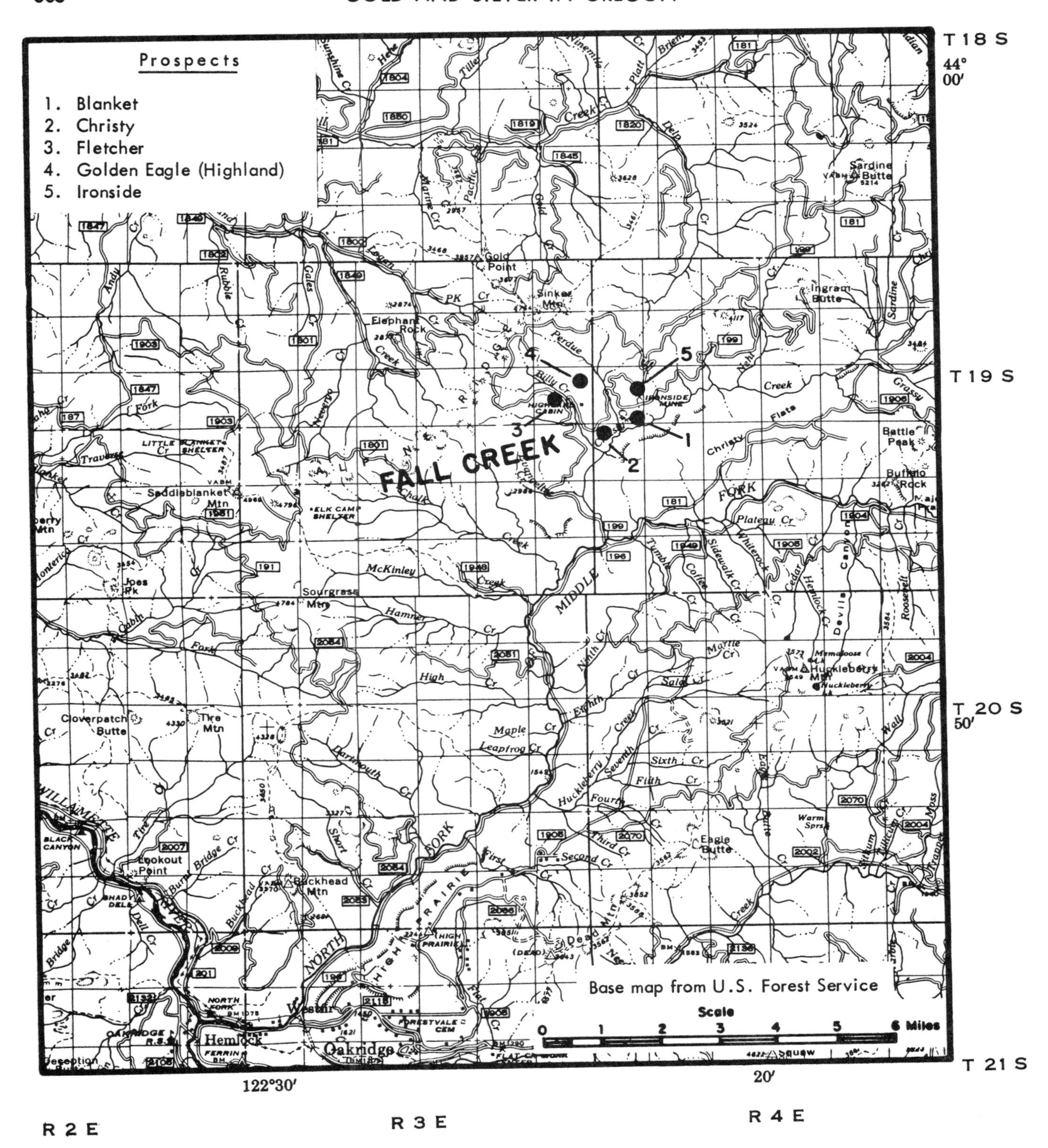

Figure 59. Index map of the Fall Creek district.

Lode Mines and Prospects of the Fall Creek District

Blanket prospect

Fall Creek District, 1

Location: Lane County, SE$\frac{1}{4}$ sec. 18, T. 19 S., R. 4 E., about 2900 feet elevation.

Development: Numerous trenches and pits.

Geology: No definite vein. Low gold content was found in altered, iron-stained tuff breccia associated with N. 10° W. fractures.

Production: Claim was being prospected in 1931. There is no record of production.

References: Callaghan and Buddington, 1938:128; Department Bulletin 14-D, 1951:85.

Christy prospect

Fall Creek District, 2

Location: Lane County, NW$\frac{1}{4}$ sec. 19, T. 19 S., R. 4 E., about 2400 feet elevation.

Development: Workings inaccessible when examined in 1931.

Geology: Altered silicified tuff breccia with some vuggy quartz and fine-grained disseminated pyrite was found on the dump of the main tunnel.

Production: Not reported.

References: Callaghan and Buddington, 1938:128-129; Department Bulletin 14-D, 1951:85.

Fletcher prospect

Fall Creek District, 3

Location: Lane County, NW$\frac{1}{4}$ sec. 13, T. 19 S., R. 3 E., about 2900 feet elevation on Billie Creek.

Development: Workings caved, amount not reported.

Geology: Veinlets and disseminated pyrite occur in light gray, silicified altered tuff.

Production: Not reported.

References: Callaghan and Buddington, 1938:129; Department Bulletin 14-D, 1951:85.

Golden Eagle (Jumbo or Highland) prospect

Fall Creek District, 4

Location: Lane County, E$\frac{1}{2}$ sec. 13, T. 19 S., R. 3 E., at about 3200 feet elevation.

Development: Three tunnels - the middle and the lower contain a total of about 650 feet of workings. Extent of the upper is not reported.

Geology: The country rocks described are various coarse- to fine-grained tuffs, agglomerates, porphyritic andesite, and a dike of augite diorite. A 200-foot-wide, mineralized altered zone trends north. Some concentration of pyrite occurs along northwest-trending fractures in the altered zone. A few high-grade assays are reported from panned concentrates.

Production: Not reported.

References: Callaghan and Buddington, 1938:129; Department Bulletin 14-D, 1951:86.

Ironside mine

Fall Creek District, 5

Location: Lane County, E$\frac{1}{2}$ sec. 18, T. 19 S., R. 4 E., at about 3120 feet elevation.

Development: There are three tunnels which have 210 feet of total workings.

Geology: Irregular streaks and masses have been mined from an area of altered and deeply weathered tuff breccia. No definite vein.

Production: The mine has been worked in a small way for several years. Ore was ground in a small five-stamp mill. Amount of production is not reported.

References: Callaghan and Buddington, 1938:129; Department Bulletin 14-D, 1951:87.

BOHEMIA DISTRICT

Location

The Bohemia district is in Townships 22 and 23 S., Ranges 1 and 2 E., in Lane County (plate 3). It is the largest and most productive mining area in the Western Cascades. The district lies about 35 miles southeast of Cottage Grove and is reached by way of the Row River road and forest roads which follow Sharps Creek and Brice Creek, making a loop through the area.

The mineralized area is roughly circular and about 5 miles in diameter. Elevations range from just under 2000 feet on Champion Creek near the north edge of the map area to 5933 feet on Fairview Peak, the highest point in the district. The area is drained by tributaries of Brice, Sharps, and Steamboat Creeks.

Geology and Mineralization

Oligocene and lower Miocene volcanic rocks having a maximum thickness of 6500 feet underlie the Bohemia area. These rocks are assigned to the Little Butte Volcanic Series by Peck and others (1964). Callaghan and Buddington (1938, p. 40) describe the geology as follows:

> "...These rocks comprise tuffs, volcanic breccias, and andesite lavas in about equal amounts, with minor lenses of coarse volcanic breccia and agglomerate and flows of rhyolite intercalated in the tuffs....These rocks commonly dip at low angles to the northeast and east, although locally the dips vary, and some east-southeast dips were observed. Several dikes of andesite varying widely in strike traverse the bedded volcanic rocks and are presumed to be closely related to them. A considerable number of small plugs, dikes, and a stock of dioritic intrusive rocks occur in a belt extending northward through the central part of the area. They are included in an area of hornfels 3½ miles long and half to three-quarters of a mile wide."

Lutton (1962) divides the volcanic rocks of the area into lower, middle, and upper units. His lower unit is composed of about 1000 feet of massive pyroclastic rocks overlain by about 300 feet of well-bedded tuffaceous shale and sandstone. His intermediate unit is composed of andesite to rhyolite domes and flows with intercalated pyroclastics which crop out in a wide, northwest-trending belt and range in thickness from 1200 to 1700 feet. The upper unit is mainly andesite and basalt flows, comprising 2000 feet of lavas with interlayered lapilli tuff at the base, overlain by 1000 feet of basic lavas.

Lutton maps three age groups of intrusive rocks. The oldest are porphyritic dacite dikes; the intermediate are basic dikes and sills; and the youngest are granitoid intrusives. The porphyritic dacite dikes strike mainly west to northwest, generally parallel to the mineralized quartz veins. Lutton believes them to be contemporaneous with his intermediate unit of andesite and rhyolite domes and probably of late Oligocene age. He suggests that the granitoid intrusives are of late Oligocene or Miocene age, and that the basic dikes and sills are of various ages, in part contemporaneous with the basic flows of his upper unit.

The veins in the area strike mainly northwest and west and the majority dip steeply southwest and south. Callaghan and Buddington (1938, p. 44) state that most of those with a northwesterly strike are in the southern half of the area, whereas those with a westerly strike are in the northern part. The general features of the mineral deposits in the district are described by Callaghan and Buddington as follows:

> "Though the Bohemia district covers a roughly circular area of about 60 square miles, the main mineralized belt occupies an area 5½ miles long and 1½ miles wide trending N. 60° W. It includes the Mayflower, Riverside, and Oregon-Colorado mines at the southeast and the Utopian, Sweepstakes and Musick at the northwest. Gold has been the principal ore mined and has been obtained largely from the oxidized parts of sulfide veins. The veins are

younger than any of the other rocks in the district, including the intrusive bodies. The vein matter in most places consists of brecciated, altered, and partly replaced country rock cemented by or containing fissure fillings of drusy or comb quartz that locally contains sulfides. In some veins there are bodies of cherty quartz with pyrite crusts along intersecting short fractures. The dominant sulfide, sphalerite, is associated with galena, chalcopyrite, and pyrite in varying amounts, and in some places with a little tetrahedrite. Galena is the dominant sulfide in the Musick vein, chalcopyrite in the Oregon-Colorado, and stibnite in the Tall Timber. Primary specularite is associated with quartz in several of the veins. The gold content of the unweathered sulfide ores is low in most places, though a few high-grade pockets have been found. There is a rough areal zonal distribution of mineral deposits in relation to intrusive rocks. Base-metal quartz shoots with variable amounts of gold and in places with specularite and dolomite are grouped in the area of most intensive igneous intrusion; and veins with generally less sulfide, more carbonate (commonly calcite), and in places stibnite occur in an area to the south, where there are fewer intrusive bodies."

Taber (1949, p. 12) described grade of the ore as follows:

"The metal content of the sulfide ores differs considerably from place to place. The gold content ranges from a trace to about 1 ounce. High-grade pockets have been found that contained as much as 10 ounces of gold per ton. The Helena mine has produced several such pockets. The silver averages about 2 ounces per ton, the lead 3 percent, the zinc 5 percent, and the copper about 1 percent. In many of the veins, zinc is the predominant sulfide metal. In a few, notably the Musick vein, lead is more abundant than zinc."

Lutton (1962) discusses the factors favorable to ore deposition. Among them he lists: 1) proximity to the center of mineralization, which he believes is probably a cupola; 2) open channels formed in the steep portions of irregular, dipping veins as the result of normal faulting; 3) concentrations of subsidiary veinlets so numerous that they form an ore body; and 4) vein intersections which give rise to ore bodies.

The distribution of veins and areas of rock alteration (shown on plate 3) and the areal mineral temperature zoning described by Callaghan and Buddington (1938, p. 48-49) may be evidence for a hidden intrusive body, or cupola, at depth beneath the central portion of the mineralized zone, as suggested by Lutton (1962, p. 137).

History and Production

Diller (1900, p. 7) reports that the Bohemia mining region was discovered by Dr. W. W. Oglesby of Junction City and Frank Brass in 1858. The name "Bohemia" originated from the nickname of James Johnson, who was popularly known as Bohemia Johnson. Johnson and George Ramsey discovered gold near the head of City Creek in 1863 and this brought in many prospectors.

The first five-stamp mill was built in 1872 on the Knott claim, now part of the Champion property. The Musick mine, the first vein found in the area, was located in 1891. The Noonday was opened in 1892. A 10-stamp mill was installed at the Champion mine in 1895 and a 20-stamp mill at the Noonday in 1896. By 1902 not less than 2000 claims had been filed in the district, although some were undoubtedly duplicate recordings. Between 1902 and 1912 West Coast Mines Co. consolidated the Champion, Helena, and Musick mines and erected a 30-stamp mill at the Champion. The Noonday was a producer between 1896 and 1908. The Vesuvius and Evening Star mines were also producers during the early 1900's.

Between 1932 and 1938 the Mahala Mines Co., the Bartels Mining Co., and the Minerals Exploration Co. produced ore with a total value of more than $400,000.

Between 1939 and 1942 Higgins and Hinsdale (H. & H. Mines, Inc.) completed some development work and erected a flotation mill and power plant at the Champion mine. These operations were suspended in 1942 before major production was attained; however, considerable development ore was milled (Harold Barton, oral communication, 1967). In 1944, F. J. Bartels acquired the mill and property from H. & H.

Mines. Bartels produced a small amount of gold from the Champion and Evening Star mines between 1945 and 1949 and milled ore from the Helena mine, which was operated by K. O. Watkins. In 1950 Watkins obtained the Champion lease and operated the Champion mill, producing about $35,000*.

In 1961 and 1962 the Office of Minerals Exploration contracted to help finance a long drift to expose the Musick vein 335 feet below the old No. 6 level. Work was done by the Emerald Empire Mining Co.

In 1964 a diamond-drilling program was announced by Federal Resources Corp. of Salt Lake City to explore the Champion, Evening Star, Musick, and other nearby properties. The program was modified to include drilling on extensions of the Helena vein and a production drift on the California-Defiance veins of the Musick; this is known as the 1000 level and work was completed at 1196 feet on May 5, 1965*.

The main production from the Bohemia district has come from the Champion, Helena, and Musick mines. A fair amount of output has also come from the Noonday, Vesuvius, and Star mines. Total production for the district is estimated to be about one million dollars. Between 1880 and 1900, according to U.S. Mint reports, Lane County (mainly the Bohemia district) produced 14,590.69 ounces of gold and 1,418.79 ounces of silver. Between 1901 and 1930, according to U.S. Bureau of Mines records, the district produced from 42,548 tons of crude ore a total of 13,694.59 ounces of gold, 8,148 ounces of silver, 14,831 pounds of copper, and 120,816 pounds of lead. Zinc was present but was considered a liability in the concentrates.

Principal Mines

The principal mines in the district are the Champion, Helena, and Musick. Since the district contains such a large number of similar veins, no attempt is made here to describe all of the prospects. Those with the more complete information and extensive development are included in the alphabetical list.

Champion mine: The Champion mine property consisted of one patented and 22 unpatented claims as reported by Taber (1949). It is located near Champion Saddle on the divide of Champion and City Creeks in the N½ sec. 13, T. 23 S., R. 1 E. The principal development is on the west- to northwest-striking Champion vein over a total distance of about 2600 feet along the strike and to an average depth of about 800 feet below the outcrops. The vein was originally developed in two separate properties. The Evening Star workings entered from the City Creek side and developed the southeastern portion of the vein. The Champion workings entered from the north side of the ridge. The two workings were eventually connected on the 600-foot level. Total development includes more than 15,000 feet of drifts and crosscuts and about 3000 feet of raises, on nine levels. Three of the main levels (600, 900, and 1200-foot) have adits.

The Champion vein varies in thickness from 1 to 8 feet and has averaged about 3 feet. Taber (1949, p. 24) describes the ore remaining in the mine as mainly sulfides similar to that of other mines in the district. The gold content is generally just under 0.5 ounce per ton. A few small, partly oxidized ore shoots remaining in the mine contain as much as 2 ounces of gold per ton, and a few shoots of oxidized ore of higher grade have been mined. Watkins (1946) states that an average of 1548 samples taken by H. & H. Mines Co. from all parts of the mine (weighted against width of sample) was as follows:

Oz. gold	Oz. silver	Percent lead	Percent zinc	Percent copper
0.555	4.21	1.72	2.15	1.71

Taber (1949, p. 24) describes the wall rocks as andesite, rhyolite tuff, and granodiorite porphyry and the vein minerals as sphalerite, pyrite, chalcopyrite, galena, hematite, and cherty to coarsely crystalline quartz. Silicification and sericitization of the wall rocks are common.

History of mining operations at the Champion is summarized by the Department (Bulletin 14-D, 1951, p. 58) as follows:

---

* Information supplied by Harold Barton, oral communication, 1967.

"The Champion vein was discovered in 1892, and in 1895 a 10-stamp mill was built on that property. In 1902 the Champion, Helena, and Musick mines were consolidated under the West Coast Mines Co. A 30-stamp mill was built at the Champion mine and it ran until 1908, partly on ore from the other properties. Only a small amount of development work was done between 1912 and 1916, and no mining was carried on between that time and 1930. During the period 1932 to 1938 approximately $100,000 was produced from the Champion by several operators, including the Mahala Mines and the Bartels Mining companies. In 1939 the property was taken over by Higgins and Hinsdale (H. & H. Mines) who built a mill and power plant and did several thousand feet of development work including most of the 1200-foot level. Operations were suspended in August 1942, however, due to high cost of materials and shortage of labor, before production had been attained. The property and mill were turned over to F. J. Bartels in 1944. From 1945 through 1949 a few cars of concentrates were shipped as well as some cars of run-of-mine gold ore." A recent photograph of the mill is shown in figure 60.

Helena mine: In 1949 the Helena mine property consisted of three patented and seven unpatented claims. Workings are on a northwest-striking vein in secs. 7 and 18, T. 23 S., R. 2 E. The vein extends across the SW$\frac{1}{4}$ sec. 7 and a short way into the SE$\frac{1}{4}$ sec. 12, T. 23 S., R. 1 E. The Helena No. 2 and other workings on Helena claims in the SE$\frac{1}{4}$ sec. 12 are on what appears to be a branching vein. The principal development is on the southeastern end of the vein, in the north edge of sec. 18, where it penetrates a large east-trending granodiorite porphyry dike.

The vein was discovered in 1896. Two ore-shoots were mined during the mining boom that lasted until 1907. Production to 1931 is estimated by Callaghan and Buddington (1938, p. 55) at $150,000. The mine has a history of several owners and operators. A 35-ton flotation mill was built in 1935. Additional production to 1949 was approximately $100,000, bringing the total to about $250,000. Small production in 1950 and 1962 would not increase this significantly. The 1964 core-drilling program disclosed characteristic sulfide mineralization both east and west of the productive zone (Harold Barton, oral communication, 1967).

Taber (1949, p. 28) describes the workings as follows:

"The mine has been developed by three principal levels and three short adit drifts. Altogether, there are about 2000 feet of drifts and crosscuts and about 500 feet of raises. All of the principal levels are connected by raises, so that the lowest (No. 7 level) may be used as a general haulage level for the mine. Except for about 100 feet of the No. 6 level, which has been stoped out, all of the drifts are open and in good condition."

The Helena vein is a strong fissure which strikes about N. 57° W. and dips 65° to 70° NE. Taber (1949, p. 27) states it is the only strong northwest-striking vein that has a northeasterly dip. It has been explored over a distance of about 3000 feet. Three ore shoots mined were discovered at the surface and mined down to about 200 feet of depth. Their length was 50 to 200 feet and thickness 3 to 8 feet, averaging 4 feet. Callaghan and Buddington (1938, p. 55) describe the sulfide ore as consisting "... chiefly of sphalerite with some pyrite, chalcopyrite, galena and a little tetrahedrite with quartz, included rock fragments, kaolin, and barite. Several stages of vein filling are shown." Grade of the ore is reported to have been quite high. Four grab samples taken by Taber from the new ore shoot at the face of No. 7 level averaged 3.15 oz./ton gold, 2.8 oz./ton silver, 4.8 percent lead, 5.8 percent zinc, and 1.2 percent copper. The paradox of relatively small amount of development with a good quality of ore is explained by Taber (1949, p. 25) as probably due to the unusually shallow depth of oxidation in the ore shoots, which limited the quantity of easy-milling ore.

Musick mine: The Musick mine is situated about 1 mile west of the Champion mine in N$\frac{1}{2}$ sec. 14, T. 23 S., R. 1 E. The Lower Musick is on Sharps Creek in NE$\frac{1}{4}$ sec. 15. The Musick Group had 14 unpatented claims situated in sec. 14 in 1949. History of the Musick is described as follows (Department Bulletin 14-D, 1951, p. 69):

"The Music vein, discovered by James C. Musick in 1891, is one of the earliest found

Figure 60. The Champion mill, a 100-ton selective flotation plant, processed ore from the Champion and Musick mines. Photograph taken in 1964 by Fred E. Miller.

Figure 61. A scene near portal of the new exploration tunnel at the Musick mine showing air-driven locomotive and ore cars. Photograph taken in 1964 by Fred E. Miller.

in the district. He organized the Bohemia Gold Mining and Milling Co., built a 5-stamp mill and operated the property until 1901. In 1902 the Musick, Helena, and Champion mines were acquired by the Oregon Securities Corporation. Ore was hauled from the Musick mine to the Champion mine by electric tram and by cable bucket tram from the portal of the Champion mine to the mill in Champion Creek gulch.

"West Coast Mines Company bought the Oregon Securities Corporation holdings in 1908. This company sold the Musick mine to L. M. Capps of Idaho in 1921. The Minerals Exploration Company leased the mine in 1935, built a 22-ton gravity concentrator mill, and produced $101,000 worth of concentrates in 1936 and 1937. About 1939, Higgins and Hinsdale Mines Company obtained a lease on the property but operations were stopped because of World War II. Kenneth O. Watkins bought the H. and H. Mines Company lease in 1944, and sold his contract in 1945 to the Tar Baby Mining Company of Salt Lake City, Utah. This company acquired the property from the L. M. Capps estate in 1946. In 1948 the mine was leased to the Helena Mines, Inc., who subleased the east end of the mine to Wyatt, Nordstrom, and Smith. This group did development work, using the main level of the Musick mine to reach their adjoining claims. In 1949 some Musick dump ore was hauled to the Champion mill for treatment."

In 1949 and 1950 ore from the Wyatt stope was milled at the Champion.

Beginning in 1961 the Emerald Empire Mining Co., with financial assistance from the O. M. E., drove a 1662-foot drift tunnel 335 feet below the No. 6 level (figure 61). The new lower adit enters from the Sharps Creek side at about 4660 elevation. Work was completed in November 1962. Results of the new exploratory work were disappointing, in that very little good ore was found in widths of greater than 30 inches in the new drift. The Musick mine has more than 7660 feet of drifts and crosscuts plus numerous stopes, raises, and winzes that explore the vein for approximately 5000 feet along the strike and to a depth of about 800 feet. The California vein, which lies north of the Musick and merges with it, has been explored by surface pits and outcrops for about 1500 feet. The vein apparently continues east and is called the War Eagle, thus covering a distance of greater than a mile (Harold Barton, oral communication, 1967). Some additional development work has been done on the nearby Mystery, Alpharetta, and Ophir veins which lie southeast of the main Musick workings. Taber (1949, p. 29) calculated that total production from the Musick may have been about $280,000. His measurements of stopes indicated at least 40,000 tons were mined and that this total would require an average of $7 per ton recovery.

The Musick vein strikes from N. 45° W. to W. and dips from 65° S. to vertical. It contains splits in the fissure and merges with the west-striking, steeply south-dipping California vein.

Geology of the Musick is similar to that at the Champion mine. The country rock is mainly rhyolite with a few exposures of andesite and tuff.

The Musick has the usual mixture of sulfides but with slightly lower gold content and higher lead than other veins in the district. The gangue is siliceous and fragments of wall rock occur mixed in the vein. Shoots in the Musick vein as much as 375 feet long and 3 to 5 feet thick have been stoped. Assays (of weathered ore) reported by Callaghan and Buddington (1938, p. 58) are as high as 1.4 ounces of gold and 2 ounces of silver per ton. A sample of partly oxidized broken ore from the stopes reported by Taber (1949, p. 30) assayed 0.38 oz. gold, 1.7 oz. silver, 1.2 percent copper, 5.6 percent lead, and 5.7 percent zinc. West of the location where the California and Musick veins merge, the resulting vein is wider (8 to 12 feet) than the main Musick vein, but it contains less ore mineralization.

Lode Mines of the Bohemia District

Champion (Evening Star) mine

Bohemia District, 1

Location: Lane County, N½ sec. 13, T. 23 S., R. 1 E. between 4400 and 5500 feet elevation.

Development: More than 15,000 feet of drifts and crosscuts and about 3000 feet of raises on 9 levels.

Geology: Vein is a fissure with brecciated country rocks cemented by vuggy quartz containing sphalerite, pyrite, chalcopyrite, galena, and hematite that is oxidized near the surface. Country rocks are rhyolite, tuff, hornfels, and granodiorite. Average vein width is about 3 feet. Strike of the Champion vein is west to about N. 45° W. and dips average about 70° SW.

Production: Mine was discovered in 1892. A 10-stamp mill was built in 1895; a 30-stamp was built in 1902 and operated until 1908. Approximately $100,000 was produced during the 1932 to 1938 period. Operations continued under new ownership until 1942, during which time a new selective flotation mill was installed. A few cars of concentrates and mine-run ore were shipped from 1945 to 1952. Partial production statistics reported by Taber, 1949, p. 23, are: 24,297 tons crude ore; 5471 ounces gold; 16,434 ounces silver; 189,583 pounds copper; 63,196 pounds lead; and 5550 pounds zinc.

References: Diller, 1900:26-27; MacDonald, 1909:83-84; Parks and Swartley, 1916:234; Smith and Ruff, 1938:42-44; Taber, 1949:22-25; Department Bull. 14-D, 1951: 57-59.

Crystal (Lizzie Bullock) mine

Bohemia District, 2

Location: Lane County, N½ sec. 11, T. 23 S., R. 1 E., about 5000 feet elevation.

Development: Lowest drift, 4580-foot level, is 400 feet long; upper drift, 4690-foot level, is 100 feet long. Other short drifts are present but workings are mostly inaccessible.

Geology: Country rocks are andesitic lavas, tuffs, and breccias. Crystal vein is quartz-cemented fault breccia with pyrite, sphalerite, chalcopyrite, and galena. Vein which strikes N. 65° W. dips steeply SW. and has been explored for 3300 feet. Reported average of 108 samples in lower workings was .03 ounce per ton gold, 2.0 ounce per ton silver, 1.1 percent copper, 2.0 percent lead, and 2.9 percent zinc. Thickness of vein was reported by Diller to be 6 feet and 7 feet.

Production: A two-stamp, water-driven mill was operated in the 1890's, during which time there was a small production, but no accurate records are reported.

References: Diller, 1900:25; Callaghan and Buddington, 1938:66-67; Taber, 1949:40-41; Department Bulletin 14-D, 1951:60-61.

El Capitan (President Group)

Bohemia District, 3

Location: Lane County, center E½, sec. 23, T. 23 S., R. 1 E., about 4000 feet elevation.

Development: Two adits total 380 feet; upper tunnel, caved in 1941; and various cuts.

Geology: Country rocks are andesite flows and tuff breccias dipping eastward that are intruded by a diorite plug to the south. Sinuous vein strikes N. 60° to 70° W., dips 80° S. to vertical, and is 15 to 30 inches thick. Vein contains vuggy quartz breccia, calcite crystals in cavities. Sulfides are pyrite, stibnite, chalcopyrite, galena, and sphalerite. A portion of the vein contains at least 25 percent stibnite. The vein has been explored for more than 1000 feet.

Production: Located about 1898. Production has been small. A two-stamp mill was reported on the property in 1930. Lane Minerals shipped a half carload of ore in 1959.

References: Parks and Swartley, 1916:55; Callaghan and Buddington, 1938:67; Taber, 1949: 50; Department Bulletin 14-D, 1951:61; H. E. L. Barton, oral communication, 1967.

Grizzly group

Bohemia District, 4

Location: Lane County, NE$\frac{1}{4}$ sec. 12, T. 23 S., R. 1 E., between 4000 and 4700 feet elevation.

Development: Three drifts, totaling about 800 feet, and several trenches explore vein for 2000 feet on strike and 600 feet vertical range.

Geology: Andesite and tuff intruded by small diorite dikes near vein which is 1 to 6 feet thick, strikes N. 60° W., dips 55° to 65° S., and contains silicified andesite, quartz, varying amounts of sphalerite, pyrite, chalcopyrite, and galena.

Production: Information lacking.

References: Stafford, 1904; Parks and Swartley, 1916:115; Callaghan and Buddington, 1938: 68; Taber, 1949:43-44; Department Bulletin 14-D, 1951:65.

Helena mine

Bohemia District, 5

Location: Lane County, N$\frac{1}{2}$ sec. 18, SW$\frac{1}{4}$ sec. 7, T. 23 S., R. 2 E. and SE$\frac{1}{4}$ sec. 12, T. 23 S., R. 1 E. from about 4500 to 5400 feet elevation on Grizzly Mountain.

Development: Three principal levels contain about 2000 feet of drifts and crosscuts and about 500 feet of raises plus stopes.

Geology: The strong fissure vein which strikes N. 57° W. and dips 65° to 70° NE. has been explored for about 3000 feet. In mined area to the south the vein averages 4 feet thick. Country rocks are andesite and a large, east-trending granodiorite porphyry dike. Vein contains sphalerite, pyrite, chalcopyrite, galena, and a little tetrahedrite in a gangue of quartz rock fragments, kaolin, and barite. Shows good gold values in oxidized zone.

Production: Vein was discovered in 1896. There is a history of several owners and operators. Total production to 1950 was approximately $250,000.

References: Diller, 1900:29-30; Parks and Swartley, 1916:234-235; Callaghan and Buddington, 1938:54-57; Taber, 1949:25-28; Department Bulletin 14-D, 1951:65-66.

Leroy group

Bohemia District, 6

Location: Lane County, W½ sec. 12, T. 23 S., R. 1 E., about 4000 feet elevation; six patented claims.

Development: Numerous cuts and tunnels have a total length of 1100 feet.

Geology: The vein strikes N. 60° to 70° W., dips 55° to 70° SW., is 5 to 6 feet wide, and occurs in a southwest-dipping granodiorite porphyry dike. It contains quartz-cemented porphyry breccia with small to moderate amounts of sphalerite, chalcopyrite, and galena and represents a large volume of low-grade material.

Production: Most of development work was done between 1900 and 1910. Production, if any, is not reported.

References: Stafford, 1904; Callaghan and Buddington, 1938:69-70; Taber, 1949:41; Department Bulletin 14-D, 1951:67-68.

Mayflower mine

Bohemia District, 7

Location: Lane County, northeast corner sec. 20, T. 23 S., R. 2 E., about 3000 to 3400 feet elevation crossing Horseheaven Creek; five patented claims.

Development: Two adits, one on each side of the creek, total length not reported. Some stopes 2 to 5 feet wide were mined to the surface.

Geology: Country rock is tuff. Vein strikes N. 70° W., dips 75° to 80° N., and consists of altered tuff with a network of drusy quartz veinlets, considerable pyrite, and little sphalerite, galena, and chalcopyrite.

Production: Work was probably done in the late 1800's and early 1900's. There is evidence of considerable ore mined but no report of production available. A small cyanide mill (first in the district) was built on the property in 1909. The mine was operated by the Kelso Gold Mining & Milling Co. during 1913-1917.

References: Stafford, 1904; Parks and Swartley, 1914:135; Callaghan and Buddington, 1938: 70-71; Taber, 1949:39-40; Department Bulletin 14-D, 1951:68.

Musick mine

Bohemia District, 8

Location: Lane County, N½ sec. 14 and NE¼ sec. 15, T. 23 S., R. 1 E., between about 4300 and 5400 feet elevation.

Development: There are about 7200 feet of drifts and crosscuts plus numerous stopes, raises, and winzes on the Musick vein. Numerous pits and short adits explore the branching California vein.

Geology: Country rocks are mainly rhyolite with some andesite and tuff. Musick vein curves, splits, and merges with the west-striking California vein. It strikes from N. 45° W. to west, dips 65° S. to vertical, and varies from 3 to 12 feet in thickness. Ore is similar to that at Champion mine, but with higher lead content. Both the California and Musick veins have been explored for more than a mile along this strike to a depth of 800 feet.

Production: Mine was discovered in 1891 and worked extensively during early part of the century by various organizations. The mine was worked during the 1930's by the Minerals Exploration Co., which built a 22-ton mill and produced $101,000 worth of concentrates. Estimated total production to 1949 has been about $280,000. An extensive exploration drift was driven during 1961-1962.

References: Diller, 1900:20-23; Stafford, 1904; Parks and Swartley, 1916:234-235; Taber, 1949:28-30; Department Bulletin 14-D, 1951:69-70.

Noonday mine

Bohemia District, 9

Location: Lane County, W½ sec. 18, T. 23 S., R. 2 E., between 4900 and 5500 feet elevation; 5 patented claims.

Development: The mine workings have three main levels with nearly 4000 feet of drifts and crosscuts and 600 feet of raises, plus extensive stopes.

Geology: The mine explores three nearly parallel veins about 100 yards apart. Development is on the central (Annie) vein, which strikes N. 45° to 70° W., and dips 75° to 80° S. It has been traced for about 1500 feet on the surface. Stoped areas were 3 to 8 feet thick and average about 5 feet thick with 1½ to 2 feet of high grade. Country rocks are mainly labradorite andesite with nearby dacite plugs. Ore (sulfide) and gangue minerals are typical of the district.

Production: Fairly complete U.S. Bureau of Mines production records from 1891 to 1945 indicate a total of 5183 ounces of gold; 2087 ounces of silver, 10,282 pounds of copper; and 19,649 pounds of lead. In 1887 the mine was second in the district to produce. Taber (1949) calculated the total production at about $100,000. Oxidized (free milling) ore was ground in a stamp mill.

References: Diller, 1900:28-29; Stafford, 1904; Callaghan and Buddington, 1938:60-62; Taber, 1949:31-33; Department Bulletin 14-D, 1951:70-71.

Oregon - Colorado mine

Bohemia District, 10

Location: Lane County, sec. 19, 29, and 30, T. 23 S., R. 2 E., between 3200 and 4000 feet elevation; 8 patented claims.

Development: The vein is explored by a lower (3300 feet elevation) adit drift 1800 feet long and an upper (3600 feet elevation) adit drift 450 feet long.

Geology: The Oregon-Colorado vein strikes about N. 50° W. and dips 60° to 65° SW. Vein filling is a quartz and chlorite-cemented breccia as much as 8 feet wide. The chief sulfides are chalcopyrite and pyrite. The vein crosses east-dipping, coarse- to fine-grained tuffs. A fine-grained andesite dike crops out at the portal of the lower tunnel. Selected ore samples reported assayed 1.0 ounce per ton gold, 3.4 ounce per ton silver, and 4.9 percent copper. Vein is explored for about 2000 feet.

Production: Most of the exploration was done by the Vesuvius Mining Co. prior to 1920. No production is reported.

References: Diller, 1900:28; Stafford, 1904; Parks and Swartley, 1916:228; Callaghan and Buddington, 1938:72-73; Taber, 1949:36-37; Department Bulletin 14-D, 1951:72.

Shotgun vein (Carlisle)

Bohemia District, 11

Location: Lane County, secs. 11 and 12, T. 23 S., R. 1 E. between 4500 and 5200 feet elevation.

Development: Four drifts and numerous cuts and pits explore the vein for 1000 feet.

Geology: The vein, which strikes northwest and dips steeply north, contains brecciated rock cemented by vuggy quartz with ore shoots containing sphalerite, galena, chalcopyrite, and less abundant pyrite. Shoots are 4 feet or more in width. Country rock is andesite cut by a narrow dacite-porphyry dike.

Production: None reported.

References: Callaghan and Buddington, 1938:75; Taber, 1949:46; Department Bulletin 14-D, 1951:75-76.

Star mine

Bohemia District, 12

Location: Lane County, NE$\frac{1}{4}$ sec. 20, T. 23 S., R. 1 E., about 3200 feet elevation.

Development: Twelve adits and several cuts explore three veins. Crosscuts and drifts total about 1300 feet (1960) plus some short raises and small stopes.

Geology: Country rocks are andesitic lavas and tuffs. Three nearly parallel fissure veins strike from N. 30° to 75° W. and dip from 45° to 80° S. They may be offset segments of one or two veins. Veins contain brecciated altered wall rocks, disseminated pyrite, white to iron-stained, frequently vuggy quartz, varying amounts of limonite in manganese oxides, and narrow clay gouge seams. Vein thickness is from a few inches to 5 feet. Oxidized zones contain free gold enrichments in confined, steeply raking ore shoots.

Production: The "Bughole" vein was discovered prior to 1910 by Pat Jennings, who reportedly produced about $30,000 and then sold property to Consolidated Mining Co. There were several other owners between 1920 and 1950, including H. E. Cully. Owners in recent years are Guy Leabo and Harry C. Miller. Total production has not been reported.

Reference: Department Bulletin 14-D, 1951:76-77.

Sultana (Miller) mine

Bohemia District, 13

Location: Lane County, S$\frac{1}{2}$ sec. 25, T. 22 S., R. 1 E., about 4000 feet elevation.

Development: Vein has been explored for more than 2500 feet along strike by trenches, a 20-foot shaft, and two long-adit drifts which together have about 2000 feet of tunnel.

Geology: Country rock is andesite. The 3- to 5-foot-wide vein strikes N. 65° to 80° W. and dips 80° S. Vein filling includes altered rock and brecciated quartz with some pyrite, chalcopyrite, sphalerite, galena, and small amount of tetrahedrite. Specimens of visible gold in sphalerite have been reported (H. E. L. Barton, oral communication, 1967).

Production: In 1916 the mine was being worked and ore being ground in a chilean mill. Since

1931 production has been approximately $10,000.

References: Parks and Swartley, 1916:114; Callaghan and Buddington, 1938:64-65; Taber, 1949:43; Department Bulletin 14-D, 1951:78; H. E. L. Barton, oral communication, 1967.

Sunset (Cape Horn vein) mine — Bohemia District, 14

Location: Lane County, north edge sec. 36, T. 22 S., R. 1 E. Vein extends west into sec. 35 and crops out between 2300 and 4500 feet elevation.

Development: Three main drifts and several small adits and prospect pits have explored the Cape Horn vein for 4000 feet horizontally in a 1000-foot vertical range. The three drifts total 1175 feet.

Geology: The vein strikes east and dips 75° to 80° S. Country rocks are andesite and altered tuff. Vein filling is banded drusy and cherty quartz, some kaolin, sphalerite, and small amounts of galena, pyrite, and chalcopyrite in ore shoots. Zinc is the principal metal reported in assays.

Production: No information.

References: Callaghan and Buddington, 1938:76-77; Taber, 1949:37-38; Department Bulletin 14-D, 1951:79.

Utopian group — Bohemia District, 15

Location: Lane County, on north boundary of sec. 3, T. 23 S., R. 1 E., extending into sec. 34, T. 22 S., R. 1 E., at about 4400 feet elevation; two patented claims.

Development: About 450 feet of drifts. One report mentions an 1800-foot drift to a depth of 900 feet.

Geology: The vein is reportedly strong and is associated with a dacite porphyry dike. The gangue is a coarse, granular, vuggy quartz with considerable disseminated sphalerite, a little galena, chalcopyrite, and pyrite.

Production: The property once belonged to the Vesuvius Mines Co. and some ore was packed on horseback to the Vesuvius mill. The railroad from Cottage Grove to Culp Creek was reportedly headed for the Utopian mine in its original planning stage.

References: Stafford, 1904; Callaghan and Buddington, 1938:78-79; Department Bulletin 14-D, 1951:80-81; H. E. L. Barton, oral communication, 1967.

Vesuvius mine — Bohemia District, 16

Location: Lane County, S½ sec. 11 and N. edge sec. 14, T. 23 S., R. 1 E. between 4800 and about 5700 feet elevation.

Development: Total length of mine workings is about 6000 feet, distributed in about 10 adits on 4 veins with major development on the Vesuvius vein.

Geology: Country rocks are andesite, some rhyolite, and interlayered tuff. The Vesuvius vein strikes N. 85° E. and dips 60° S. It forms an ore shoot at its intersection with the Jasper vein, which strikes N. 67° W. and dips steeply south. The shorter Stocks-Harlow vein is about parallel to the Vesuvius and lies about 600 feet north of it. The Storey vein, situated about a third of a mile east of the Vesuvius vein, strikes N. 56° W. and dips 70° to 75° S. Stope width on the Vesuvius-Jasper ore shoot is 3 to 5 feet. Character of the vein filling and mineralization is said to be typical of the area.

Production: Veins were discovered about 1895. Oxide ore was milled in a five-stamp mill. The bulk of mining and development work was done before 1923. Estimates of production range from $5,000 to $15,000.

References: Diller, 1900:24-25; Stafford, 1904; Parks and Swartley, 1916:227-228; Callaghan and Buddington, 1938:62-63; Taber, 1949:34-36; Department Bulletin 14-D, 1951:81.

War Eagle (Wall Street, Gilbertson) group — Bohemia District, 17

Location: Lane County, N½ secs. 13 and 14, T. 23 S., R. 1 E., between 4100 and 5200 feet elevation.

Development: Many cuts, pits, and tunnels. A 600-foot tunnel is on the No. 1 vein, and the No. 2 vein has been prospected for more than 2000 feet. Total footage of underground development is not reported.

Geology: Country rocks are tuffs and andesites. Four veins, the War Eagle No. 1, No. 2, Morning Glory, and Cross vein have been prospected. No. 1 strikes about N. 75° W. and dips 80° S.; No. 2 strikes about N. 55° W. and dips nearly vertically. They intersect. The Morning Glory vein strikes N. 85° W. and the Cross vein strikes N. 15° E. No. 1 and No. 2 veins have each been traced for nearly half a mile. The War Eagle No. 1 appears to extend west to include the California vein of the Musick group and east to the Bertha claim of the Champion group. The No. 1 vein is 3 to 4 feet thick with brecciated andesite, quartz, and disseminated sulfides (sphalerite, pyrite, galena, and chalcopyrite). No. 2 vein is from 2½ to 8 feet thick and similar in composition. Some of the oxidized ore yields good gold-silver assays. The Cross vein cuts through the Champion diorite dike. It contains quartz-cemented breccia and thin layers of specularite.

Production: Not reported.

References: Diller, 1900:124; Callaghan and Buddington, 1938:79-80; Taber, 1949:44; Department Bulletin 14-D, 1951:82.

OTHER DEPOSITS IN WESTERN CASCADES

A few widely separated mineralized localities are known in the Western Cascades south of the Bohemia district. Chief among these are the Zinc mine on the South Umpqua River in Douglas County, the Buzzard or Al Sarena mine near the head of Elk Creek, and the Barron mine about 8 miles east of Ashland. Both of the latter are in Jackson County (figure 55). These three areas of mineralization are associated with small intrusive bodies of probable Miocene age. The host rocks are Tertiary volcanics, mainly tuffs with associated basalt and andesite flows. The rocks in the vicinity of the Zinc mine are assigned to the Eocene Colestin Formation by Peck and others (1964). Those underlying the Al Sarena and Barron mines are mapped as Oligocene to lower Miocene volcanics by Wells and Peck (1961). Each of the areas is described by Callaghan and Buddington (1938, p. 130-136). Mineralization in these areas consists of clay alteration, disseminated pyrite, sphalerite, chalcopyrite, and galena. In addition, arsenopyrite is reported at the Al Sarena and Barron mines.

The Zinc mine has been explored by short adits on both sides of the river. There is no record of production. The northwest-striking vein lies between two dikes of augite diorite, each about 150 feet wide and 200 feet apart (Callaghan and Buddington, 1938).

The Al Sarena (Buzzard mine) workings are said by Callaghan and Buddington (1938) to be in excess of 3200 feet of drifts and crosscuts and 1000 feet of raises and winzes. Production from 1909 to 1918 was reported to be nearly $24,000, chiefly in gold with some silver and lead. Placer gold found in Elk Creek was traced upstream and the deposit was discovered in 1897 by Peter and Mark Applegate. The area shows extensive alteration with disseminated sulfides, mainly pyrite. The ore is reported to occur in gouge seams and consists of streaks and lenses of sphalerite with smaller amounts of pyrite and galena with very little quartz.

The Barron mine had workings that totaled about 1000 feet, including raises, winzes, and stoped areas, when reported by Callaghan and Buddington (1938), and the workings were partly caved. The vein lies in a northwest-striking 50-foot-wide zone of altered rock which dips steeply northeast. Callaghan and Buddington (1938, p. 135) describe the vein as follows:

> "...The vein, as shown on the lower level, consists of a series of branching and intersecting fractures, some of which are filled with gouge, some with fragments of altered rock, and some with altered rock cemented by cherty quartz, which in places contains sulfides. Comb quartz is inconspicuous. The vein is over 10 feet wide at the crosscut but pinches to 1 or 2 feet both to the northwest and to the southeast. This is essentially the lower limit of the ore shoot that has been partly stoped. An open cut reveals 40 feet of altered rock between the main vein and one lying to the west. Sulfides exposed in the drift occur in small stringers and consist chiefly of sphalerite with a little galena, chalcopyrite, pyrite, and arsenopyrite. Winchell (1914) mentions in addition stibnite, malachite, wire silver, realgar, and probable pyrargyrite. Altered rock consisting chiefly of clay minerals and a little sericite and carbonate, cherty quartz, calcite, and a little barite occurs with the sulfides. Most of the gold has been obtained from the leached and iron-stained vein matter, and leaching has extended to the main level, though it has not been complete."

Early history of the property has been lost. Production from 1917 to 1931 is reported to have been about $9000. Records of smelter shipments indicate that the gold-to-silver ratio is about 1:25. Some ore reportedly remained in the mine in 1931. No mining has been done of the property since reported on by Callaghan and Buddington.

Other Mines of the Western Cascades

Al Sarena (Buzzard) mine
Other Areas, Western Cascades, 2

Location: Jackson County, sec. 29, T. 31 S., R. 2 E., 3500 feet elevation.

Development: Drifts, crosscuts, raises, and winzes total about 4200 feet.

Geology: Altered Tertiary volcanic rocks are mineralized with disseminated pyrite and some sphalerite, galena, chalcopyrite, and arsenopyrite. Narrow enrichments occur in fault zones.

Production: Deposit was discovered in 1897. It produced $24,000 during 1909-1918. Equipment included a 100-ton flotation mill. Reported 35,000 tons of ore blocked out.

References: Callaghan and Buddington, 1938:131-132; Department Bulletin 14-C (Jackson), 1943:195-197.

Barron mine
Other Areas, Western Cascades, 3

Location: Jackson County, N½ sec. 23, T. 39 S., R. 2 E., at 3450 feet elevation.

Development: Total workings about 1000 feet.

Geology: Host rocks are hydrothermally altered lavas and tuffs. Sulfides including sphalerite, galena, chalcopyrite, pyrite, arsenopyrite, and stibnite occur with malachite, wire silver, and realgar in a gangue of clay, sericite, calcite, cherty quartz, and a little barite. Gold:silver ratio is about 1:25. The vein is 1 to 10 feet wide in a broader altered shear zone striking northwest and dipping steeply northeast.

Production: Early history lost. Production between 1917 and 1931 was about $9000. A 10-stamp mill was used.

References: Winchell, 1914:123; Parks and Swartley, 1916:25; Callaghan and Buddington, 1938:134-135; Department Bulletin 14-C (Jackson), 1943:26-28.

Zinc mine
Other Areas, Western Cascades, 1

Location: Douglas County, secs. 23 and 24, T. 29 S., R. 1 W., 1500 feet elevation on South Umpqua River and about 2000 feet elevation on Zinc Creek Road.

Development: Drifts penetrate both banks of the river. There is roughly 2300 feet of workings in 17 adits, most of which are very short.

Geology: Country rocks are tuff breccia cut by dikes of augite diorite and andesite. Mineralized zones of disseminated pyrite and clay alteration are common. A few lenses of sphalerite and associated galena and chalcopyrite have been found. Some calcite and marcasite are associated.

Production: Discovered in 1910. Some ore is reported to have been shipped in the 1920's, but no production is credited.

References: Callaghan and Buddington, 1938:130; Department Bulletin 14-C, vol. 1, 1940: 130; Department mine file report, 1940.

Bibliography

Allen, R.M., Jr., 1948, Geology and mineralization of the Morning mine and adjacent region, Grant County, Oregon: Oregon Dept. Geology and Mineral Industries Bull. 39.
Ashley, R.P., 1966, Metamorphic petrology and structure of the Burnt River Canyon area, northeastern Oregon: Stanford Univ. doctoral dissertation, 193 p.
Baldwin, E.M., 1964, Geology of Oregon: Eugene, Ore., Univ. Oregon Book Store, 2nd ed.
Bales, W.E., 1951, Geology of the lower Brice Creek area, Lane County, Oregon: Univ. Oregon master's thesis.
Bostwick, D.A., and Koch, G.S., Jr., 1962, Permian and Triassic rocks of northeastern Oregon: Geol. Soc. America Bull., v. 73, no. 3, p. 419-422.
Brooks, H.C., and Williams, G.A., (in progress): Geology of the Mineral quadrangle, Oregon-Idaho.
Brown, C.E., and Thayer, T.P., 1966, Geologic map of the Canyon City quadrangle, northeastern Oregon: U.S. Geol. Survey Misc. Geol. Inv. Map 1-447.
Brown, R.W., 1956, New items in Cretaceous and Tertiary floras of the western United States: Jour. Wash. Acad. Sci., v. 46, p. 106-108.
Browne, J.R., 1867, 1869, Reports on the mineral resources of the United States and Territories west of the Rocky Mountains: U.S. Treas. Dept.
Buddington, A.F., and Callaghan, E., 1936, Dioritic intrusive rocks and contact metamorphism in the Cascade Range of Oregon: Am. Jour. Sci., 5th ser., v. 31, p. 421-449.
Bur h, A., 1942, Development of metal mining in Oregon: Oregon Hist. Quarterly, v.43, no. 2, P. 105-128.
Butler, G.M., and Mitchell, G.J., 1916, Preliminary survey of the geology and mineral resources of Curry County, Oregon: Oregon Bur. Mines and Geology, Mineral Res. of Oregon, v. 2, no. 2.
Callaghan, E., and Buddington, A.F., 1938, Metalliferous mineral deposits of the Cascade Range in Oregon: U.S. Geol. Survey Bull. 893, 141 p.
Collier, A.J., 1914, The geology and mineral resources of the John Day region: Oregon Bur. Mines and Geology, Mineral Res. of Oregon, v. 1, no. 3.
Dickinson, W.R., and Vigrass, L.W., 1965, Geology of the Suplee-Izee area, Crook, Grant, and Harney Counties, Oregon: Oregon Dept. Geology and Mineral Ind. Bull. 58.
Diller, J.S., 1896, A geological reconnaissance in northwestern Oregon: U.S. Geol. Survey 17th Ann. Rept., part 1, p. 441-520.
______. 1898, Description of the Roseburg quadrangle, Oregon: U.S. Geol. Survey Geologic Atlas of the United States, Roseburg Folio no. 49.
______. 1900, The Bohemia mining region of western Oregon, with notes on the Blue River mining region and on the structure and age of the Cascade Range: U.S. Geol. Survey 20th Ann. Rept., part 3, p. 1-36.
______. 1902, Topographic development of the Klamath Mountains: U.S. Geol. Survey Bull. 196.
______. 1903, Description of the Port Orford quadrangle, Oregon: U.S. Geol. Survey Geologic Atlas of the United States, Port Orford Folio No. 89.
______. 1914, Mineral resources of southwestern Oregon: U.S. Geol. Survey Bull. 546, 147 p.
Diller, J.S., and Kay, G.F., 1908, Mines of the Riddles quadrangle, Oregon: U.S. Geol. Survey Bull. 340, p. 134-152.
______. 1909, Mineral resources of the Grants Pass quadrangle and bordering districts: U.S. Geol. Survey Bull. 380-A, p. 48-79.
______. 1924, Description of the Riddle quadrangle, Oregon: U.S. Geol. Survey Geologic Atlas of the United States, Riddle Folio No. 218.
Dole, H.M., and Baldwin, E.M., 1947, A reconnaissance between the Almeda and Silver Peak mines of southwestern Oregon: The ORE BIN, v. 9, no. 12, p. 95-100.
Dott, R.H., Jr., 1961, Permo-Triassic diastrophism in the western Cordilleran region: Am. Jour. Sci., v. 259, p. 561-582.
______. 1965, Mesozoic-Cenozoic tectonic history of the southwest Oregon coast in relation to Cordilleran orogenesis: Jour. Geophys. Research, v. 70, no. 18, p. 4687-4707.
______. 1966, Late Jurassic unconformity exposed in southwestern Oregon: The ORE BIN, v. 28, no. 5, p. 85-97.
Emmons, W.H., 1933, Recent progress in studies of supergene enrichment: Ore deposits of the Western States (Lindgren Volume): A.I.M.E., p. 386-418.
Engineering and Mining Journal, 1929, Dismantling and rebuilding a gold dredge: Engineering and Mining Journal, v. 128, p. 736-737.
Gilluly, J., 1931, Copper deposits near Keating, Oregon: U.S. Geol. Survey Bull. 830-A.
______. 1933, Replacement origin of the albite granite near Sparta, Oregon: U.S. Geol. Survey Prof. Paper 175-C, p. 1-81.
______. 1937, Geology and mineral resources of the Baker quadrangle, Oregon: U.S. Geol. Survey Bull. 879.
Gilluly, J., Reed, J.C., and Park, C.F., Jr., Some mining districts of eastern Oregon: U.S. Geol. Survey Bull. 846-A.
Goodspeed, G.E., 1939, Geology of the gold quartz veins of Cornucopia: A.I.M.E. Tech. Pub. 1035.
______. 1956, New data concerning the geology at Cornucopia, Oregon: International Geological Congress, 20th Session, Mexico City, 1956.
Grant, U.S., and Cady, G.H., 1914, Preliminary report on the general and economic geology of the Baker district of eastern Oregon: Oregon Bur. Mines and Geology, Mineral Res. of Oregon, v. 1, no. 6, p. 129-161.
Grants Pass Courier, 1904, Grants Pass, Oregon July 28, 1904.
Griggs, A.B., 1945, Chromite-bearing sands of the southern part of the coast of Oregon: U.S. Geol. Survey Bull. 945-E, p. 113-150.
Haines, F.D., Jr., and Smith, V.S., 1964, Gold on Sterling Creek; a century of placer mining: Medford, Ore. Grandee Printing Center, Inc., 104 p.
Hewett, D.F., 1931, Zonal relations of the lodes of the Sumpter quadrangle: A.I.M.E. Trans., v. 5, p. 305-346.
Hornor, R.R., 1918, Notes on black sand deposits of southern Oregon and northern California: U.S. Bur. Mines Tech. Paper 196, p. 10
Howard, J.K., and Dott, R.H., Jr., 1961, Geology of Cape Sebastian State Park and its regional relationships: The ORE BIN, v. 23, no. 8, p. 75-81.
Imlay, R.W., 1961, Late Jurassic ammonites from the western Sierra Nevada, California, U.S. Geol. Survey Prof. Paper 374-D.
Imlay, R.W., DOle, H.M., Wells, F.G., and Peck, D.L., 1959, Relations of certain Jurassic and Lower Cretaceous formations in southwestern Oregon: American Assn. Petroleum Geologists, Bull., v. 43, no. 12, p. 2270-2785;.
Irwin, W.P., 1964, Late Mesozoic orogenies in the ultramafic belts of northwestern California and southwestern Oregon: U.S. Geol. Survey Prof. Paper 501-C, p. C1-C9.
Kimbal, J.P., 1902, Bohemia mining district: Engineering and Mining Jour., June, 1902, p. 889-891.
Kleweno, W.P., Jr., and Jeffords, R.M., 1961, Devonian rocks in the Suplee area of central Oregon [abs.]: Geol. Soc. America Spec. Paper 68, p. 34-35; and Oregon Dept. Geology and Mineal Ind. The ORE BIN, v. 23, no. 5, p. 50.
Koch, G.S., Jr., 1959, Lode mines of the central part of the Granite mining district, Grant County, Oregon: Oregon Dept. Geology and Mineral Industries Bull. 49, 49 p.
Koch, J.G., 1966, Late Mesozoic stratigraphy and tectonic history, Port Orford-Gold Beach area, southwestern Oregon coast: American Assn. Petroleum Geologists Bull, v. 50, no. 1, p. 25-71.
Libbey, F.W., 1963, Lest we forget: The ORE BIN, v. 25, no. 6, p. 93-109.
______. 1967, The Almeda mine, Josephine County, Oregon: Oregon Dept. Geology and Mineral Ind. Short Paper 24, 53 p.
Lindgren, W., 1901, The gold belt of the Blue Mountains of Oregon: U.S. Geol. Survey 22nd Ann. Rept., Pt. 2, p. 551-776.
______. 1933, Ore deposits of the western states: A.I.M.E. Committee on the Lindgren Volume, eds.
Livingston, D.C., 1925, A geologic reconnaissance of the Mineral and Cuddy Mountains districts, Idaho: Idaho Bur. Mines and Geology Pamph. No. 13.
Lorain, S.H., 1938, Gold mining and milling in northeastern Oregon: U.S. Bur. Mines Inf. Circ. 7015.
Lowry, W.D., (in press), Geology of the Ironside Mountain quadrangle, Oregon: Oregon Dept. Geology and Mineral Ind. Bulletin.
Lutton, R.J., 1962, Geology of the Bohemia mining district, Lane County, Oregon: Univ. Arizona doctoral dissertation.
MacDonald, D.F., 1909, Notes on the Bohemia mining district, Oregon: U.S. Geol. Survey Bull. 380, p. 80-84.
McDaniel, D.L., and Marshall, C.L., 1911, Report on Treasure mine: Univ. Oregon thesis.
Merriam, C.W., and Berthiaume, S.A., 1943, Late Paleozoic formations of central Oregon: Geol. Soc. America Bull., v. 54, no. 2, p. 145-171.
Nelson, R.E., 1959, Facts and yarns of the Bohemia gold mines: (privately printed), 30 p.

Editor's Note: In an earlier printing, Page No. 324 was blank. Blank pages have been eliminated in the present book thus that page number does not appear.

Nolf, B., 1967, Geology of the sedimentary rocks surrounding the Wallowa batholith: Princeton Univ. doctoral dissertation.
Nolf, B., and Taubeneck, W.H., 1963, Permo-Triassic stratigraphic revisions, Wallowa Mountains, northeastern Oregon [abs.]: Geol. Soc. America Spec. Paper 73, p. 92-93.
Oregon (State of) Dept. Geology and Mineral Industries, 1939, Oregon Metal Mines Handbook, Bulletin 14-A, northeastern Oregon - east half [Baker, Union, and Wallowa Counties], 125 p.
______. 1940, Oregon Metal Mines Handbook, Bulletin 14-C, southwestern Oregon, vol. 1 [Coos, Curry, and Douglas Counties], 133 p.
______. 1941, Oregon Metal Mines Handbook, Bulletin 14-B, northeastern Oregon - west half [Grant, Morrow, and Umatilla Counties], 157 p.
______. 1942, Oregon Metal Mines Handbook, Bulletin 14-C, southwestern Oregon, vol. 2, sec. 1, [Josephine County], 229 p.
______. 1943, Oregon Metal Mines Handbook, Bulletin 14-C, southwestern Oregon, vol. 2, sec. 2 [Jackson County], 208 p.
______. 1951, Oregon Metal Mines Handbook, Bulletin 14-D, northwestern Oregon, 116 p.
Oregon Observer, 1904, July 2 and July 9, 1904.
Pardee, J.T., 1909, Faulting and vein structure in the Cracker Creek gold district, Baker County, Oregon: U.S. Geol. Survey Bull. 380, p. 85-93.
______. 1910, Placer gravels of the Sumpter and Granite districts, eastern Oregon: U.S. Geol. Survey Bull. 430, p. 59-66.
______. 1934, Beach placers of the Oregon coast: U.S. Geol. Survey Circ. 8, 41 p.
Pardee, J.T., and Hewett, D.F., Geology and mineral resources of the Sumpter quadrangle: Oregon Bur. Mines and Geology, Mineral Res. or Oregon, v.1, no. 6, p. 1-128.
Pardee, J.T., Hewett, D.G., and others, 1941, Preliminary geologic map of the Sumpter quadrangle, Oregon: Oregon Dept. Geology and Mineral Ind. map, with text. Scale 1:96,000.
Parks, H.M., and Swartley, A.M., 1916, Handbook of the mining industry of Oregon: Oregon Bu. Mines and Geology, Mineral Res. of Oregon, v. 2, no. 4, 306 p.
Peck, D.L. Griggs, A.B., Schlicker, H.G., Wells, F.G., and Dole, H.M., 1964, Geology of the central and northern part of the Western Cascade Range in Oregon: U.S. Geol. Survey Prof. Paper 449, 56 p.
Peck, D.L. Imlay, R.W., and Popenoe, W.P., 1956, Upper Cretaceous rocks of parts of southwestern Oregon and northern California: American Assn. of Petroleum Geologists Bull., v. 40, no. 8, p. 1968-1984.
Prostka, H.J., 1962, Geology of the Sparta quadrangle, Oregon: Oregon Dept. Geology and Mineral Ind. map, with text. Scale 1:62,500.
______. 1963, The geology of the sparta quadrangle, Oregon: Johns Hopkins Univ. doctoral dissertation, unpub., 236 p.
______. 1967, Preliminary geologic map of the Durkee quadrangle, Oregon: Oregon Dept. Geology and Mineral Ind. Geol. Map Series, No. 3.
Ramp, Len, 1960, Gold placer mining in southwestern Oregon: The ORE BIN, v. 22, no. 8, p. 75-79.
Raymond, R.W., 1870-1877, Statistics of the mines and mining in (also mineral resources of) the states and territories west of the Rocky Mountains: U.S. Treas. Dept. ann. repts.
Rogue River Courier (names changed to Grants Pass Courier).
Rosenberg, F.J., 1941, Amalgamated mine: private report in files of F.J. Rosenberg, mining engineer, Portland, Oregon.
Ross, C.P., 1938, The geology of part of the Wallowa Mountains: Oregon Dept. Geology and Mineral Ind. Bull. 3, 74 p.
Shenon, P.J., 1933-a, Copper deposits in the Squaw Creek and Silver Peak districts and at the Almeda mine, southwestern Oregon, with notes on the Pennell and Farmer and Banfield prospects: U.S. Geol. Survey Circ. 2, 35 p.
______. 1933-b, Geology of the Roberton, Humdinger, and Robert E. gold mines, southwestern Oregon: U.S. Geol. Survey Bull. 830-B, p. 33-55.
______. 1933-c, Geology and ore deposits of the Takilma-Waldo district, Oregon: U.S. Geol. Survey Bull. 846-B, p. 33-194.
Smith, W.D., and Allen, J.E., 1941, Geology and physiography of the northern Wallowa Mountains, Oregon: Oregon Dept. Geology and Mineral Ind. Bull. 12, 64 p.
Smith, W.D., and Ruff, L.L., 1938, The geology and mineral resources of Lane County, Oregon: Oregon Dept. Geology and Mineral Industries Bull. 11, 65 p.
Spreen, C.A. 1939, A history of placer gold mining in Oregon, 1850-1870: Oregon Univ. master's thesis, unpub.
Stafford, O.F., 1904, The mineral resources and mineral industry of Oregon for 1903: Oregon Univ. Bull., v. 1, no. 4.
Stowell, G.E., 1921, Report on the geology and ore deposits of the Quartzville district, Oregon: Oregon Bur. Mines and Geology (unpublished rept. in files of Oregon Dept. Geology and Mineral Ind.).
Swartley, A.M., 1914, Ore deposits in northwestern Oregon: Oregon Bur. Mines and Geology Mineral Res. of Oregon, v. 1, no. 8, 229 p.
Taber, J.W., 1949, A reconnaissance of lode mines and prospects in the Bohemia mining district, Lane and Douglas Counties, Oregon: U.S. Bur. Mines Inf. Circ. 7512.
Taliaferro, N.L., 1942, Geologic history and correlation of the Jurassic of southwestern Oregon and California: Geol. Soc. America Bull., v. 53, no. 1, p. 71-112.
Taubeneck, W.H., 1957, Geology of the Elkhorn Mountains, northeastern Oregon: Bald Mountain batholith: Geol. Soc. America Bull., v. 68, no. 2, p. 181-238.
______. 1958, Wallowa batholith, Wallowa Mountains, northeastern Oregon [abs.]: Geol. Soc. America Bull., v. 69, no. 12, pt. 2, p. 1650.
______. 1963, Wallowa Mountains uplift, northeastern Oregon: Geol. Soc. America Spec. Paper 73, 385 p.
______. 1964, Cornucopia stock, Wallowa Mountains, northeastern Oregon: field relationships: Geol. Soc. America Bull., v. 75, no. 11, p. 1093-1116.
______. 1967, Petrology of Cornucopia tonalite unit, Cornucopia stock, Wallowa Mountains, northeastern Oregon: Geol. Soc. America Spec. Paper 91, 56 p.
Thayer, T.P., 1939, Geology of the Salem Hills and the North Santiam River basin, Oregon: Oregon Dept. Geology and Mineal Ind. Bull. 15, 40 p.
______. 1956, Preliminary geologic maps of the Aldrich Mountain, Mount Vernon, and John Day quadrangles: U.S. Geol. Survey Maps MF 49, 50, and 51.
______. 1963, The Canyon Mountain complex, Oregon, and the alpine mafic magma stem: U.S. Geol. Survey Prof. Paper 475-C, art. 81, p. C82-C85.
Thayer, T.P., and Brown, C.E., 1964, Pre-Tertiary orogenic and plutonic intrusive activity in central and northeastern Oregon: Geol. Soc. American Bull., v. 75, no. 12, p. 1255-1262.
Vallier, T.L., 1967, Geology of part of the Snake River canyon and adjacent areas in northeastern Oregon and western Idaho: Oregon State Univ. doctoral dissertation, 267 p.
Walling, A.C., 1884, History of Southern Oregon: Portland, Oregon.
Watkins, Kenneth, 1946, Unpublished report on several properties in the Bohemia district.
Wells, F.G., 1933, Notes on the Chieftan and Continental mines, Douglas County, Oregon: U.S. Geol. Survey Bull. 830-B, p. 57-62.
______. 1956, Geology of the Medford quadrangle, Oregon-California: U.S. Geol. Survey Geol. Quad. Map GQ-89.
Wells, F.G., and others, 1940, Preliminary geologic map of the Grants Pass quadrangle, Oregon: Oregon Dept. Geology and Mineral Industries map.
Wells, F.G., Hotz, P.E., and Cater, F.W., Jr., 1949, Preliminary description of the geology of the Kerby quadrangle, Oregon: Oregon Dept. Geology and Mineral Ind. Bull. 40, 23 p.
Wells, F.G., and Peck, D.M., 1961, Geologic map of Oregon west of the 121st meridian: U.S. Geol. Survey (in cooperation with Oregon Dept. Geology and Mineral Ind.) Misc. Geol. Invest. Map 1-325.
Wells, F.G., and Walker, G.W., 1953, Geology of the Galice quadrangle, Oregon: U.S. Geol. Survey Geol. Quad. Map GQ-25.
Wetherell, C.E., 1960, Geology of part of the southeastern Wallowa Mountains, northeastern Oregon: Oregon State Univ. master's thesis, 208 p., unpub.
White, D.J., and Wolfe, H.D., 1950, Report of reconnaissance of the area from Panther Butte to Tellurium Peak, Douglas County, Oregon: The ORE BIN, v. 12, no. 12, p. 71-76.
Winchell, A.N., 1914, Petrology and mineral resources of Jackson and Josephine Counties, Oregon: Oregon Bur. Mines and Geology Mineral Res. of Oregon, v. 1, no. 5 265 p.
Wolff, E.N., 1965, Geology of the northern half of the Caviness quadrangle, Oregon: Univ. Oregon doctoral dissertation, 200 p., unpub.
Youngberg, E.A., 1947, Mines and prospects of the Mount Reuben mining district, Josephine County, Oregon: Oregon Dept. Geology and Mineral Ind. Bull. 34, 35 p.

INDEX OF MINES

See separate Index for Placers on page 331

LODE MINES AND PROSPECTS

PLACERS

—Photography by June Hose, Folsom, Calif.
Bert Webber at the site of operations of Yuba Dredge No. 21 in July 1994. The story of this dredge and others is in his book *Dredging For Gold*.

About the Editor

Bert Webber is a Research Photojournalist who likes to look for subjects that have not been earlier presented in books or, if there has been such a book he is able to find material that would make for a better one. His subjects are all non-fiction, are mostly about Oregon and the Oregon Trail and some unique matters of World War II in the Pacific. These books are aimed for use by librarians and teachers as source material, are usually loaded with pictures, and all are written at *Reader's Digest* level for popular acceptance.

He learned newspaper-style camera work and photo lab techniques in the U. S. Army Signal Corps in World War-II. He served in Alaska, Scotland, England, Belgium and twice in France making still and motion pictures. After the war, he was a commercial photographer and had affiliation with a number of newspapers for his pictures and news articles.

Webber graduated from Whitworth College in Journalism and in Library Science. Later, he earned the Master of Library Science by studying at Portland State University and the University of Portland. He was a teacher and school librarian but retired from that profession in 1970 to pursue full-time writing.

Bert Webber has hundreds of newspaper and magazine articles and over fifty books. He is listed in *Who's Who in the West, Who's Who in the World* and in *Contemporary Authors.* He was awarded the "Decree of Merit" in *Men of Achievement* at the International Biographical Centre, Cambridge, England.

For a hobby, he plays baritone horn in the Southern Oregon Symphonic Band where he also serves on the band's Board of Control. Webber is an Elder in the Presbyterian Church.

Bert and his wife, Margie, a retired Registered Nurse, who is co-author of some of the books, have four grown children and eight grandchildren. They live in Oregon's Rogue River Valley just a few miles from where that Jacksonville street caved in to reveal an old gold mine underneath (page *i*-B). □